A Short Course in Discrete Mathematics

A Short Course in Discrete Mathematics

Edward A. Bender
S. Gill Williamson
University of California, San Diego

Dover Publications, Inc.
Mineola, New York

Bibliographical Note

A Short Course in Discrete Mathematics is a new work, first published in book
form by Dover Publications, Inc., in 2005.

Library of Congress Cataloging-in-Publication Data

Bender, Edward A., 1942–
    A short course in discrete mathematics / Edward A. Bender, S. Gill
Williamson.
        p. cm.
    Includes indexes.
    ISBN-13: 978-0-486-43946-4 (pbk.)
    ISBN-10: 0-486-43946-1 (pbk.)
    1. Mathematics. 2. Computer science—Mathematics. I. Williamson, S. Gill
(Stanley Gill) II. Title.

QA39.3.B453 2005
510—dc22

                                                          2004063429

Manufactured in the United States by Courier Corporation
43946105   2014
www.doverpublications.com

Preface

Discrete mathematics is an essential tool in almost all subareas of computer science. Interesting and challenging problems in discrete mathematics arise in programming languages, computer architecture, networking, distributed systems, database systems, AI, theoretical computer science, and other areas.

The course. The University of California, San Diego, has a lower-division two-quarter course sequence in discrete mathematics that includes Boolean arithmetic, combinatorics, elementary logic, induction, graph theory and finite probability. These courses are core undergraduate requirements for majors in Computer Science, Computer Engineering, and Mathematics-Computer Science. This text, *A Short Course in Discrete Mathematics*, was developed for the first quarter and *Mathematics for Algorithm and System Analysis* was developed for the second quarter.

This book consists of six units of study (Boolean Functions and Computer Arithmetic; Logic; Number Theory and Cryptography; Sets and Functions; Equivalence and Order; and Induction, Sequences and Series), each divided into two sections. Each section contains a representative selection of problems. These vary from basic to more difficult, including proofs for study by mathematics students or honors students.

The review questions. "Multiple Choice Questions for Review" appear at the end of each unit. The explanatory material in this book is directed towards giving students the mathematical language and sophistication to recognize and articulate the ideas behind these questions and to answer questions that are similar in concept and difficulty. Many variations of these questions have been successfully worked on exams by most beginning students using this book at UCSD.

Students who master the ideas and mathematical language needed to understand these review questions gain the ability to formulate, in the neutral language of mathematics, problems that arise in various applications of computer science. This skill greatly facilitates their ability to discuss problems in discrete mathematics with other computer scientists and with mathematicians.

Table of Contents

> Asterisks (stars) are used in the text to mark more
> difficult material that is not needed in later sections.

relation, modular arithmetic, modular addition, modular multiplication, floor function, ceiling function, diagonalization proofs

---

Unit SF: Sets and Functions

Unit EO: Equivalence and Order

Boolean Functions and Computer Arithmetic

Section 1: Boolean Functions

We recall the concept of a function and some of the terminology.

Definition 1 (Function)   *If A and B are sets, a function from A to B is a rule that tells us how to find a unique $b \in B$ for each $a \in A$. We write $f(a) = b$ and say that f maps a to b. We also say the value of f at a is b.*

   We write $f : A \to B$ to indicate that f is a function from A to B. We call the set A the domain of f and the set B the range or, equivalently, codomain of f.

   To specify a function completely you must give its domain, range and rule.

   In this section, we'll study a special class of functions called "boolean functions." General properties of functions are studied in Unit SF.

   A *Boolean function* is a function f from the Cartesian product $\times^n\{0,1\}$ to $\{0,1\}$. Alternatively, we write $f : \times^n\{0,1\} \to \{0,1\}$. The set $\times^n\{0,1\}$, by definition, the set of all n-tuples $(x_1, \cdots, x_n)$ where each x_i is either 0 or 1, is called the *domain* of f. The set $\{0,1\}$ is called the *codomain* (or, sometimes, *range*) of f. The Cartesian product $\times^n\{0,1\}$ is also written $\{0,1\}^n$. This corresponds to writing the product of n copies of y as y^n.

Example 1 (Tabular representation of Boolean functions)   One way to represent a function whose domain is finite is with a table. Each element x of the domain has a row of the table listing the domain element x and the corresponding function value $f(x)$. For example, the two tables

p	q	f
0	0	0
0	1	1
1	0	0
1	1	1

p	q	r	g
0	0	0	1
0	0	1	1
0	1	0	0
0	1	1	0
1	0	0	0
1	0	1	0
1	1	0	1
1	1	1	1

define Boolean functions $f : \times^2\{0,1\} \to \{0,1\}$ and $g : \times^3\{0,1\} \to \{0,1\}$. In the case of f, the first two values of each row represent the argument of f and the third entry in the same row represents the value of f at that argument. We have $f(0,0) = 0$, $f(0,1) = 1$, $f(1,0) = 0$, and $f(1,1) = 1$. For g, the first three values of each row represent an element of $\times^3\{0,1\}$ (a triple (p,q,r) of 0's or 1's) and the fourth entry of the row represents the value of $g(p,q,r)$. Thus, $g(0,0,0) = 1$, $g(0,0,1) = 1$, $g(0,1,0) = 0$, etc.

Boolean Functions and Computer Arithmetic

Notice that we have used p, q and r for variables instead of x, y and z, which are usually the choice for variable names in algebra. We can choose any names we please for variables. Names such as p, q and r are commonly used in the study of Boolean functions. $\square$

The tabular form of a Boolean function is often called a *truth table* because of the connection between Boolean functions and logic, which we will study later.

Example 2 (The number of Boolean functions) How many Boolean functions are there with domain $\times^3\{0,1\}$? More generally, how many Boolean functions are there with domain $\times^n\{0,1\}$? We do this in two steps.

First, how many elements are there in the domain $\times^n\{0,1\}$? We use the notation $|S|$ to denote the number of elements of S (also called the cardinality of S). Thus we are asking for the value of $|\times^n\{0,1\}|$. We can select an element of the domain by making n choices $x_1, x_2, \ldots, x_n$ where $x_i \in \{0,1\}$ for $i = 1, 2, \ldots, n$. Thus there are two choices for x_1 and two choices for x_2 and so on. The total number of elements in the domain is thus $2 \times 2 \times \cdots \times 2 = 2^n$. In other notation, $|\times^n\{0,1\}| = |\{0,1\}|^n$. If you have trouble seeing why this is true, the footnote may help.[1]

Second, we must construct the Boolean function. In constructing a Boolean function $h : \times^3\{0,1\} \to \{0,1\}$, there are two possible choices that we could assign to $h(p,q,r)$ for each $(p,q,r) \in \times^3\{0,1\}$. We saw in the previous paragraph that there are 2^3 elements (p,q,r) in $\times^3\{0,1\}$. Thus the total number of Boolean functions $h : \times^3\{0,1\} \to \{0,1\}$ is $2^{(2^3)}$. In general, the number of Boolean functions $h : \times^n\{0,1\} \to \{0,1\}$ is $2^{(2^n)}$. $\square$

In this section we are concerned about a particular way of representing Boolean functions in terms of certain more "primitive" representations. An example that is familiar from high school mathematics is the representation of polynomial and rational functions, such as $1 + x$, $1 + x^2$, $(2 + x^2 - x^3)/(1 + x^2)$, starting with the constant functions c and the identity function x. All of these functions are created by adding, multiplying, subtracting, and/or dividing the simple starting functions.

Example 3 (The simplest Boolean functions: constants, identity, not) The simplest functions are the constant functions. Since we have only two constants, 0 and 1, there are two constant Boolean functions.

The next simplest Boolean functions are those that depend only only a single variable, but are not constant. Since there are $2^{2^1} = 2^2 = 4$ one-variable Boolean functions and since there are two constants, there are $4 - 2 = 2$ nonconstant, one-variable Boolean functions. One is the identity function $f(p) = p$ for all $p \in \{0,1\}$. The other one-variable function is "**not** p." For the moment, let's call this function $n(p)$. The definition is simple: $n(p) = 0$ if $p = 1$, $n(p) = 1$ if $p = 0$. This function is usually denoted by $\sim p$ rather than $n(p)$. Thus,

---

[1] How many elements are there in the Cartesian product $S \times T$ of a set with s elements and a set with t? Imagine an s by t array. Can you see why each entry in the array can be thought of as an element of $S \times T$? If so, we are almost done. We have shown that $|S \times T| = |S| \cdot |T|$. Apply this over and over again to $\{0,1\} \times \{0,1\} \times \cdots \times \{0,1\}$.

we would write $\sim 0 = 1$ and $\sim 1 = 0$. The symbol "$\sim$" is called the "unary operator **not**." *Unary* means it is a function that has one variable. The minus sign in ordinary arithmetic is a familiar example of a unary operator. □

Example 4 (Two-variable Boolean functions: and, or, exculsive or) We know there are $2^{2^2} = 2^4 = 16$ two-variable Boolean functions $f(p,q)$. From the previous example, 2 of these are constant, 2 of them are not constant and depend only on p, and 2 of them are not constant and depend only on q. This leaves $16 - 6 = 10$ functions that depend on both variables. We certainly don't want to give names to all of them! In this example, we define three commonly-used, two-variable Boolean functions.

The first function we define is "p **and** q." Again, for the moment, let's call this function (a function of two Boolean variables) $a(p,q)$. By definition, $a(p,q) = 0$ unless both $p = 1$ and $q = 1$, in which case $a(p,q) = 1$. The function $a(p,q)$ is denoted by $p \wedge q$. We write $0 \wedge 0 = 0$, $0 \wedge 1 = 0$, $1 \wedge 0 = 0$, and $1 \wedge 1 = 1$.

The next function is "p **or** q." Again, for the moment, let call this function (a function of two Boolean variables) $o(p,q)$. By definition, $o(p,q) = 1$ unless both $p = 0$ and $q = 0$, in which case $o(p,q) = 0$. The function $o(p,q)$ is denoted by $p \vee q$. We write $0 \vee 0 = 0$, $0 \vee 1 = 1$, $1 \vee 0 = 1$, and $1 \vee 1 = 1$.

The last function is the *exclusive or* function. It is written $\mathbf{xor}(p,q)$ or $p \oplus q$. By definition $p \oplus q = 0$ if $p = q$ and $p \oplus q = 1$ if $p \neq q$.

Here are our functions in tabular form:

p	q	$p \wedge q$		p	q	$p \vee q$		p	q	$p \oplus q$
0	0	0		0	0	0		0	0	0
0	1	0		0	1	1		0	1	1
1	0	0		1	0	1		1	0	1
1	1	1		1	1	1		1	1	0

The symbols $\vee$, $\wedge$, and $\oplus$ are called "binary operators" because they involve two variables (like addition and multiplication in ordinary arithmetic). □

Example 5 (Boolean functions and logic) People often call Boolean variables such as p and q *statement variables*. Why? Because Boolean functions are related to logic. In logic we think of p and q as statements, 0 as "false," and 1 as "true."

Suppose p stands for "I have classes tomorrow" and q stands for "I will stay home tomorrow." Let's look at our basic Boolean functions.

- $\sim p$ stands for "**not** (I have classes tomorrow)," which can be written in more normal English as "I do not have classes tomorrow." As mentioned earlier, 0 is thought of as "false" and 1 is thought of as "true." According to our definition of the function $\sim$, if p is true, then $\sim p$ is false. This is also true about our statements: If the statement "I have classes tomorrow" is true, then the statement "I do not have classes tomorrow" is false.

- $p \wedge q$ stands for the statement, "I have classes tomorrow and I will stay home tomorrow." You should verify that the definition of $p \wedge q$ agrees with our usual interpretation of "and": $p \wedge q$ is true if and only if both p and q are true. The statement $p \wedge q$ is also read "p but q", especially if q is surprising as in "I have classes tomorrow, but I will stay home tomorrow."

- $p \vee q$ stands for the statement, "Either I have classes tomorrow or I will stay home tomorrow." Unfortunately "or" is ambiguous in English. If I have classes *and also* stay home, some people may think the statement is not true. Others may think that it is true. Our definition of **or** is *not* ambiguous: $p \vee q$ is true if either p or q or both are true.

- $p \oplus q$ stands for the statement "Either I have classes tomorrow or I will stay home tomorrow, but not both." Why is this? The value of $p \oplus q$ is 1 (true) if and only if one of p and q is 1 (true) and the other is 0 (false).

When converting ordinary language into symbolic form, it is important to be aware of ambiguities so that the statements can be converted correctly. What we have been discussing is called *propositional logic*. We will explore the connection between Boolean functions and logic in the unit on logic. □

Because of the close connection with logic, the tabular form of a Boolean function is often called a *truth table*.

Why are the Boolean functions represented by $\sim$, $\wedge$, $\vee$ and $\oplus$ important? In the previous example, we have seen a hint of their importance in logic. They are also important because they can be used to define more complex Boolean functions, in the same way that the basic operations of arithmetic can be used to define more complex algebraic functions.

Example 6 (Defining more complex Boolean functions) We could try to define a Boolean function using $\sim$, $\wedge$, and $\vee$ by stating that

$$f(p,q) = p \wedge \sim q \vee \sim p \wedge q. \qquad \text{(ambiguous)}$$

The problem with this "definition" is that it is not clear what the order of application of these operators should be. It is conventional to give top priority to the $\sim$ operator. Thus the expression used to define f can be clarified a bit: $f(p,q) = p \wedge (\sim q) \vee (\sim p) \wedge q$. In addition, the order in which the $\wedge$ and $\vee$ are performed must be specified by grouping them in the definition of the function f. Thus, as an example, we could group them

$$f(p,q) = (p \wedge (\sim q)) \vee ((\sim p) \wedge q). \qquad \text{(not ambiguous)}$$

Now the function f is clearly defined. You should make sure that you use enough parentheses. Using too few may result in an ambiguous function definition. Using more than needed does not change a function. For example, the usage of $\sim$ is defined by the precedence rules, so we could just write

$$f(p,q) = (p \wedge \sim q) \vee (\sim p \wedge q). \qquad \text{(still not ambiguous)}$$

If you're unsure about precedence rules, be safe and use extra parentheses!

Using the formula for f we can compute values and give f in tabular form:

p	q	f
0	0	0
0	1	1
1	0	1
1	1	0

This is the exclusive or function. We have proved $p \oplus q = (p \wedge \sim q) \vee (\sim p \wedge q)$. $\square$

In the previous example, we found that $p \oplus q$ could be written as an equivalent Boolean function using $\sim$, $\wedge$ and $\vee$. This is a particular example of a much more general result:

Theorem 1 (Representing functions) *Suppose $n > 0$ and $f : \times^n\{0,1\} \rightarrow \{0,1\}$. There is a function g using only $\sim$, $\vee$ and $\wedge$ that is equal to f. In fact, we can use just $\sim$ and $\wedge$ or, if we prefer, we can use just $\sim$ and $\vee$.*

Proof: We'll illustrate how to do this with the function

p	q	r	f
0	0	0	0
0	0	1	1
0	1	0	1
0	1	1	0
1	0	0	0
1	0	1	0
1	1	0	0
1	1	1	1

Look at the rows in the table where the $f = 1$. The first such is $(p,q,r) = (0,0,1)$. Note that $(\sim p) \wedge ((\sim q) \wedge r)$ equals 1 when $(p,q,r) = (0,0,1)$ and equals 0 otherwise.[2] Similarly, for the rows $(p,q,r) = (0,1,0)$ and $(p,q,r) = (1,1,1)$, we have $(\sim p) \wedge (q \wedge (\sim r))$ and $p \wedge (q \wedge r)$. You should be able to see why the function

$$g(p,q,r) = \Big((\sim p) \wedge ((\sim q) \wedge r)\Big) \vee \Big((\sim p) \wedge (q \wedge (\sim r))\Big) \vee \Big(p \wedge (q \wedge r)\Big)$$

is equal to f.

It should be clear how to do this in general: Construct an appropriate "anding" for each row of the table where the function equals 1. Then "or" all these andings together. This proves the first claim.

In this paragraph we take a break to discuss some terminology. An expression of this sort, namely an "or" of "ands" of variables and their negations, is called *disjunctive normal form*. Where did "disjunctive" come from? An "or" is sometimes called a *disjunction*. If we

---

[2] We don't need all these parentheses. The value of $\sim p \wedge \sim q \wedge r$ is unambiguous. We're just playing it safe!

Boolean Functions and Computer Arithmetic

replace the roles of "and" and "or," we obtain *conjunctive normal form*, so called because "and" is called a *conjunction*. End of terminology discussion and back to the proof.

One might object that we haven't taken care of the case when the function is a contradiction since then there are no rows where the function is 1. This is taken care of by 0 or, if you prefer, by $p \wedge \sim p$.

How can we get rid of $\vee$? We claim that

$$P \vee Q \vee R \vee \cdots \vee T \quad \text{and} \quad \sim\bigl((\sim P) \wedge (\sim Q) \wedge (\sim R) \wedge \cdots \wedge (\sim T)\bigr)$$

are equal. Why? The only way the first expression can be 0 is for *all* of $P, Q, \ldots, T$ to be 0. The only way the second expression can be 0 is for the expression inside the large parentheses to be 1. The only way this can happen is for *all* of $\sim P, \sim Q, \ldots, \sim T$ to be 1 — which is the same as *all* of $P, Q, \ldots, T$ being 0. Applying this to g with

$$P = (\sim p) \wedge ((\sim q) \wedge r), \quad Q = (\sim p) \wedge (q \wedge (\sim r)), \quad R = p \wedge (q \wedge r),$$

and none of the terms $\ldots, T$, we have the equivalent function

$$\sim\Bigl(\sim\bigl((\sim p) \wedge ((\sim q) \wedge r)\bigr) \wedge \sim\bigl((\sim p) \wedge (q \wedge (\sim r))\bigr) \wedge \sim\bigl(p \wedge (q \wedge r)\bigr)\Bigr).$$

You should see that this works in general.

What about getting rid of $\wedge$ instead of $\vee$? A similar trick can be used, working from the "inside" of the formula instead of from the "outside." We leave it to your inventiveness to find the trick. ∎

In the last half of the previous proof, we used a trick that let us replace $\vee$ with $\wedge$. In fact, this trick is used often enough that we should call it a rule.[3] It is very useful to have a catalog of simple rules to help in deciding whether or not functions are equal, without having to always construct tables of functions. Here is such a catalog. In each case the standard name of the rule is given first, followed by the rules as applied first to $\wedge$ and then to $\vee$.

Theorem 2 (Algebraic rules for Boolean functions)   *Each rule states that two different-looking Boolean functions are equal. That is, they look different but have the same table.*

Associative Rules:	$(p \wedge q) \wedge r = p \wedge (q \wedge r)$	$(p \vee q) \vee r = p \vee (q \vee r)$
Distributive Rules:	$p \wedge (q \vee r) = (p \wedge q) \vee (p \wedge r)$	$p \vee (q \wedge r) = (p \vee q) \wedge (p \vee r)$
Idempotent Rules:	$p \wedge p = p$	$p \vee p = p$
Double Negation:	$\sim\sim p = p$	
DeMorgan's Rules:	$\sim(p \wedge q) = \sim p \vee \sim q$	$\sim(p \vee q) = \sim p \wedge \sim q$
Commutative Rules:	$p \wedge q = q \wedge p$	$p \vee q = q \vee p$
Absorption Rules:	$p \vee (p \wedge q) = p$	$p \wedge (p \vee q) = p$
Bound Rules:	$p \wedge 0 = 0 \quad p \wedge 1 = p$	$p \vee 1 = 1 \quad p \vee 0 = p$
Negation Rules:	$p \wedge (\sim p) = 0$	$p \vee (\sim p) = 1$

---

[3] It is sometimes said that a rule or method is a trick that is used more than once.

These rules are "algebraic" rules for working with $\wedge$, $\vee$, and $\sim$. You should memorize them as you use them. They are used just like rules in ordinary algebra: whenever you see an expression on one side of the equal sign, you can replace it by the expression on the other side. Each of the rules can be proved by constructing tables for the functions on each side of the equal sign and verifying that those tables give the same function values.

Truth tables are similar to the tabular method for proving set identities (see Section 1 of Unit SF). The algebraic rules for Boolean functions are almost identical to the rules for sets in Section 1 of Unit SF. When two apparently very different situations (sets and Boolean functions in this case) are similar, one should look for an explanation. We provide an explanation in Unit Lo.

One useful consequence of this connection is that we can use "Venn diagrams" to prove identities for Boolean functions. (If you are not familiar with Venn diagrams, read the first four pages of Unit SF.) How does this work? Think of the variables p, q and so on as sets. Then make the translations:

$$\wedge \text{ to } \cap, \qquad \vee \text{ to } \cup, \qquad \sim \text{ to complement.}$$

The universal set is 1 and the empty set is 0. For example, you can prove the first of DeMorgan's Rules for Boolean functions, by showing that $(P \cap Q)^c$ and $P^c \cup Q^c$ give the same regions in the Venn diagram for two sets P and Q. Try it.

Example 7 (Manipulating functions) We want to simplify the function

$$(\sim(p \wedge \sim q)) \wedge (p \vee q).$$

Here are our calculations:

$$
\begin{aligned}
(\sim(p \wedge \sim q)) \wedge (p \vee q) &= (\sim p \vee \sim\sim q) \wedge (p \vee q) &&\text{(DeMorgan's rule)} \\
&= (\sim p \vee q) \wedge (p \vee q) &&\text{(double negation)} \\
&= ((\sim p \vee q) \wedge p) \vee ((\sim p \vee q) \wedge q) &&\text{(distributive rule)} \\
&= (p \wedge (\sim p \vee q)) \vee (q \wedge (q \vee \sim p)) &&\text{(commutative rule 3 times)} \\
&= (p \wedge (\sim p \vee q)) \vee q &&\text{(absorbtion rule)} \\
&= ((p \wedge \sim p) \vee (p \wedge q)) \vee q &&\text{(distributive rule)} \\
&= (0 \vee (p \wedge q)) \vee q &&\text{(negation rule)} \\
&= (p \wedge q) \vee q &&\text{(bound rule)} \\
&= q &&\text{(commutative and absorbtion rules)}
\end{aligned}
$$

Except for the last step, we gave each step in detail. In actual calculations, you can combine steps as we did in the last step. How you decide to manipulate things can make a big difference. For example,

$$
\begin{aligned}
(\sim(p \wedge \sim q)) \wedge (p \vee q) &= (\sim p \vee \sim\sim q) \wedge (p \vee q) &&\text{(DeMorgan's rule)} \\
&= (\sim p \vee q) \wedge (p \vee q) &&\text{(double negation)} \\
&= (q \vee \sim p) \wedge (q \vee p) &&\text{(commutative rule)} \\
&= q \vee (\sim p \wedge p) &&\text{(distributive rule)} \\
&= q \vee 0 &&\text{(negation rule)} \\
&= q &&\text{(bound rule)}
\end{aligned}
$$

Boolean Functions and Computer Arithmetic

The same thing happens in algebra; however, you are more likely to do things the shorter way in algebra because you are more familiar with those manipulations. There is a trade off between taking time to try finding a shorter way and simply going ahead. This is a problem faced by designers of "symbolic manipulation" packages such as Maple® and Mathematica®. ∎

Exercises for Section 1

1.1. Let f = "she is out of work" and s = "she is spending more." Write the following statements in symbolic form:

   (a) She is out of work but she is spending more.

   (b) Neither is she out of work nor is she spending more.

1.2. Let r = "she registered to vote" and v = "she voted." Write the following statements in symbolic form: She registered to vote but she did not vote.

1.3. Make a truth table for $\sim\!\big((p \wedge q) \vee \sim\!(p \vee q)\big)$.

1.4. Make a truth table for $\sim\!p \wedge (q \vee \sim\!r)$.

1.5. Make a truth table for $\big(p \vee (\sim\!p \vee q)\big) \wedge \sim\!(q \wedge \sim\!r)$.

1.6. Using DeMorgan's rule, state the negation of the statement: "Mary is a musician and she plays chess."

1.7. Using DeMorgan's rule, state the negation of the statement: "The car is out of gas or the fuel line is plugged."

1.8. Show that $p \vee (p \wedge q) = p$ follows from the idempotent rule, distributive rule, and the absorption rule $p \wedge (p \vee q) = p$.

1.9. Is the function $(p \wedge q) \vee r$ equal to the function $p \wedge (q \vee r)$?

1.10. Is the function $(p \vee q) \vee (p \wedge r)$ equal to the function $(p \vee q) \wedge r$?

1.11. Is the function $\big((\sim\!p \vee q) \wedge (p \vee \sim\!r)\big) \wedge (\sim\!p \vee \sim\!q)$ equal to the function $\sim\!(p \vee r)$?

1.12. Is the function $(r \vee p) \wedge \big(\sim\!r \vee (p \wedge q)\big) \wedge (r \vee q)$ equal to the function $p \wedge q$?

1.13. Is the function $\sim\!(p \vee \sim\!q) \vee (\sim\!p \wedge \sim\!q)$ equal to the function $\sim\!p$?

1.14. Is the function $\sim\!\big((\sim\!p \wedge q) \vee (\sim\!p \wedge \sim\!q)\big) \vee (p \wedge q)$ equal to the function $\sim\!p$?

1.15. Is the function $\Big(p \wedge \big(\sim\!(\sim\!p \vee q)\big)\Big) \vee (p \wedge q)$ equal to the function $p \vee q$?

Section 2: Number Systems and Computer Arithmetic

The number system we are most familiar with is the "base 10" system. In that system, an "n-digit" number is represented by a sequence of "digits," $d_{n-1} \cdots d_1 d_0$. For example, 243598102 is a 9-digit number base 10. The numerical value of the number $d_{n-1} \cdots d_1 d_0$ is $d_{n-1}10^{n-1} + d_{n-2}10^{n-2} + \cdots + d_1 10^1 + d_0 10^0$. We are familiar, from elementary school, with various tedious algorithms (treated with mystical reverence by the "back-to-basics" educational advocates) for adding, subtracting, multiplying, and dividing numbers in the base 10 system using pencil and paper calculations. Although these algorithms are sometimes useful, mostly we use our hand calculators or computers to do these calculations. Here, for example, is the almost instantaneous result of asking a computer program to compute 500!, the product of the first 500 positive integers, and represent the answer in base 10.

```
12201368259911100687012387854230469262535743428031928421924135883858453
73153881997605496447502203281863013616477148203584163378722078177200480
78520515932928547790757193933060377296085908627042917454788242491272634
43056701732707694610628023104526442188787894657547771498634943677810376
44274033827365397471386477878495438489595537990423241061271326984327745
71554630997720278101456108118837370953101635632443298702956389662891165
89747695720879269288712817800702651745077684107196243903943225364226052
34945850129918571501248706961568141625359056693423813008856224924689156
41267756544818865065938479517753608940057452389403357984763639449053130
62323749066445048824665075946735862074637925184200459369692981022263971
95259719094521782333175693458150855233282076282002340262690789834245171
20062077146409794561161276291459512372299133401695523638509428855920187
27433795173014586357570828355780158735432768888680120399882384702151467
60544540766353598417443048012893831389688163948746965881750450692636533
81750554781286400000000000000000000000000000000000000000000000000000000
0000000000000000000000000000000000000000000000000000000000000000000000000
0000000000000000
```

Example 8 (Base 256) There is no reason why we have to use base 10. For certain applications, base 256 is used. In base 10 we have ten familiar symbols to use for the digits, namely

$$0, 1, 2, 3, 4, 5, 6, 7, 8, 9 \,.$$

What do we use in base 256 for the symbols used for the digits of numbers? One obvious choice is to use

$$[0], [1], \ldots, [254], [255] \,.$$

We have written these digit symbols with square braces to avoid confusion when digit symbols are concatenated to form numbers. In the base-256 system, an "n-digit" number, as in base 10, is represented by a sequence $d_{n-1} \cdots d_1 d_0$ where each d_i is a digit symbol (from $[0], [1], \ldots, [254], [255]$). For example, $[12][43][251][198][210][122][2][0][0][0]$ is a 10-digit number base 256. The numerical value of the number $d_{n-1} \cdots d_1 d_0$ is $d_{n-1}256^{n-1} + d_{n-2}256^{n-2} + \cdots + d_1 256^1 + d_0 256^0$. The symbols d_i are used in two different

Boolean Functions and Computer Arithmetic

ways here, but in practice there should be no confusion. For example, the 10-digit base-256 number [12][43][251][198][210][122][2][0][0][0] has numerical value

$$12 \cdot 256^9 + 43 \cdot 256^8 + 251 \cdot 256^7 + 198 \cdot 256^6 + 210 \cdot 256^5 + 122 \cdot 256^4 + 2 \cdot 256^3.$$

In base 10, this number is 5747975020917562330316. $\square$

Example 9 (Number systems base b) In fact, a number system can be defined for any integer base $b > 1$. We need unique symbols for the *digit symbols*, say $[0], [1], \ldots, [b-1]$ or, more generally, $D_0, \ldots, D_{b-1}$. The digit symbol D_i is said to have *index* or *rank* i in the list of digit symbols for the base-b number system. If d is a digit symbol, we use $\iota(d)$ ("iota of d") to represent its rank in the list of digit symbols. Thus, $\iota(D_k) = k$ or $\iota([k]) = k$ in the lists of digit symbols just described. Then, an n-digit positive number can be represented uniquely as $d_{n-1} \cdots d_1 d_0$ where each d_i is one of the digit symbols $D_0, \ldots, D_{b-1}$ and $\iota(d_{n-1}) > 0$. The numerical value of this number is $\iota(d_{n-1})b^{n-1} + \iota(d_{n-2})b^{n-2} + \cdots + \iota(d_1)b^1 + \iota(d_0)b^0$. In base b, using digit symbols $[0], [1], \ldots, [b-1]$, the nonnegative numbers, ordered according to numerical value, start off $[0], [1], [2], \ldots, [b-1]$. These are the 1-digit numbers, base b. Next come the 2-digit numbers

$$[1][0], \ldots, [1][b-1], \ [2][0], \ldots, [2][b-1], \ldots, \ [b-1][0], \ldots, [b-1][b-1].$$

If you wanted to do paper and pencil calculations for addition, subtraction, multiplication or long division in base b, you could use the "back-to-basics" algorithms with base-b digit symbols instead of base-10 symbols. We will study such algorithms for $b = 2$ later in this section.

The largest n-digit number in base b is

$$[b-1][b-1]\ldots[b-1] = (b-1)b^{n-1} + \cdots + (b-1) = (b-1)\frac{b^n - 1}{b-1} = b^n - 1.$$

If we add [1] to it, we get

$$[b-1][b-1]\ldots[b-1] + [1] = [1][0]\cdots[0],$$

which is the smallest $(n+1)$-digit number and has value b^n. $\square$

Example 10 (Transforming numbers between bases) Given a number, how can we write it in base b?

It would be nice to begin with an example in base 10, but there's no familiar way to specify a number without using base 10. We can fake it a bit by writing the number in words. Let's write twenty-one hundred seven in base 10.

- If we divide by 10, we obtain two hundred ten and a remainder of seven. Thus our rightmost digit is [7]. Now start again with two hundred ten.

- If we divide by 10, we obtain twenty-one and a remainder of zero. Thus our rightmost digit is [0]. Now start again with twenty-one.

- If we divide by 10, we obtain two and a remainder of one. Thus our rightmost digit is [1]. Now start again with two.

- If we divide by 10, we obtain zero and a remainder of two. Thus our rightmost digit is [2]. Since we've reached zero, we're done.

Putting the rightmost digits together in order, we have [2][1][0][7], or 2107 in the usual notation for base-10 numbers.

Now suppose our number N is written in the familiar way (base 10) and we want to write it in base b. We proceed in the same manner as in the previous paragraph, dividing by b each time instead of by 10.

Here is an example for base 256, using digit symbols $[0], [1], \ldots, [255]$. Suppose we are given the decimal (base-10) number 3865988647 and want to write it in base 256.

- 38659886477/256 is 15101518 with a remainder of 39. Thus our rightmost digit is [39]. Now start again with 15101518.

- 15101518/256 is 58990 with a remainder of 78. Thus our rightmost digit is [78]. Now start again with 58990.

- 58990/256 is 230 with a remainder of 110. Thus our rightmost digit is [110]. Now start again with 230.

- 230/256 is 0 with a remainder of 230. Thus our rightmost digit is [230]. Since we've reached 0, we're done.

Thus the base-10 number 3865988647 equals [230] [110] [78] [39] in base 256. We can check that this is correct:

$$[230][110][78][39] = 230 \cdot 256^3 + 110 \cdot 256^2 + 78 \cdot 256 + 39 = 38659886477. \qquad \square$$

Computer Arithmetic

In the remainder of this section, we will study base 2, 8, and 16 numbers. Instead of bases 2, 8 and 16, people also say

binary for base 2, *octal* for base 8 and *hexadecimal* for base 16.

You should practice converting some base-10 numbers of your choosing into their binary, octal and hexadecimal equivalents.

Computers, for the most part, work directly with base $b = 2$. The symbols for the digits are usually taken to be 0, 1. Thus, $10011 = 2^4 + 2^1 + 2^0$. In base ten, 10011 is denoted by 19. If the base needs to be mentioned explicitly in the mathematics, one usually writes it as a subscript: $10011_2 = 19_{10}$. Base 2 is inconvenient for people because it leads to long strings of digits. (Try writing a small number like 2345 in base 2.) As we shall see, converting between base 2 and bases 8 and 16 is quite simple. As a result, when we need to deal with base 2 in studying a computer program, we often write it in base 8 or base 16 instead to avoid long strings of digits.

Boolean Functions and Computer Arithmetic

Example 11 (Octal and hexadecimal number systems) Given a binary number such as 1101100111010101110, one can start at the right and group the digits into groups of three as follows: 1|101|100|111|010|101|110. If each group of three is now replaced by its decimal equivalent, one obtains 1547256. This sequence of digits is the base-8 representation of the given binary number, using the base-8 digit symbols 0, 1, 2, 3, 4, 5, 6, 7. In summary, $1101100111010101110_2 = 1547256_8$. The process is easily reversed. For example, $7237631_8 = 111010011111110011001_2$. To understand why this works in general, consider the following computation for the latter example:

$$7237631_8 = 7 \cdot 8^6 + 2 \cdot 8^5 + 3 \cdot 8^4 + 7 \cdot 8^3 + 6 \cdot 8^2 + 3 \cdot 8^1 + 1 \cdot 8^0$$
$$= (2^2 + 2^1 + 2^0)2^{18} + (2^1)2^{15} + (2^1 + 2^0)2^{12} + (2^2 + 2^1 + 2^0)2^9$$
$$+ (2^2 + 2^1)2^6 + (2^1 + 2^0)2^3 + (2^0)2^0 .$$

In this calculation, we have replaced each power of 8, 8^k, by $(2^3)^k = 2^{3k}$. Each base-8 digit is replaced by its corresponding expansion in terms of powers of 2. By using the distributive law on each of the summands in the latter expression (e.g., $(2^2 + 2^1 + 2^0)2^{18} = 2^{20} + 2^{19} + 2^{18}$) we get the defining expansion as powers of two of the binary number $111010011111110011001_2 = 7237631_8$.

Base-16 numbers relate to base-2 numbers in a manner directly analogous to the way base-8 numbers relate to base-2 numbers. One problem is to decide on the symbols used to represent the 16 digits needed for the base-16 number system. One could use $[0], [1], [2], \ldots, [10], [11], [12], [13], [14], [15]$. Thus, $[11][8][10]$ would represent the number $11 \cdot 16^2 + 8 \cdot 16^1 + 10 \cdot 16^0$. Although this is sometimes done, it is more common to use A, B, C, D, E, and F in place of $[10], [11], [12], [13], [14], [15]$. In place of $[0], [1], [2], \ldots, [9]$, the ordinary base-10 digit symbols are used, $0, 1, 2, 3, 4, 5, 6, 7, 8, 9$. Thus $[11][8][10]$ would be written B8A. To transform a base-2 number such as 1101100111010101110 to a base-16 number, first group the digits, right to left, in groups of four: 110|1100|1110|1010|1110. Then change each group to base 10, using the symbols A, B, C, D, E, and F for the base-10 numbers 10 through 15. For 110|1100|1110|1010|1110 the result is 6|C|E|A|E which we write 6CEAE. Going in the reverse direction is just as easy, namely, replace each base-16 digit by its base two form.

To transform numbers from base-10 representation to bases 2, 8, or 16, transform to any one of these bases, say base 2, and then transform between 2, 8, and 16 using the algorithms described above. Going in the reverse, from bases 2, 8, or 16 to base 10, just use the definition of the base expansion directly (e.g., $6CEAE = 6 \cdot 16^4 + 12 \cdot 16^3 + 14 \cdot 16^2 + 10 \cdot 16^1 + 14 \cdot 16^0$). □

Example 12 (Binary arithmetic) Here we show an addition, multiplication, and subtraction using the familiar base-10 algorithms, followed in each case by a similar base-2 computation.

```
                                   5 8 4        1 1 1        3  9  9   5  9       0  1  1  1
   3 4 7 8 3      1 1 1 0        x  3 0 7     x 1 0 1      4 ∅ ∅ 0 ∅ ∅ 2      1 ∦ ∅ ∅ ∅ 0
 + 6 1 3 5 8    + 1 0 1 1          4 0 8 8      1 1 1      - 3 8 7 6 5 1 3    - 1 0 0 0 0 1
   9 6 1 4 1    1 1 0 0 1        1 7 5 2        1 1 1        1 2 4 0 8 9        0 0 1 1 1 1
                                1 7 9 2 8 8  1 0 0 0 1 1
```

Section 2: Number Systems and Computer Arithmetic

You should study each example and understand the similarities and the differences. The base-2 addition corresponds to $14_{10} + 11_{10}$, the base-2 multiplication corresponds to $7_{10} \times 5_{10}$, and the base-2 subtraction corresponds to $48_{10} - 33_{10}$.

The familiar long division algorithm from elementary school works fine for base-2 numbers. It is actually easier in base 2. Here is an example of long division carried out using the standard algorithm and a similar calculation using base-2 numbers (corresponding to 554_{10} divided by 9_{10} to get 61_{10} with remainder 5_{10}).

```
              9 9 9 7 6                         1 1 1 1 0 1
  4 2 5 | 4 2 4 9 0 0 0 1        1 0 0 1 | 1 0 0 0 1 0 1 0 1 0
          3 8 2 5                          1 0 0 1
          _____                          _____
          4 2 4 0                          1 0 0 0 0
          3 8 2 5                          1 0 0 1
          _____                          _____
          4 1 5 0                            1 1 1 1
          3 8 2 5                            1 0 0 1
          _____                            _____
          3 2 5 0                            1 1 0 0
          2 9 7 5                            1 0 0 1
          _____                            _____
          2 7 5 1                              1 1 1 0
          2 5 5 0                              1 0 0 1
          _____                              _____
            2 0 1                                1 0 1
```

The layout of the long division algorithm is designed for pencil and paper calculation and is obviously not relevant to computer based calculations. The long division algorithm shown above that seeks to divide 42490001 (the "dividend") by 425 (the "divisor"), scans the dividend 42490001 from left to right to find the shortest sequence of digits, in this case 4249, that represents an integer greater than or equal to 425. This gives us our starting point. Since 425 has three digits, the shortest sequence will always have either three or four digits (that is the whole point of why this algorithm is used for elementary school kids). The first step is to divide 425 into this number 4249, thus representing the dividend as

$$42490001 = 9 \times 425 \times 10^4 + 4240001.$$

We then move over one digit in the remainder. Thus we divide 425 into 4240. As a result we get

$$42490001 = 9 \times 425 \times 10^4 + 9 \times 425 \times 10^3 + 415001.$$

The next step, applied to 415001, gives

$$42490001 = 9 \times 425 \times 10^4 + 9 \times 425 \times 10^3 + 9 \times 425 \times 10^2 + 32501.$$

The next step, applied to 32501, gives

$$42490001 = 9 \times 425 \times 10^4 + 9 \times 425 \times 10^3 + 9 \times 425 \times 10^2 + 7 \times 425 \times 10^1 + 2751.$$

The final step, applied to 2751 gives

$$42490001 = 9 \times 425 \times 10^4 + 9 \times 425 \times 10^3 + 9 \times 425 \times 10^2 + 7 \times 425 \times 10^1 + 6 \times 425 + 201.$$

This latter expression is, using standard base-10 notation,

$$42490001 = 99976 \times 425 + 201.$$

[13]

BF-13

Boolean Functions and Computer Arithmetic

Each step could be carried out without being able to divide! For example, to divide 4249 by 425, simply subtract 425 from 4249 as many times as possible. It can be done nine times and the remainder is 415, the result after the nine subtractions. Thus $4249 = 9 \times 425 + 415$.

For base-2 division, avoiding division is easy since the number of subtractions we'll be able to do at each step is either one or zero. If you like to program, you may find it interesting to program this algorithm for base-2 numbers. The required divisions at each stage, for base-2 numbers, can be carried out by subtraction, as done in the above example. □

The important feature of binary arithmetic on a computer is that, at the most primitive hardware level, the size of the register is fixed. We have to understand binary addition with the constraint of fixed register size. Also, we need to discuss negative numbers.

Example 13 (Binary addition and register size) Suppose that the register has places for n binary bits. We number the bits right-to-left from 0 to $n - 1$. Here is an example with $n = 16$.

bit position	15	14	13	12	11	10	9	8	7	6	5	4	3	2	1	0
bit value	0	1	0	1	1	1	0	0	0	1	1	1	0	1	0	1

The bit with position 0 (rightmost bit) is called the "least significant" bit and the bit with position 15 is called the "most significant" bit. In the example just given, the binary number has the value

$$2^{14} + 2^{12} + 2^{11} + 2^{10} + 2^6 + 2^5 + 2^4 + 2^2 + 2^0 = 23669 \text{ (base-10)}.$$

We call arithmetic done using an n-bit register "n-bit binary arithmetic".

The largest number that can be represented with 16 binary bits is

$$1111111111111111 = \sum_{i=0}^{15} 2^i = 2^{16} - 1.$$

If you are using a 16-bit register to add binary numbers, everything is fine unless you add two numbers whose sum is greater than $2^{16} - 1$. If that happens, too bad! It is an "overflow" and, if you are lucky, you are informed of the problem without having the program mistakenly continue on, perhaps to start World War III. If we get an overflow we reconsider what we are doing and try to avoid it by being more clever. For modern computers, the register size is big enough (at least 32 bits) so that we have lots of room to work with basic integers. We can use these register operations to build software systems that can work with integers of arbitrary size by using as many registers as needed to store the binary form of a number. Such packages are found in software systems that do "multiple precision arithmetic," an example being GNU MP. □

So far, we have only mentioned positive numbers. So, subtracting a smaller number from a larger one on a 16-bit register should work just fine. But what if we want to allow negative numbers? We could use + and − just as we do with base-10 numbers. For example -101_2 would be -5 and $+101_2$ would be $+5$, which we often write simply as 5. Alternatively, we can use a bit somewhere to keep track of the sign, say 0 for $+$ and 1 for $-$. Following our usual convention about sign placement, we could use the leftmost bit, the bit in position 15. In our 16-bit register, -5 would then be 1000000000000101 and $+5$ would be 0000000000000101, which is the way we were storing 5 in our register. Notice that the size of our numbers is now restricted to 15 bits because one bit is used for the sign. Thus $|x| < 2^{15}$.

Next we need to work out rules for adding and subtracting when numbers may be positive or negative. This is the obvious way to do things, but there's an easier way. Surprisingly, this easier way exists because of the problem we ran into in the previous example!

Example 14 (Negative numbers and register size) Forget about negative numbers for a minute. Imagine adding 1 to the contents of a 16-bit register. It's clear what to do unless the register contains 1111111111111111. When we add 1, the result is 0000000000000000 with a 1 to carry and no place to put it — our overflow problem. Since we have no place to put it, let's just throw it away! Since 0000000000000000 is zero, we have the equation

1111111111111111 plus 1 equals 0.

Since we know that $(-1) + 1 = 0$, it looks like we should think of 1111111111111111 as being -1. Wait, the binary number 1111111111111111 equals $2^{16} - 1$. What happened?

Think of telling time, but forget about the hours and just talk about minutes. The time can be k minutes after the hour for $k = 0, 1, \ldots 59$. Sixty minutes after the hour is the start of the next hour, and since we aren't keeping track of what the hour is, that's simply zero minutes after the hour. Add one minute to 59 minutes after the hour and we're back to zero — just liking adding 1 to 1111111111111111. Where are the negative numbers? Fifty-nine minutes *after* the hour is also one minute *before* the (next) hour. So "after" is positive and "before" is negative. Well then, is it 59 or -1? That's up to you. If we want to be "fair" and have about as many negative minutes as positive ones, we could go up to 29 or 30 minutes after (positive) or before (negative) the hour.

You can picture all this on the dial of an old fashioned (analog) watch or clock. When the minute hand is before 6, we count minutes after the hour (clockwise) and get a positive number. When the minute hand is after 6, we count minutes before the hour (counter clockwise) and get a negative number. When the minute hand is on the 6, we have either $+30$ or -30, depending on what we decide. Our large numbers (more than 30 minutes after the hour) are now thought of as negative numbers (before the hour). Minutes before and after the hour are related by

(minutes before the hour) + (minutes after the hour) = 60.

"One hour" in our register has 2^{16} minutes and 30 minutes corresponds to 2^{15}. If you are familiar with modular arithmetic, you may recognize that the clock is doing arithmetic modulo 60 and our register is doing it modulo 2^{16}.

Just as with the clock, numbers that look large in the register are to be thought of as negative numbers. We can simply look at the leftmost bit to see if a number is large. If the

Boolean Functions and Computer Arithmetic

leftmost bit is 1, the number is thought of as negative. This is different from the obvious idea; for example, consider -1:

$$\text{“obvious” } -1 = 1000000000000001 \qquad \text{new } -1 = 1111111111111111.$$

Recall that the new value is -1 (one before zero) because when we add 1 to it and throw away the carry we get 0000000000000000. $\Box$

The idea we just introduced is called *two's complement* notation. It won't do us any good unless we understand how to do arithmetic with such numbers and how to convert a negative number like -5 into the bits in a register. That's the subject of the next example. As we'll see, the arithmetic is easy.

Example 15 (Two's complement arithmetic) Let's suppose we have an n-long register and let $x < 2^{n-1}$ be a positive number stored in the register. Since $x < 2^{n-1}$, the leftmost bit of x is zero. What should be stored in the register to represent $-x$? In the previous example we said it was the number we needed to add to a register containing x to get zero, after throwing away the carry bit that we had no place to put. If we were doing ordinary grade school arithmetic, the carry bit would have given us 2^n, one larger than $2^n - 1$, the register filled with ones. Let's restate what we just concluded:

The positive number that should be thought of as $-x$
is
the number y such that $x + y = 2^n$.

This is just the before and after minutes rule for a clock.

In other words, our representation of $-x$ is the number $y = 2^n - x$. The number $2^n - x$ is called the *two's complement* of x. More generally, if some number z, positive or negative, is stored in a register, we find $-z$ by computing the two's complement of the value stored for z.

Let's look at an example of computing the two's complement. We take $n = 16$ and $x = 0100100011001000$. We want to compute $2^n - x$. The lefthand calculation in the following figure is the obvious way to do the calculation.

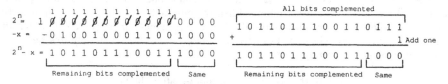

The observation at the bottom of the lefthand calculation gives a simple short-cut rule for computing the n-bit two's complement of x directly in terms of the digits of x. Scan the digits of x from right to left until the first 1 is encountered. Complement all digits of x to the left of that first 1. Some texts describe the "short cut" process of taking the two's complement as complementing all bits of x and adding 1 to the result. As shown in the right-hand calculation in the above figure, this leads to the same thing. Use whichever short-cut method you prefer.

Section 2: Number Systems and Computer Arithmetic

How do we add two numbers? For the time being, forget about overflow. Add the way we have already learned to do it, throwing away the carry bit. This gives the correct answer regardless of whether the numbers are positive or negative or one of each as long as the numbers are not too big in size. (We'll be specific about "too big" later.) Why is this?

First, let's try an example. Suppose have a 16-long register and want to add -5 and -8. You should compute the two's complements of 5 and 8, add the results and check that the answer is the two's complement of 13. Here's how we can see that without doing all the work. The two's complements are $2^{16} - 5$ and $2^{16} - 8$. Adding them as numbers we get $2^{16} + 2^{16} - 13$. Adding them in register arithmetic, one 2^{16} will be thrown away as a carry bit. (This has to happen because our register answer must be between 0 and $2^{16} - 1$.) Thus we get $2^{16} - 13$. Clearly this works in general. If we were to add -5 and 8, we would get $2^{16} - 5 + 8 = 2^{16} + 3$. The 2^{16} would be thrown away as a carry bit, giving us 3.

What about numbers that are "too big" (overflow)?

- If both numbers are positive and the answer looks negative, there was an "overflow." This will happen if we have $0 < x, y < 2^{15}$ but $x + y \geq 2^{15}$ because when we add there is a carry into the sign bit, making it 1 which is supposed to indicate a negative number.

- If both numbers are negative and the answer looks positive, there was overflow.

If you think about it and look at examples, you should be able to convince yourself that this is the only way overflow can happen. For example, if one number is positive and the other negative, there cannot be any overflow. This gives us a simple rule:

In two's complement arithmetic, there is overflow if and only if
x and y have the same sign bit and $x + y$ has a different sign bit.

What about subtracting two numbers, say x and y? You can either do the subtraction in the usual way, or take the n-bit two's complement of y and add (ignoring any value carried into the $(n + 1)^{\text{th}}$ bit). Both methods are illustrated in the following calculation ($n = 16$). You should write the binary numbers as ordinary base-10 numbers (including signs) and check the calculation.

```
  0 1  1 1 1   0 1  1 1 1
  1̸ 0̸ 0̸ 0̸ 0̸ 0 1̸ 0̸ 0̸ 0̸ 0̸ 0̸ 0 1 1 1         1 0 0 0 0 0 1 0 0 0 0 0 0 1 1 1
- 0 1 0 0 1 1 0 0 1 1 0 0 1 0 0 0       + 1 0 1 1 0 0 1 1 0 0 1 1 1 0 0 0
  ───────────────────────────────         ───────────────────────────────
  0 0 1 1 0 1 0 1 0 0 1 1 1 1 1 1         0 0 1 1 0 1 0 1 0 0 1 1 1 1 1 1
```

Thus, representing the negative of a number using two's complement, reduces subtraction to addition.

What about multiplying two numbers? Multiplication is really just shifting and adding: When we multiplied 111 by 101 we added 111 to the result of shifting 111 to the left by two bits. What if bits get shifted off the left end of the register? Shifting x by k bits is the same as adding 2^k copies of x together, so that's okay. Thus, multiplication works regardless of whether the numbers are positive or negative. Overflow is somewhat trickier and we won't discuss it.

What about dividing two numbers? When division is understood properly, it works okay. We won't go into that. You might like to think about it. □

Boolean Functions and Computer Arithmetic

In the previous example, we saw how doing two's complement arithmetic reduces to addition. To complete our discussion, we look at how to build a circuit for binary addition. Computer circuits implement Boolean functions. We will discuss how to do addition using **and, or** and **xor**.

Example 16 (A circuit for binary addition) Suppose we want to add the two binary numbers $a_{n-1}a_{n-2}\cdots a_1a_0$ and $b_{n-1}b_{n-2}\cdots b_1b_0$. Let the answer be $s_ns_{n-1}\cdots s_1s_0$. (The leading digits of these numbers might be zero.) Here is how we did it in Example 12.

(0) Add the binary digits a_0 and b_0 to obtain the two digit sum c_0s_0. The digit c_0 is the "carry" digit.

(1) Add the binary digits a_1, b_1 and c_0 to obtain the two digit sum c_1s_1.

$\cdots$

($n-2$) Add the binary digits a_{n-2}, b_{n-2} and c_{n-3} to obtain the two digit sum $c_{n-2}s_{n-2}$.

($n-1$) Add the binary digits a_{n-1}, b_{n-1} and c_{n-2} to obtain the two digit sum s_ns_{n-1}.

We discuss the circuit for this in two steps. First, adding two binary digits — called a "half adder." Then, adding three binary digits — called a "full adder."

Let p, q, s and c be the binary digits in $p+q = cs$. Thus c is the carry digit. Considering the four possibilities for (p, q), we get the following table for the Boolean functions s and c:

p	q	s	c
0	0	0	0
0	1	1	0
1	0	1	0
1	1	0	1

Note that $s = p \oplus q$, the "exclusive or" function, and $c = p \wedge q$, the "and" function. We can visualize these two Boolean functions in one circuit diagram as follows:

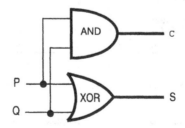

This circuit with two Boolean variables as inputs and two as outputs is sometimes referred to as a "half adder." The functions **and** and **xor** in the circuit are referred to as *logic gates* or just *gates*. You should imagine the values of p and q entering at the points labeled P and Q and moving along the wires. A $\bullet$ indicates a branch — the value moves along both wires leading out from the $\bullet$. Thus the value of p enters both gates and the value of q enters both gates. The values of c and s emerge at the points labeled C and S. These points can be used as inputs to another circuit, if desired.

The half adder circuit can be used to compute $a_0 + b_0 = c_0 s_0$ in Step (0). Now we need to add three binary digits for Steps (1) through $(n-1)$. In other words, we want to find c and s so that $p + q + r = cs$. This can be done in steps:

$$p + q = c's' \quad \text{then} \quad s' + r = c''s \quad \text{then} \quad c' + c'' = c.$$

Why doesn't the last step give a two digit answer? It's impossible. Suppose there were two digits c^*c. We are adding three digits, p, q and r. By the above calculations, the answer (in binary) is c^*cs. The largest possible answer, obtained when $p = q = r = 1$ is 11, which has only two digits. Thus $c^* = 0$. Actually, this is nothing more than the fact that when we do addition the carry is never more than a single digit.

We can do the calculations described above with three half adders, one for each addition. Since the third half adder cannot produce a carry, we'll throw away that portion of the circuit. Here's the result, with the two complete half adders in dashed boxes:

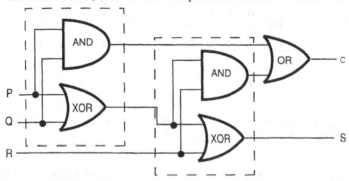

Wait! The figure has a misprint — an **or** where there should be an **xor**. In fact, either one works. Why is that? Here's the table for computing $c' + c''$:

c'	c''	c
0	0	0
0	1	1
1	0	1
1	1	*

The table has no entry for $c' = c'' = 1$ because that never occurs. (Remember the discussion in the previous paragraph about the carry being a single digit.) Since it never happens, we can define this entry any way we choose. Choosing 0 gives $c = c' \oplus c''$. Choosing 1 gives $c = c' \vee c''$.

Our device for adding two binary numbers can now be built. Use a half adder for Step (0). Next, take the carry from this half adder, a_1 and b_1 as inputs to a full adder. This is Step (1). Next, take the carry from this adder, a_2 and b_2 as inputs to a full adder. This is Step (2). Continue in this manner.

How long does our circuit take to add two n-bit registers? Suppose a single gate takes time T. Then a half adder takes time T because the two gates do their calculations at the same time. You should be able to see that our full adder takes time $3T$. We use $n-1$ full adders and one half adder for an n-bit register. Thus the total time is $(n-1)3T + T = (3n-2)T$. $\square$

[19]

Exercises for Section 2

2.1. For each of the following, construct a Boolean function equal to the function S defined by the given truth table. Make your function as simple as you can. Then design a circuit for the function.

(a)

P	Q	R	S
0	0	0	0
0	0	1	0
0	1	0	1
0	1	1	0
1	0	0	1
1	0	1	1
1	1	0	0
1	1	1	0

(b)

P	Q	R	S
0	0	0	0
0	0	1	1
0	1	0	0
0	1	1	1
1	0	0	0
1	0	1	0
1	1	0	1
1	1	1	0

2.2. Design a circuit that represents the Boolean function S where $S(P, Q, R) = 0$ if and only if $(P, Q, R) = (0, 0, 0)$ or $(P, Q, R) = (1, 1, 1)$.

2.3. Design a circuit to represent the response of the lights in a room to the light switches under each of the following conditions.

(a) There are two switches, moving either switch to the opposite position turns the room lights on if off and off if on.

(b) There are three switches, moving any switch to the opposite position turns the room lights on if off and off if on.

2.4. Show that the following two circuits represent the same Boolean function.

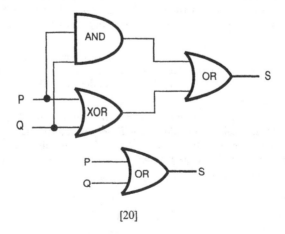

2.5. Show that the following two circuits represent the same Boolean function.

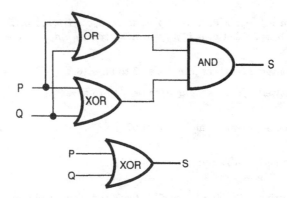

2.6. Show that the Boolean function $(\sim P \wedge \sim Q) \vee (P \oplus Q)$ equals the Boolean function computed by the following circuit with just two logic gates (NOT and AND):

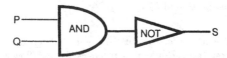

2.7. Find a circuit with at most three logic gates, each of which is allowed to have at most two inputs, that is equal to the Boolean function defined by the following truth table.

P	Q	R	S
0	0	0	0
0	0	1	0
0	1	0	0
0	1	1	0
1	0	0	1
1	0	1	1
1	1	0	0
1	1	1	1

2.8. Compute the difference of the two base two numbers: $1110100_2 - 10111_2$.

2.9. Convert 10110111110000101_2 from binary to hexadecimal (i.e., base 16) and octal (i.e., base 8).

2.10. Convert the following as indicated.

(a) Convert 61502_8 to decimal.

   (b) Convert $EB7C5_{16}$ to octal.

2.11. Let $b, m, l, f, -, z, k, a, n, y, e, x, j, w, d, v, o, u, c, g, t, p, h, s, q, i, r$ be the digit symbol list for base 27. Let $n, m, k, j, f, s, q, h, z, p, c, x, y, e, d, w$ be the digit symbol list for base 16.

   (a) Convert $hi - there$ from base 27 to base 16.

   (b) Convert $cfemxysnnjnq$ from base 16 to base 27.

2.12. Find the 8-bit two's complement of 67_{10}.

2.13. Find the 8-bit two's complement of 108_{10}.

2.14. The number 10001001_2 is the 8-bit two's complement of a number k. What is the decimal representation of k?

2.15. The number 10111010_2 is the 8-bit two's complement of a number k. What is the decimal representation of k?

2.16. Using base-2 arithmetic, compute $79 - 43$. Then compute it using 8-bit two's-complement registers. Remember to check for overflow.

2.17. Using base-2 arithmetic, compute $-15 - 46$. Then compute it using 8-bit two's-complement registers. Also compute $46 + 46 + 46$ Remember to check for overflow.

2.18. We have defined and learned how to use the idea of two's complement for n-bit binary numbers. What about "n-digit ten's complement" for base ten arithmetic? Define the appropriate notion of ten's complement and show, by example, how to compute with it in a way that is analogous to computing with two's complement.

Multiple Choice Questions for Review

In each case there is one correct answer (given at the end of the problem set). Try to work the problem first without looking at the answer. Understand both why the correct answer is correct and why the other answers are wrong.

1. Let
   m = "Juan is a math major,"
   c = "Juan is a computer science major,"
   g = "Juan's girlfriend is a literature major,"
   h = "Juan's girlfriend has read Hamlet," and
   t = "Juan's girlfriend has read The Tempest."
   Which of the following expresses the statement "Juan is a computer science major and a math major, but his girlfriend is a literature major who hasn't read both The Tempest and Hamlet."

   (a) $c \wedge m \wedge (g \vee (\sim h \vee \sim t))$

   (b) $c \wedge m \wedge g \wedge (\sim h \wedge \sim t)$

   (c) $c \wedge m \wedge g \wedge (\sim h \vee \sim t)$

   (d) $c \wedge m \wedge (g \vee (\sim h \wedge \sim t))$

   (e) $c \wedge m \wedge g \wedge (h \vee t)$

2. The function $((p \vee (r \vee q)) \wedge \sim(\sim q \wedge \sim r)$ is equal to the function

   (a) $q \vee r$

   (b) $((p \vee r) \vee q)) \wedge (p \vee r)$

   (c) $(p \wedge q) \vee (p \wedge r)$

   (d) $(p \vee q) \wedge \sim(p \vee r)$

   (e) $(p \wedge r) \vee (p \wedge q)$

3. The truth table for $(p \vee q) \vee (p \wedge r)$ is the same as the truth table for

   (a) $(p \vee q) \wedge (p \vee r)$

   (b) $(p \vee q) \wedge r$

   (c) $(p \vee q) \wedge (p \wedge r)$

   (d) $p \vee q$

   (e) $(p \wedge q) \vee p$

4. The Boolean function $[\sim(\sim p \wedge q) \wedge \sim(\sim p \wedge \sim q)] \vee (p \wedge r)$ is equal to the Boolean function

   (a) q (b) $p \wedge r$ (c) $p \vee q$ (d) r (e) p

5. Which of the following functions is the constant 1 function?

   (a) $\sim p \vee (p \wedge q)$

[23]

(b) $(p \wedge q) \vee (\sim p \vee (p \wedge \sim q))$

(c) $(p \wedge \sim q) \wedge (\sim p \vee q)$

(d) $((\sim p \wedge q) \wedge (q \wedge r)) \wedge \sim q$

(e) $(\sim p \vee q) \vee (p \wedge q)$

6. Consider the statement, "Either $-2 \leq x \leq -1$ or $1 \leq x \leq 2$." The negation of this statement is

   (a) $x < -2$ or $2 < x$ or $-1 < x < 1$

   (b) $x < -2$ or $2 < x$

   (c) $-1 < x < 1$

   (d) $-2 < x < 2$

   (e) $x \leq -2$ or $2 \leq x$ or $-1 < x < 1$

7. The truth table for a Boolean expression is specified by the correspondence $(P, Q, R) \rightarrow S$ where $(0,0,0) \rightarrow 0$, $(0,0,1) \rightarrow 1$, $(0,1,0) \rightarrow 0$, $(0,1,1) \rightarrow 1$, $(1,0,0) \rightarrow 0$, $(1,0,1) \rightarrow 0$, $(1,1,0) \rightarrow 0$, $(1,1,1) \rightarrow 1$. A Boolean expression having this truth table is

   (a) $[(\sim P \wedge \sim Q) \vee Q] \vee R$

   (b) $[(\sim P \wedge \sim Q) \wedge Q] \wedge R$

   (c) $[(\sim P \wedge \sim Q) \vee \sim Q] \wedge R$

   (d) $[(\sim P \wedge \sim Q) \vee Q] \wedge R$

   (e) $[(\sim P \vee \sim Q) \wedge Q] \wedge R$

8. Which of the following statements is **FALSE**:

   (a) $(P \wedge Q) \vee (\sim P \wedge Q) \vee (P \wedge \sim Q)$ is equal to $\sim Q \wedge \sim P$

   (b) $(P \wedge Q) \vee (\sim P \wedge Q) \vee (P \wedge \sim Q)$ is equal to $Q \vee P$

   (c) $(P \wedge Q) \vee (\sim P \wedge Q) \vee (P \wedge \sim Q)$ is equal to $Q \vee (P \wedge \sim Q)$

   (d) $(P \wedge Q) \vee (\sim P \wedge Q) \vee (P \wedge \sim Q)$ is equal to $[(P \vee \sim P) \wedge Q] \vee (P \wedge \sim Q)$

   (e) $(P \wedge Q) \vee (\sim P \wedge Q) \vee (P \wedge \sim Q)$ is equal to $P \vee (Q \wedge \sim P)$.

9. To show that the circuit corresponding to the Boolean expression $(P \wedge Q) \vee (\sim P \wedge Q) \vee (\sim P \wedge \sim Q)$ can be represented using two logical gates, one shows that this Boolean expression is equal to $\sim P \vee Q$. The circuit corresponding to $(P \wedge Q \wedge R) \vee (\sim P \wedge Q \wedge R) \vee (\sim P \wedge (\sim Q \vee \sim R)$ computes the same function as the circuit corresponding to

   (a) $(P \wedge Q) \vee \sim R$

   (b) $P \vee (Q \wedge R)$

   (c) $\sim P \vee (Q \wedge R)$

   (d) $(P \wedge \sim Q) \vee R$

   (e) $\sim P \vee Q \vee R$

10. Using binary arithmetic, a number y is computed by taking the n-bit two's complement of $x - c$. If n is eleven, $x = 10100001001_2$ and $c = 10101_2$ then $y =$

(a) 0110000111_2

(b) 0110000110_2

(c) 0110001110_2

(d) 0100011110_2

(e) 0110000000_2

11. In binary, the sixteen-bit two's complement of the hexadecimal number $DEAF_{16}$ is

    (a) 0010000101010111_2

    (b) 1101111010101111_2

    (c) 0010000101010011_2

    (d) 0010000101010001_2

    (e) 0010000101000001_2

12. In octal, the twelve-bit two's complement of the hexadecimal number $2AF_{16}$ is

    (a) 6522_8

    (b) 6251_8

    (c) 5261_8

    (d) 6512_8

    (e) 6521_8

Answers: **1** (c), **2** (a), **3** (d), **4** (e), **5** (b), **6** (a), **7** (d), **8** (a), **9** (c), **10** (b), **11** (d), **12** (e).

Unit Lo

Logic

Logic is the tool for reasoning about the truth and falsity of statements. There are two main directions in which logic develops.

- The first is the depth to which we explore the structure of statements. The study of the basic level of structure is called propositional logic. First order predicate logic, which is often called just predicate logic, studies structure on a deeper level.

- The second direction is the nature of truth. For example, one may talk about statements that are usually true or true at certain times. We study only the simplest situation: a statement is either always true or it is considered false.

"True" and "false" could be replaced by 1 and 0 (or any other two symbols) in our discussions. Using 1 and 0 relates logic to Boolean functions. In fact, *propositional logic* is the study of Boolean functions, where 1 plays the role of "true" and 0 the role of "false." As we saw in Unit BF, Boolean functions can be thought of as computer circuits. Thus, propositional logic, Boolean functions, and computer circuits are different ways of interpreting the same thing.

Propositional logic is not sufficient for all our logic needs. Mathematics requires predicate logic. This and other logics are employed in the design of expert systems, robots and artificial intelligence.

Section 1: Propositional Logic

If it is not fresh in your mind, you should review the material in the first section of Unit BF (Boolean Functions). In that section we were wearing our "arithmetic hat." Now we are wearing our "logic hat" and so refer to things differently:

Arithmetic Hat	Logic Hat
0 and 1, respectively	false and true, respectively
Boolean variable	statement variable
form of function	statement form
value of function	truth value of statement (form)
equality of function (forms)	equivalence of statement forms

We should explain some of these terms a bit more.

- In English, statement variables have structure — verbs, subjects, prepositional phrases, and so on. In propositional logic, we don't see the structure. You're used to that because variables in high school algebra don't have any structure; they just stand for (unknown) numbers.

Logic

- A function can be written in many ways. For example, $xy + x$, $x + yx$, $x(y + 1)$ and $(x + z)y + x - yz$ are all ways of writing the same function. Logicians refer to the particular way a function is written as a *statement form*.

You may wonder why we're concerned with statement *forms* since we're not concerned with function forms in other areas of mathematics but just their values. That is a misconception. We **are concerned** with function forms in algebra. It's just that you're so used to the equality of different forms that you've forgotten that. Knowing that certain forms represent the same function allow us to manipulate formulas. For example, the commutative ($ab = ba$ and $a + b = b + a$) and distributive ($a(b + c) = ab + ac$) laws allow us to manipulate the function forms $xy + x$, $x + yx$ and $x(y + 1)$ to show that they all have the same value; that is, they all represent the same function. As soon as the equality of the function forms is less familiar, you're aware of their importance. For example $(a^u)^v = a^{uv}$, $\sin(2x) = 2 \sin x \cos x$ and $\mathrm{d}(e^x)/\mathrm{d}x = e^x$.

Since some of you may still be confused, let's restate this. For our purposes, we shall say that two statement forms are *different as statement forms*, or simply *different* if they "look different." They are the same if they "look the same." This is not very precise, but is good enough. Thus, for example, $p \vee q$ and $q \vee p$ look different and so are different statement forms. We say that two statement forms are *logically equivalent* (or simply *equivalent*) if they have the same truth table. The statement forms $p \vee q$ and $q \vee p$ are equivalent (have same truth table). Likewise, $(p \wedge q) \vee r$ and $(p \vee r) \wedge (q \vee r)$ are different statement forms that are equivalent, as may be seen by doing a truth table for each form and comparing them. We are familiar with these ideas from high school algebra. For example, $x(y + z)$ and $xy + xz$ look different but are equivalent functions.

Sometimes we'll let our logic hat slip and use Boolean function terminology. In particular, we'll often use 0 instead of "false" and 1 instead of "true."

The constant functions are particularly important and are given special names.

Definition 1 (Tautology, contradiction) *A statement form that represents the constant 1 function is called a tautology. In other words, the statement form is true for all truth values of the statement variables. A statement form that represents the constant 0 function is called a contradiction. In other words, the statement form is false for all truth values of the statement variables.*

Recall the some of the basic functions studied in Unit BF: **not**, **and** and **or**, denoted by $\sim$, $\wedge$ and $\vee$, respectively. We defined these three functions by giving their values in tabular form, which is called a *truth table* just as it is for Boolean functions in Unit BF. In that unit, definitions were as follows, where we have replaced 0 and 1 by F and T to emphasize "false" and "true;" however, we'll usually use 0 and 1.

p	$\sim p$
F	T
T	F

p	q	$p \wedge q$
F	F	F
F	T	F
T	F	F
T	T	T

p	q	$p \vee q$
F	F	F
F	T	T
T	F	T
T	T	T

p	q	p "equals" q
F	F	T
F	T	F
T	F	F
T	T	T

We said there were three functions, but there is a fourth table. Besides, p "equals" q isn't a function—is it? What happened? The statement p "equals" q is either true of false. Thus, we can think of "equals" as a function with domain $\{F,T\}^2$ and range $\{F,T\}$. In symbols, "equals" : $\{F,T\}^2 \to \{F,T\}$. In what follows, we'll replace "equals" with the symbol "$\Leftrightarrow$" (equivalence) which is usually used in logic. We use the more familiar "$=$" for assigning meaning and values. Thus

- $q =$ "the sky is blue" assigns an English meaning to q.

- $q = p \vee r$ says that q "means" $p \vee r$; that is, we should replace q by the statement form $p \vee r$.

- $p = 1$ means we are assigning the value 1 (true) to p.

Since propositional logic can be viewed as the study of Boolean functions, the techniques we developed for proving results about Boolean functions (Venn diagrams, truth tables and algebraic) can also be used in propositional logic. For convenience, we recall the theorem for manipulating Boolean statements:

Theorem 1 (Algebraic rules for statement forms) *Each rule states that two different statement forms are equivalent. That is, they look different but have the same truth table.*

Associative Rules:	$(p \wedge q) \wedge r \Leftrightarrow p \wedge (q \wedge r)$	$(p \vee q) \vee r \Leftrightarrow p \vee (q \vee r)$
Distributive Rules:	$p \wedge (q \vee r) \Leftrightarrow (p \wedge q) \vee (p \wedge r)$	$p \vee (q \wedge r) \Leftrightarrow (p \vee q) \wedge (p \vee r)$
Idempotent Rules:	$p \wedge p \Leftrightarrow p$	$p \vee p \Leftrightarrow p$
Double Negation:	$\sim\sim p \Leftrightarrow p$	
DeMorgan's Rules:	$\sim(p \wedge q) \Leftrightarrow \sim p \vee \sim q$	$\sim(p \vee q) \Leftrightarrow \sim p \wedge \sim q$
Commutative Rules:	$p \wedge q \Leftrightarrow q \wedge p$	$p \vee q \Leftrightarrow q \vee p$
Absorption Rules:	$p \vee (p \wedge q) \Leftrightarrow p$	$p \wedge (p \vee q) \Leftrightarrow p$
Bound Rules:	$p \wedge 0 \Leftrightarrow 0 \qquad p \wedge 1 \Leftrightarrow p$	$p \vee 1 \Leftrightarrow 1 \qquad p \vee 0 \Leftrightarrow p$
Negation Rules:	$p \wedge (\sim p) \Leftrightarrow 0$	$p \vee (\sim p) \Leftrightarrow 1$

Truth tables and algebraic rules are practically the same as the tabular method and algebraic rules for sets discussed in Section 1 of Unit SF. The next example explains why this is so. You may want to read the first four pages of Unit SF now.

Example 1 (Logic and Sets) We've already pointed out that propositional logic and Boolean arithmetic can be viewed as different aspects of the same thing. In this example, we show that basic manipulation of sets are also related.

Suppose we are studying some sets, say P, Q and R. Let the corresponding lower case letters p, q and r stand for the statement that x belongs to the set. For example p is the statement "$x \in P$".

Consider the distributive rule for sets:

$$P \cap (Q \cup R) = (P \cap Q) \cup (P \cap R).$$

It is equivalent to saying that

$$x \in P \cap (Q \cup R) \qquad \text{if and only if} \qquad x \in (P \cap Q) \cup (P \cap R)$$

Logic

for all x in the universal set. What does $x \in P \cap (Q \cup R)$ mean? It means $x \in P$ and $x \in (Q \cup R)$. What does $x \in (Q \cup R)$ mean? It means $x \in Q$ or $x \in R$. Putting this all together and using our logic notation, $x \in P \cap (Q \cup R)$ means $p \wedge (q \vee r)$. Similarly $x \in (P \cap Q) \cup (P \cap R)$ means $(p \wedge q) \vee (p \wedge r)$. Thus the set form of the distributive rule is the same as

$$p \wedge (q \vee r) \Leftrightarrow (p \wedge q) \vee (p \wedge r),$$

the distributive rule for logic.

It should be obvious how things are translated between set notation and logic notation and why we get the same algebraic rules. A practical consequence of this is that we can use Venn diagrams to prove logic statements just as we used them in Unit SF.

Why is the tabular method for proving set identities like a truth table? The answer is simple, consider P, Q and R again. There are exactly eight possibilities for the location of x, corresponding to the eight regions of the Venn diagram. For example, if $x \in P$, $x \notin Q$ and $x \in R$, then the corresponding row in the tabular method for sets begins "Yes No Yes" and the truth table for p, q and r begins $T\ F\ T$. Thus the way to translate between the two methods is Yes$\leftrightarrow T$ and No$\leftrightarrow F$. $\Box$

Implication

"If it is raining, then the sidewalk is wet." This is a simple example of an *implication* statement. Some other forms are "The sidewalk is wet whenever it is raining" and "If the sidewalk isn't wet, then it isn't raining." We want to include implication in propositional logic since statements of the form "If X, then Y" play an important part in logical reasoning. To do so, we must face two problems:

- It is not clear how we should view "If X, then Y" when X is false. For example, what should we think if it isn't raining?

- Due to the variety and ambiguity of English, translation into Boolean statements may not be clear.

In the remainder of this section, we investigate carefully the relationship between English language assertions and Boolean functions (Boolean statement forms) associated with implication.

Let $r =$ "it is raining" and let $w =$ "the sidewalk is wet." In symbolic notation, we use $r \Rightarrow w$ to stand for the statement "If it is raining, then the sidewalk is wet." Usually, when such a statement is made we are primarily concerned with the situation when r is true. For the study of logic, we must be concerned with all situations, so we need to know how to think about $r \Rightarrow w$ when r is false. If it is not raining, the sidewalk may be wet (it rained earlier, the sprinklers are on, etc.) or the sidewalk may not be wet. Thus when r is false, we have no reason to disbelieve the statement $r \Rightarrow w$. Of course, we have no reason to believe it either, so we are free to choose whatever we want for the truth value in this case. We take the generous approach and call $r \Rightarrow w$ true when r is false. (Actually we're not being generous—this is the standard interpretation.)

Let's put all this into a definition.

Definition 2 (Implication)   *We define $p \Rightarrow q$ to be the Boolean function, called impli-cation or p implies q or the "conditional of q by p." As a Boolean function, $p \Rightarrow q$ has the following truth table:*

p	q	$p \Rightarrow q$
0	0	1
0	1	1
1	0	0
1	1	1

The expression $p \Rightarrow q$ is a Boolean statement form. It is equivalent to the statement form $\sim p \vee q$, as can be seen by comparing truth tables:

p	q	$\sim p \vee q$
0	0	1
0	1	1
1	0	0
1	1	1

Note that the Boolean function $f(p,q)$ defined by $p \Rightarrow q$ has value 1 when $p = 0$, independent of the value of q.

We've been a bit sloppy: we've said $r =$ "It is raining" and also, by the definition, $r = 1$. Clearly "It is raining" is not the same as "1." What's going on? Since we are studying the truth values of statements, we should have said

$$r = \text{the truth value of "It is raining."}$$

We'll continue with the common practice of abusing the terminology by omitting the words "the truth value of."

Example 2 (Implication, rain and Venn diagrams)   Let R be the set of all times when it is raining and let W be the set of all times when the sidewalk is wet. What does our earlier example $r \Rightarrow w$ say about the sets R and W? Suppose t is a time; that is, an element of our universal set of all times.

- If $t \in R$, then it is raining at time t and so, by $r \Rightarrow w$, $t \in W$. Thus $R \subseteq W$.

- If $t \notin R$, then it is not raining at time t. In this case, $r \Rightarrow w$ gives us no information about whether or not t is in W. Why is that? When r is false, $r \Rightarrow w$ is true regardless of whether or not w is true; that is, whether or not t is in W.

What happened to the 0 case of $r \Rightarrow w$ in the definition of implication? That is the case $t \in R$ (because r is true) and $t \notin W$ (because w is false). Since the definition of implication says this is false, it says that this cannot happen. This is a consequence of $R \subseteq W$.

This is the way we represent $p \Rightarrow q$ with Venn diagrams: There is a set P where p is true and another set Q where q is true and we insist that $P \subseteq Q$.

Logic

How can we show that $p \Rightarrow q$ is not true for specific p and q? (For example, "If it is raining, then my car won't start.") We must find an instance where p is true and q is false because this is the only time $p \Rightarrow q$ is false. (It's raining, but my car starts.) You'll see more of this later. ◻

Example 3 (Statement forms associated with implication) Here is a table that defines the basic statement forms associated with the conversational use of implication.

p	q	$p \Rightarrow q$	$\sim q \Rightarrow \sim p$	$q \Rightarrow p$	$\sim p \Rightarrow \sim q$	$p \Leftrightarrow q$
0	0	1	1	1	1	1
0	1	1	1	0	0	0
1	0	0	0	1	1	0
1	1	1	1	1	1	1

Starting with the statement form $p \Rightarrow q$, the statement form $\sim q \Rightarrow \sim p$ is called the *contrapositive* of $p \Rightarrow q$, the statement form $q \Rightarrow p$ is called the *converse* of $p \Rightarrow q$, and the statement form $\sim p \Rightarrow \sim q$ is called the *inverse* of $p \Rightarrow q$. Note that, although the statement and its contrapositive are different statement forms (they look different) they are equivalent (i.e., the same as Boolean functions). Likewise, the converse and the inverse are equivalent. But, and this is important, the statement and its converse are **not** equivalent. The final statement form $p \Leftrightarrow q$ is called *double implication* or "biconditional" and is equivalent to $(p \Rightarrow q) \wedge (q \Rightarrow p)$.

Here, in tabular form, are some additional facts related to implication.

p	q	$p \Rightarrow q$	$\sim p \vee q$	$\sim(p \Rightarrow q)$	$p \wedge \sim q$
0	0	1	1	0	0
0	1	1	1	0	0
1	0	0	0	1	1
1	1	1	1	0	0

The statement forms $p \Rightarrow q$ and $\sim p \vee q$ are equivalent as are $\sim(p \Rightarrow q)$ and $p \wedge \sim q$. ◻

Example 4 (Right triangles and the Pythagorean theorem) Throughout this example, suppose $0 < a \le b \le c$ are some fixed numbers. Take $R =$ "The triangle with side lengths a, b, c is a right triangle" and $P =$ "$a^2 + b^2 = c^2$." We know from high school that "If R then P." As a Boolean statement form we may write $R \Rightarrow P$. If you proved this fact by starting with a right triangle and using a geometric argument to show that $a^2 + b^2 = c^2$, then the statement form $R \Rightarrow P$ represented the state of your knowledge at that point in time. You then probably went on to learn the law of cosines: $a^2 + b^2 - 2\,ab\,\cos(\theta) = c^2$. Using that, you can easily see that the converse $P \Rightarrow R$ is true. Now you can represent the state of your knowledge by $R \Leftrightarrow P$.

The statement form $R \Rightarrow P$ is equivalent to $\sim R \vee P$. Either a triangle is not a right triangle or it satisfies $a^2 + b^2 = c^2$.

Start with the statement, "If the triangle with side lengths a, b, c is a right triangle, then $a^2 + b^2 = c^2$."

- The contrapositive of that statement is "If $a^2 + b^2 \neq c^2$, then the triangle with side lengths a, b, c is not a right triangle."

- The converse is, if $a^2 + b^2 = c^2$, then the triangle with side lengths a, b, c is a right triangle."

- The inverse is, "If the triangle with side lengths a, b, c is a not right triangle, then $a^2 + b^2 \neq c^2$." $\square$

Example 5 (The many English forms for $p \Rightarrow q$) In this example we'll discuss most of the ways implication is written in English. Pay careful attention to when we use the phrase "statement form" and when we use the phrase "Boolean function." Be sure to read the last part of the example where we discuss the distinction between statement form and Boolean function further.

- *if ... then*: The basic English form, "If p then q," is understood to stand for the statement form $p \Rightarrow q$. Note that the "if" is associated with p. Alternatively to this, one sees "q if p." Again, the "if" is associated with p, so this stands for the statement form $p \Rightarrow q$. Thus, "If it's raining, then it's cloudy" is interpreted as the same statement form as "It's cloudy if it's raining." Both stand for the statement form "raining" $\Rightarrow$ "cloudy."

- *only if*: Sometimes we say "p only if q," as in "I'll go to the party only if I finish studying." Some people would paraphrase this as, "If I don't finish studying, then I won't go to the party." In other words "p only if q" is translated into the statement form $\sim q \Rightarrow \sim p$. This statement form is equivalent, as a Boolean function, to the statement form $p \Rightarrow q$, because an implication form is equivalent to its contrapositive form. In other words, "p only if q", however it is interpreted as a statement form, is equivalent as a Boolean function to "If I go to the party, then I finished studying." Thus the phrase "p only if q" can be translated *as a Boolean function* into either one of the equivalent statement forms $\sim q \Rightarrow \sim p$ or $p \Rightarrow q$, whichever is most convenient for the discussion at hand.

- *if and only if*: The biconditional, $p \Leftrightarrow q$ is sometimes stated as "p if and only if q" and written "p iff q".

- *sufficient*: The expression, "p is sufficient for q"(or "p is a sufficient condition for q") is usually translated into the statement form $p \Rightarrow q$. Some students find it helpful to (silently to themselves) expand this phrase to "p is sufficient *to force q to happen*." Then it is easier to remember that this means $p \Rightarrow q$. Instead of saying "p is sufficient for q", one sometimes says "a sufficient condition for q is p."

- *necessary*: The statement "p is necessary for q" usually stands for the statement form $\sim p \Rightarrow \sim q$. Some students (again silently to themselves) expand this to "p is a necessary consequence of q." They find this easier to associate with the equivalent (as a Boolean function) form $q \Rightarrow p$. Instead of saying "q is necessary for p", one says "a necessary condition for p is q."

- *necessary and sufficient*: Combining the two previous bulleted items, we see that "p is necessary and sufficient for q" is equivalent to $p \Leftrightarrow q$, the biconditional. Notice that we simply combined "necessary" and "sufficient", just as we combined "if" and "only if" earlier to get the biconditional.

- *unless*: Another possible source of confusion is the term "unless." To say "p unless q" is, formally, to specify the statement form $\sim q \Rightarrow p$. The most common usage of "unless" in English is something like, "The building is safe, unless the fire alarm is ringing." Formally, this means, "If the fire alarm is not ringing, then the building is safe." Think of a night watchman sitting in his office with the fire alarm on the wall. Since the alarm isn't ringing he relaxes, maybe even takes a nap. His assumption is that "If the fire alarm is not ringing then all is well, the building is safe, I can relax." If you asked him, "Why are you sleeping?" he might reply, "The building is safe unless the fire alarm is ringing."

  An equivalent Boolean function is $\sim p \Rightarrow q$, the contrapositive of $\sim q \Rightarrow p$. Thus we have "If the building is not safe, then the fire alarm is ringing." Note the symmetry in translating "p unless q" into either of the equivalent forms $\sim p \Rightarrow q$ or $\sim q \Rightarrow p$. In translating "p unless q" into a Boolean function, simply apply "$\sim$" to one of p or q and have that imply the other without applying "$\sim$".

Let's review the role of the concepts of a "statement form" and a "Boolean function" in the above discussion.

(a) Generally, when we are translating an English description of an implication into symbolic form, we are concerned most of all with obtaining the correct Boolean function. With a little practice you will find this easy to do.

(b) In the rare case when we are being pedantic and want to know if some statement form is the contrapositive, converse, or inverse, of an implication described in English, then we need to associate a precise *statement form* with the English sentence. Our policy will be to always give you that statement form when you need to know it for the discussion or question. The one exception to this policy is the case of "if p then q" (or "q if p") which we always associate with the statement form $p \Rightarrow q$.

Thus, in most cases, as with other English usages, all you will need to be able to do is translate "if p then q" into an equivalent form as a Boolean function: $p \Rightarrow q$, $\sim q \Rightarrow \sim p$, $\sim p \vee q$, etc. $\quad\square$

Example 6 (A way of translating English implications) Of course, you can memorize the rules from the previous example (and that may be a good idea), but what if you forget or if you run into something new? Suppose we see a sentence that relates two phrases A and B; for example, "If A then B" or "A requires B." Suppose we also realize that an implication is involved. How can we determine whether to write $A \Rightarrow B$ or $B \Rightarrow A$ or some other implication?

Here's a trick: The truth table for $p \Rightarrow q$ has only one row which is false and that occurs when p is true and q is false. Take your sentence and figure out how to make it false and set things up so that it corresponds to (True)$\Rightarrow$(False).

Let's do some examples.

What about "A requires B?" Consider "Fishes require water." This is false if something is a fish and does not require water. In general "A requires B" is false when A is true and B is false. Thus, we have $A \Rightarrow B$.

What about "A is necessary for B?" Consider "Enrollment is necessary for credit." This is false if I receive credit even though I am not enrolled. In other words "A is necessary for B" is false when B is true and A is false. Thus we can write it as $B \Rightarrow A$.

What about "*A* unless *B*?" Consider "I will flunk unless I study." This is false when I don't flunk and I don't study. Thus, it is false when *A* and *B* are both false. Since we need (True)⇒(False), we need to negate something. One possibility is ~*A* ⇒ *B* and another is ~*B* ⇒ *A*. Which is correct? They both are — one is the contrapositive of the other. However, they sound different in English. Compare "If I passed, then I studied" and "If I don't study, then I won't pass." The first is celebration after the fact and the second is a warning about what I should do.

What about "*A* or *B*?" Wait! There's no implication here. In logic all that matters is the truth table. Any statement form involving two variables that is false in only one of the four cases can be written as an implication. "*A* or *B*" is false only when both *A* and *B* are false. Thus ~*A* ⇒ *B*. You've actually seen this before: we learned that $p \Rightarrow q$ and $\sim p \vee q$ are equivalent, so set $p = \sim A$ and $q = B$. □

Exercises for Section 1

1.1. In Example 1 we noted that algebraic operations in propositional logic, set theory and Boolean functions can be viewed as different aspects of the same thing. What logic and set operations correspond to the exclusive or operation for Boolean functions?

1.2. Let h = "he is happily married," and w = "he is wealthy," and s = "he is smart." Write the following statements in symbolic form:

(a) He is happily married and wealthy but not smart.

(b) He is not wealthy, but he is happily married and smart.

(c) He is neither happily married, nor wealthy, nor smart.

1.3. Let n = "Nancy will major in computer science" and k = "Karen will major in computer science." Write the following statement in symbolic form: Either Nancy will major in computer science or Karen will major in computer science, but not both.

1.4. Let p = "computer is out of memory flag, COM, is zero (i.e., off)," and q = "disk error has occurred flag, DEO, is zero," and r = "ZIP disk hasn't enough memory left flag, ZIP, is zero." Write the following statements in symbolic form:

(a) COM is off and DEO is off and ZIP is off.

(b) COM is off but DEO is on.

(c) There is enough memory in the computer; however, either a disk error has occurred or the ZIP disk is out of memory.

(d) The computer is out of memory and no disk error has occurred, but the ZIP disk is out of memory.

(e) Either the computer is out of memory or both COM == 0 and DEO == 0.

1.5. Is the statement form $(p \wedge q) \vee (\sim p \vee (p \wedge \sim q)) \vee r$ a tautology, contradiction, or neither?

1.6. Is the statement form $(p \wedge \sim q) \wedge (\sim p \vee q) \wedge r$ a tautology, contradiction, or neither?

1.7. Is the statement form $((\sim p \wedge q) \wedge (q \vee r)) \wedge \sim q \wedge r$ a tautology, contradiction, or neither?

1.8. Construct a truth table for $p \vee (\sim p \wedge q) \Rightarrow q$.

1.9. Construct a truth table for $p \vee (\sim p \wedge q) \Rightarrow \sim q$.

1.10. Construct a truth table for $(p \Rightarrow q) \Rightarrow (q \Rightarrow p)$.

1.11. Write negations of the following statements in English that is as easily understood as possible.

(a) If P is a pentagon then P is a polygon.

(b) If Tom is Ann's father, then Jim is Ann's uncle and Sue is her aunt and Mary is her cousin.

1.12. Write the converses and inverses for the statements in the previous exercise.

1.13. Why is the assertion, "There is some statement $p \Rightarrow q$ that is not equivalent to its contrapositive," equivalent to the statement, "There is some statement $p \Rightarrow q$ whose converse is not equivalent to its inverse?" (Note: both statements are false.)

1.14. Write the contrapositives for the statements in Exercise 1.11.

1.15. Write the contrapositive of the statement "Dennis won't enter the America's Cup unless he is sure of victory." Use the interpretation of "p unless q" as the statement form $\sim p \Rightarrow q$.

1.16. You were told by your high school principal that you will "graduate with honor" (call that H) only if you either "make the honor roll each semester" (M) or "complete all language requirements" (C), and if, in addition, you "get straight A's in biology" (B) and "letter in at least one athletic activity" (A). You lettered in track, got straight A's in all your science classes (including biology), and completed all language requirements, but at graduation you were not given any honors. Did your high school principal lie to you?

1.17. Write two different statement forms using "if" and "then" that are equivalent to the following: "Learning to program in C is a necessary condition for learning to program in C++. "

1.18. Given $(\sim p \vee q) \Rightarrow (r \vee \sim q)$, rewrite it as a statement form using only $\sim$ and $\wedge$.

1.19. Given $\Big((p \Rightarrow (q \Rightarrow r)) \Leftrightarrow ((p \wedge q) \Rightarrow r) \Big) \wedge \sim p \wedge \sim q \wedge \sim r$, rewrite it using only $\sim$ and $\vee$.

1.20. Start with the statement form "Getting up when the alarm rings is a sufficient condition for me to get to work on time." Rewrite it in an equivalent if- then form.

1.21. Start with the statement form, "Having sides of length 3, 4, and 5 is a sufficient condition for this triangle to be a right triangle." Rewrite it in an equivalent if-then form.

1.22. Start with the statement form, "Doing all of the programming assignments is a necessary condition for Jane to pass her Java course." Rewrite this statement in an equivalent if-then form.

1.23. "If the program is running then there is at least 250K of RAM." Which of the following are equivalent to this statement?

(a) If there is at least 250K of RAM then the program is running.

(b) If there is less than 250K of RAM then the program is not running.

(c) The program will run only if there is at least 250K of RAM.

(d) If the program is not running then there is less than 250K of RAM.

(e) A necessary condition for the program to run is that there are at least 250K of RAM.

(f) A sufficient condition for the program to run is that there is at least 250K of RAM.

Section 2: Predicate Logic

We have been studying statements that are either true or false. But, consider the statement
"$x^2 > 1$." In order to decide if this statement is true of false, we need to know the numerical
value of x. If $x = 1.1$, then "$x^2 > 1$" is true. If $x = 0.9$, then "$x^2 > 1$" is false. The best
way to think of this is to regard the statement "$x^2 > 1$" as a function $S(x) = $ "$x^2 > 1$." If
we take this point of view, we need to specify the domain of S. First suppose the domain
of S is $\mathbb{R}$, the set of all real numbers. The codomain (or range) of S, by our description
just given, is a set of statements that are either true or false (e.g., $S(0.9) = $ "$0.9^2 > 1$",
$S(2.3) = $ "$2.3^2 > 1$"). The function S is an example of a *predicate*.

Definition 3 (Predicate and truth set) *A predicate is any function whose codomain
is statements that are either true or false. There are two things to be careful about:*

- *The codomain is statements **not** the truth value of the statements.*

- *The domain is arbitrary — different predicates can have different domains.*

*The truth set of a predicate S with domain D is the set of those $x \in D$ for which $S(x)$ is
true. It is written*

$$\{x \in D \mid S(x) \text{ is true}\} \quad \text{or simply} \quad \{x \mid S(x)\}.$$

Note that $S(0.9) = $ "$0.9^2 > 1$" is a correct statement consistent with the way we
commonly use functional notation. But $S(0.9) = $ FALSE or $S(0.9) = 0$ is not a correct
statement even though "$0.9^2 > 1$" is false. This is because the codomain of S is a set of
statements, not the set $\{0, 1\}$. Instead of "$S(0.9) = 0$" we should say "$S(0.9)$ is false." Likewise, we say "$S(1.1)$ is true." These are sometimes shortened to "$\sim S(0.9)$" and "$S(1.1)$."

The expression $\{x \mid S(x)\}$ may look strange, but it is consistent with the usual use of
the notation. If the domain is known, there is no need to mention it and $\{x \mid \ldots\}$ means
the set of those x for which $\ldots$ is true. The truth set of the predicate $S(x) = $ "$x^2 > 1$"
with domain $\mathbb{R}$ is the set $\{x \mid x > 1\} \cup \{x \mid x < -1\}$.

With some domains, it is more natural to think of a predicate as a function of more
than one variable. For example, the domain may be $\mathbb{R} \times \mathbb{R}$ and the predicate may be
"$P(x,y) = \left((x > y > 0) \Rightarrow (x^2 > y^2)\right)$." Notice that $P(x,y)$ is true for all $x,y \in \mathbb{R}$.
In other words "For all $x,y \in \mathbb{R}, S(x,y)$" is true. This sort of statement is the essence of
predicate logic, so we introduce some terminology.

Definition 4 (Quantifiers) *The phrase "for all" is called a universal quantifier and
is written $\forall$ ("A" rotated $180°$). If $S(x)$ is a predicate and the set D is contained in the
domain of x, the statement "$\forall x \in D, S(x)$" is read "for all $x \in D, S(x)$ is true," or just
"for all $x \in D, S(x)$." The statement "$\forall x \in D, S(x)$" is true if and only if $S(x)$ is true for
every $x \in D$; otherwise the statement "$\forall x \in D, S(x)$" is false. If the value of D is clear,
we may write simply $\forall x\, S(x)$.*
  The phrase "for some" is called an existential quantifier and is written $\exists$ ("E" rotated

180°). If $S(x)$ is a predicate and the set D is contained in the domain of x, the statement "$\exists x \in D, \; S(x)$" is read "for some $x \in D$, $S(x)$ is true," or just "for some $x \in D$, $S(x)$." It is also read "there exists $x \in D$ such that $S(x)$." The statement "$\exists x \in D, \; S(x)$" is true if and only if $S(x)$ is true for at least one $x \in D$; otherwise the statement "$\exists x \in D, \; S(x)$" is false. If the value of D is clear, we may write simply $\exists x \; S(x)$.

In terms of truth sets:

- "$\forall x \in D, \; S(x)$" is equivalent to saying that the truth set of $S(x)$ contains the set D.

- "$\exists x \in D, \; S(x)$" is equivalent to saying that the truth set of $S(x)$ contains at least one element of the set D

One can view much of mathematics as an attempt to understand the truth sets of certain predicates. For example, can you describe the truth set of the predicate $S(b,c) =$ "x^2+bx+c has no real roots"? You can answer this if you know that "the roots of $x^2 + bx + c$ are $(-b \pm \sqrt{b^2 - 4c})/2$" is true for all $(b,c) \in \mathbb{R} \times \mathbb{R}$ and that "$(\sqrt{d} \in \mathbb{R}) \Leftrightarrow (d \geq 0)$" is true for all $d \in \mathbb{R}$. The answer is $\{(b,c) \mid b^2 < 4c\}$.

To work with the notation and also introduce ideas we will need later, we'll look at some examples from elementary number theory. The word "elementary" here means easy to state, not, necessarily, easy to solve. To make it easier to specify domains, we need some notation.

Definition 5 (Notation for sets of numbers) *Recall that $\mathbb{R}$ denotes the real numbers, $\mathbb{Z}$ denotes the integers, and $\mathbb{Q}$ denotes the rational numbers (ratios of integers). In addition, $\mathbb{N}$ denotes the nonnegative integers (the "natural numbers"), $\mathbb{N}^+$ denotes the nonzero natural numbers (positive integers), and $\mathbb{P}$ denotes the primes. A natural number n is prime if $n \geq 2$ and the only divisors of n are n and 1. An integer $n \geq 2$ that is not prime is composite.*

The number 2 is the smallest prime and the only even prime. The other primes less than 20 are $3, 5, 7, 11, 13, 17, 19$.

Example 7 (Goldbach's conjecture) A mathematician named Christian Goldbach (1690–1764), noticed that $4 = 2+2$, $6 = 3+3$, $8 = 3+5$, $10 = 5+5$, $12 = 5+7$, $14 = 7+7$, $16 = 5+11$, etc., making him think that every even number greater than or equal to 4 can be written as the sum of two primes. We can state this in our notation:

$$\forall\, n \in \mathbb{N}, \; \Big((n \geq 4) \wedge (n \text{ even})\Big) \Rightarrow \Big(\exists\, p, q \in \mathbb{P}, \; n = p+q\Big).$$

Goldbach made this conjecture in 1742 in a letter to Euler (1701-1783). Of course, it can be written in various other ways; for example,

$$\forall\, n \geq 2, \; \exists\, p, q \in \mathbb{P}, \; 2n = p+q,$$

where it is understood from the context that n must be an integer and not something like $\sqrt{5}$ or π.

[39]

Lo-13

Logic

Sadly (for mathematicians, since few others are interested) it is unknown whether or not Goldbach's conjecture is true. At least we have learned how to make the assertion, if not how to prove or disprove it. However, something is known for odd numbers: It is known that

$$\exists\, K \in \mathbb{N},\ \forall\, n \geq K,\ \Big((n \text{ odd}) \Rightarrow \exists\, p, q, r \in \mathbb{P},\ n = p + q + r\Big)$$

is true. This can be stated as "every sufficiently large odd number is the sum of three primes." (The "sufficiently large" is due to "$n \geq K$.") This was proved by Ivan Vinogradov (1891–1983) in 1937. $\square$

Example 8 (Sets and logic again) In Example 1 we saw how set identities could be thought of in terms of propositional logic. We can also phrase this in predicate logic terms.

Let U be the universal set. For every set A, B, C and so on that is being considered, introduce the predicates $A(x), B(x), C(x)$ and so on. Define $A(x)$ to be true if and only if $x \in A$ and do likewise for the other predicates.[1] A statement about sets is now equivalent to the corresponding statement about predicates with a universal quantifier. For example, $\sim(A \cup B) = (\sim A \cap \sim B)$ is true if and only if

$$\forall\, x \in U,\ \Big(\sim(A(x) \vee B(x)) \Leftrightarrow (\sim A(x) \wedge \sim B(x))\Big).$$

Why is that? The logic statement asserts that $x \in \sim(A \cup B)$ if and only if $x \in (\sim A \cap \sim B)$. This is essentially the element method of proof. $\square$

Example 9 (Quantifiers and negation) Let $R(x) =$ "$x + 2$ is prime" be a predicate. The statement "$\forall\, n \in \mathbb{P},\ R(n)$" is an example of the universal quantifier "for all" applied to this predicate. Another way to say the same thing is "$\forall(n \in \mathbb{P}),\ (n+2 \in \mathbb{P})$." We have used parentheses to make it easier to see that the predicate is $n + 2 \in \mathbb{P}$ and that $n \in \mathbb{P}$ belongs with the quantifier. You should practice inserting parentheses in what follows to make it easier to read.

Using the normal English meanings of the statements, you should be able to see that the negation of these statements is "$\exists\, n \in \mathbb{P},\ \sim R(n)$," which can be written "$\exists\, n \in \mathbb{P},\ n + 2 \notin \mathbb{P}$." Both negation statement forms mean the same thing. The symbol "$\notin$" is the negation of $\in$. Since $\in$ stands for "is in" or "is an element of," $\notin$ stands for "is not in" or "is not an element of."

In this case, "$\forall\, n \in \mathbb{P},\ R(n)$" is false and "$\exists\, n \in \mathbb{P},\ \sim R(n)$" is true since $7 \in \mathbb{P}$ and $9 \notin \mathbb{P}$. When $\exists$ is read "there exists," the symbol $\ni$ is sometimes used for "such that." Thus we can write either "$\exists\, n \in \mathbb{P},\ n + 2 \notin \mathbb{P}$" or "$\exists\, n \in \mathbb{P} \ni n + 2 \notin \mathbb{P}$." $\square$

The negation of a "for all" to get a "for some" in the previous example is an application of the following theorem for moving negation through quantifiers. You should be able to

---

[1] If you remember the definition of "characteristic function," you should be able to see that $A(x)$ is simply the characteristic function for the set A.

see that the theorem is true by translating the notation into ordinary English. We omit the formal proof

Theorem 2 (Negating quantifiers)   *Let D be a set and let $P(x)$ be a predicate that is defined for $x \in D$. Then*

$$\sim\!\Big(\forall(x \in D),\ P(x)\Big) \ \Leftrightarrow\ \Big(\exists(x \in D),\ \sim\!P(x)\Big)$$

and

$$\sim\!\Big(\exists(x \in D),\ P(x)\Big) \ \Leftrightarrow\ \Big(\forall(x \in D),\ \sim\!P(x)\Big)$$

Example 10 (You can't buy it here)   A grocery store chain has the disclaimer

ALL ITEMS NOT AVAILABLE AT ALL STORES.

in its weekly flyer of specials. What did they say and how could they have said what they meant?

Let I be the set of items referred to and S the set of stores. Let $A(i,s)$ be the predicate indicating that item i is available at store s. To translate the statement, we need to know how NOT should be applied. If the interpretation is

ALL ITEMS NOT (AVAILABLE AT ALL STORES),

then we can rewrite the statement as $\forall i \in I,\ \sim\!\Big(\forall s \in S,\ A(i,s)\Big)$, which our theorem tells us is equivalent to $\forall i \in I,\ \exists s \in S,\ \sim\!A(i,s)$. In English this says that, for every item in the flyer, the company has at least one store where you won't be able to get it. That's not a good way to run a business, so our choice of parentheses must be wrong.

The other possibility is

ALL ITEMS (NOT AVAILABLE) AT ALL STORES,

which translates as $\forall i \in I, \forall s \in S,\ \sim\!A(i,s)$. This is even worse! In English it says no matter what item you look for and no matter what store you look in, the item won't be available.

It seems fairly obvious that what they want to say is $\sim\!\forall i \in I,\ \forall s \in S,\ A(i,s)$. In other words, it is not the case that all items are available at all stores. This is rather awkward. Moving the negation through the quantifiers, we obtain $\exists i \in I,\ \exists s \in S,\ \sim A(i,s)$, which can be written as

SOME ITEMS ARE UNAVAILABLE AT SOME STORES.

Notice that we have written "UNAVAILABLE" instead of "NOT AVAILABLE" to avoid the problem of where to put parentheses that we considered in the two previous paragraphs.   □

Logic

Example 11 (Twin primes) Let $S(x) =$ "x and $x+2$ are prime" be a predicate.[2] If $S(x)$ is true, we call x and $x+2$ twin primes. We could rewrite $S(x)$ as "$(x \in \mathbb{P}) \wedge (x+2 \in \mathbb{P})$."

The Twin Prime conjecture asserts that there are infinitely many twin primes. How can we express this in our notation since we do not have the phrase "infinitely many?" Here is a precise way of stating the Twin Prime conjecture:

"For all $m \in \mathbb{N}$, there exists $n \in \mathbb{N}$ such that $n \geq m$ and $S(n)$."

Using the symbols we've just learned, we can rewrite it as

$$\text{"} \forall \, m \in \mathbb{N} \, \Big(\exists \, n \in \mathbb{N} \ni \big((n \geq m) \wedge S(n) \big) \Big). \text{"}$$

We often combine $\exists \, n \in \mathbb{N}$ and $n \geq m$ and often omit the $\ni$:

$$\text{"} \forall \, m \in \mathbb{N}, \ \exists \, n \geq m, \ S(n). \text{"}$$

If you are puzzled why this states that there are infinitely many primes n such that $n+2$ is also a prime, it helps to look at the negation. The negation of the statement is

$$\sim\!\Big(\forall \, m \in \mathbb{N}, \ \exists \, n \geq m, \ S(n) \Big) \quad \Leftrightarrow \quad \exists \, m \in \mathbb{N}, \ \sim\!\Big(\exists \, n \geq m, \ S(n) \Big)$$
$$\Leftrightarrow \quad \exists \, m \in \mathbb{N}, \ \forall \, n \geq m, \ \sim\!S(n).$$

Note that we applied Theorem 2 twice: the first time to move $\sim$ inside "$\forall \, m \in \mathbb{N}$" and the second time to move $\sim$ inside "$\exists \, n \geq m$."

Let's look at our negative statement $\exists \, m \in \mathbb{N}, \ \forall \, n \geq m, \ \sim\!S(n)$. If there were only finitely many primes p such that $p+2$ is also prime, we could take m to be bigger than the largest such p, and the negative would be proved. On the other hand, if there were infinitely many twin primes, no matter how we chose m, there would be larger twin primes (i.e., larger n so that $S(n)$ is true) and so the negative would not be true. $\square$

Example 12 (Fermat numbers) A Fermat number (Pierre de Fermat, 1601–1665) is an integer of the form $F_n = 2^{2^n} + 1$ for $n \in \mathbb{N}$. The first five Fermat numbers are $F_0 = 3$, $F_1 = 5$, $F_2 = 17$, $F_3 = 257$, $F_4 = 65537$, and $F_5 = 4294967297$. In 1640 Fermat conjectured that F_n is a prime for every n. Let $S(n) =$ "$F_n \in \mathbb{P}$." If we want to assert that all F_n are primes, we would say, "$\forall \, n \in \mathbb{N}, \ S(n)$," or "$\forall \, n \in \mathbb{N}, \ F_n \in \mathbb{P}$." The negation of this statement is, "$\exists \, n \in \mathbb{N}, \ F_n \notin \mathbb{P}$." The negation is true, since $F_5 = 4294967297 = 641 \times 6700417$. F_5 is the first composite (i.e. non-prime) Fermat number. It is easy to factor F_5 with modern computers, but it was hard when people computed by hand. Thus Fermat was led to the false conjecture "$\forall n \in \mathbb{N}, \ S(n)$." Are there infinitely many Fermat primes (i.e., Fermat numbers that are prime)? If we thought so, we would conjecture "$\forall \, m \in \mathbb{N}, \ \exists \, n \geq m, \ S(n)$." If we thought not, we would conjecture "$\exists \, m \in \mathbb{N}, \ \forall \, n \geq m, \ \sim\!S(n)$." No one knows which assertion is correct. It is known that F_6 through F_{20} are, like F_5, composite.

---

[2] This is not the same as the predicate R in Example 9. Explain why. Express $S(x)$ in terms of R.

High school math students who take geometry often suffer through "straight edge and compass" constructions of various geometric figures. The ancient Greeks figured out that any regular polygon with $3 \cdot 2^k$ sides, $k \in \mathbb{N}$, or with $5 \cdot 2^k$ sides, $k \in \mathbb{N}$ could be constructed with straight edge and compass. This led them to wonder (in Greek, of course) "For which n can a regular polygon with n sides be constructed with straight edge and compass?" Let $P(n)$ be the predicate "a regular n-sided polygon can be constructed with straight edge and compass." We have just said that the Greeks proved $\forall \, k \in \mathbb{N}$, $\bigl(P(3 \cdot 2^k) \wedge P(5 \cdot 2^k)\bigr)$ and they wondered what the truth set of $P(n)$ is. In 1796, Karl Friedrich Gauss (1777–1855), then 18 years old, proved that it is possible to construct a polygon with $m \cdot 2^k$ sides ($k \in \mathbb{N}$ and m odd) using ruler and compass if and only if m is a product of distinct Fermat primes, including $m = 1$, the empty product. (In other words, he found the truth set for $P(n)$.) Thus, such constructions are known to be possible for $m = 1$, 3, 5, 17, 257, 65535, 3×5, 5×17, and so on up to $3 \times 5 \times 17 \times 257 \times 65535 = 4294967295$. If more Fermat primes are found, we can add to this list. It should be noted that, although only 18, Gauss worked very hard at math. $\square$

Example 13 (Mersenne primes and perfect numbers) A number of the form $M_p = 2^p - 1$, where $p \in \mathbb{P}$, is called a *Mersenne number* after Marin Marsenne (1588–1648). If M_p is prime, then it is called a *Mersenne prime*. $M_2 = 3$, $M_3 = 7$, $M_5 = 31$ and $M_7 = 127$ are Mersenne primes. But, $M_{11} = 23 \times 89$ is not a prime. The first thirty-one values of p for which M_p is prime are 2, 3, 5, 7, 13, 17, 19, 31, 61, 89, 107, 127, 521, 607, 1279, 2203, 2281, 3217, 4253, 4423, 9689, 9941, 11213, 19937, 21701, 23209, 44497, 86243, 110503, 132049, 216091. (The 25th and 26th Mersenne primes, corresponding to $p = 21701$ and $p = 23209$, were discovered in 1978 by two high school students, Laura Nickel and Curt Noll.) It is not known whether or not the statement "$\forall \, k \in \mathbb{N}$, $\exists \, p \in \mathbb{P}$, $\bigl((p \geq k) \wedge (M_p \in \mathbb{P})\bigr)$" is true or false. (What does this statement assert about the number of Mersenne primes?)

A mathematician named Leonhard Euler (1701–1783) studied numbers called *perfect numbers* and found a remarkable connection between them and Mersenne primes in 1770. A perfect number $n \in \mathbb{N}^+$ is a number that is equal to the sum of all of its factors (other than itself). The smallest perfect number is $6 = 3 + 2 + 1$. The next smallest is $28 = 14 + 7 + 4 + 2 + 1$. It is known that

$$\forall \, n \in \mathbb{N}^+, \; \Bigl(\exists \, p \in \mathbb{P}, \; \bigl((n = 2^{p-1} M_p) \wedge (M_p \in \mathbb{P})\bigr) \; \Leftrightarrow \; (n \text{ is even and perfect})\Bigr).$$

Euclid knew the "if-then" part (the left side implies the right). Euler proved the reverse implication — a gap of about two millennia! Thus $2^{2-1} M_2 = 6$, $2^{3-1} M_3 = 28$, $2^{5-1} M_5 = 496$ and $2^{7-1} M_7 = 8128$ are the first four even perfect numbers.

One could make the statement, "$\forall \, k \in \mathbb{N}$, $\exists \, n > k$, (n is even and perfect)." Is that statement known to be true or false? What do you think and why?

One could also make the statement, "$\exists \, n \in \mathbb{N}^+$, ($n$ is odd and perfect)." No one knows whether this statement is true or false. It is known that there are no odd perfect numbers less than 10^{160}, in other words "$\sim \exists \, n < 10^{160}$, ($n$ is odd and perfect)" is true. $\square$

Example 14 (Fermat's Last Theorem) In 1637, Pierre de Fermat wrote, in French, "I have discovered a truly remarkable proof which this margin is too small to contain." Proof of what? He wrote it in his private shorthand notation which we have learned to decode. In our notation his claim was

$$\forall\, n \in \mathbb{N}^+ \left\{ \left(\exists\, (x,y,z) \in \mathbb{N}^+ \times \mathbb{N}^+ \times \mathbb{N}^+,\ (x^n + y^n = z^n) \right) \Leftrightarrow (n \le 2) \right\}.$$

This is known as *Fermat's Last Theorem*. If you have been watching TV in the last few years you will know that this "marginal statement" has finally been proven true by Andrew Wiles, after more than 350 years of attempts by many mathematicians.

You may wonder why Fermat never wrote up his proof and why it took so long to rediscover his proof. Fermat claimed to have proved many things without writing down the proofs. (Don't try this on a test!) He also made many conjectures. All his claims, except the "last theorem" were proven some time ago. All his conjectures are false. What was Fermat's proof of his "last theorem?" Most mathematicians believe that his proof was incorrect. One of the techniques that Fermat used is known as "infinite descent." It can be used for the cases $n = 3$ and $n = 4$, but cannot be used in general. Some people believe Fermat assumed it would work in general because it worked in these two cases. $\square$

This concludes our "number theory" examples. They were chosen to show you how to work with the notation of predicate logic. They were also chosen to introduce you to some famous problems in number theory. There are important applications of number theory to computer science, but not, so far as we know, applications of these particular examples. They are just hard problems and, as such, intellectual challenges.

At this moment the two Voyager spacecraft (Voyager I and Voyager II) are speeding away from the solar system at the rate of a million miles per day. They were launched in 1977. About 2010 they will arrive at the heliopause and enter interstellar space. Aboard each is a recording containing greetings in 59 languages, a whale noise, 12 minutes of sound — including the smack of a kiss, a baby's cry, an EEG of a young women in love, 116 pictures of this and that, 90 minutes of music including a Navajo chant, a Japanese *shakuhachi* piece, a Pygmy girl's initiation song, a Peruvian wedding song, Bach, Beethoven, Mozart, Stravinsky, Louis Armstrong, Blind Willie Johnson, and, last but not least, Chuck Berry singing "Johnny B. Goode." If any aliens find that stuff, they may think we are nuts sending such material and keep right on going. Much better would have been short crisp questions, easily translated,[3] asking questions such as, "Are there any odd perfect numbers?" "Is Goldbach's conjecture true?" The ability to answer such questions should be a matter of intellectual pride to any culture. At the time Voyager was launched, Fermat's Last Theorem was the "obvious" candidate question. Ironically, we now know the answer without any help from aliens.

---

[3] Questions in physics would also be of interest, but more background concepts may be needed than are needed for simple number theory problems.

Example 15 (Algebraic rules for predicate logic) In propositional logic we have Theorem 1 to help us manipulate statement forms. What about analogous rules for predicate logic?

As long as we are not trying to pull things through quantifiers, the same rules apply. For example

$$P(x) \vee (Q(y) \wedge R(x,y)) \quad \Leftrightarrow \quad (P(x) \vee Q(y)) \wedge (P(x) \vee R(x,y)).$$

Why is this? For each particular choice of x and y, the predicates become statement variables and so we are back in propositional logic.

What happens when quantifiers are involved? Theorem 2 tells us how to move $\sim$ through quantifiers. Sometimes we can move quantifiers through $\vee$ and $\wedge$, and sometimes not:

$$
\begin{array}{ll}
\text{True:} & \forall\, x \in D,\ (P(x) \wedge Q(x)) \quad \Leftrightarrow \quad (\forall\, x \in D,\ P(x)) \wedge (\forall\, x \in D,\ Q(x)) \\
\text{False:} & \exists\, x \in D,\ (P(x) \wedge Q(x)) \quad \Leftrightarrow \quad (\exists\, x \in D,\ P(x)) \wedge (\exists\, x \in D,\ Q(x)) \\
\text{False:} & \forall\, x \in D,\ (P(x) \vee Q(x)) \quad \Leftrightarrow \quad (\forall\, x \in D,\ P(x)) \vee (\forall\, x \in D,\ Q(x)) \\
\text{True:} & \exists\, x \in D,\ (P(x) \vee Q(x)) \quad \Leftrightarrow \quad (\exists\, x \in D,\ P(x)) \vee (\exists\, x \in D,\ Q(x))
\end{array}
$$

In the exercises, you will be asked to explain this.

It should be clear from this that manipulating quantifiers is trickier than the manipulations of propositional logic. Given the problems with algebraic manipulation, how does one go about proving statements in predicate logic? Since the variables often have infinite domains (as in our number theory examples), we can't construct truth tables because the would have to have an infinite number of rows. Proving things in predicate logic can be difficult.

Now you know why this section has less manipulation and proofs than the section on Boolean functions in Unit BF. $\square$

Exercises for Section 2

The following exercises will give you basic practice in predicate logic.

2.1. Each statement below should be rewritten in the form "$\forall\, \cdots\, x,\ \cdots$."

    (a) Every real number is negative, zero, or positive.

    (b) No computer scientists are unemployed.

2.2. Start with the statement, "$\forall n \in \mathbb{N}$, if n^2 is even then n is even." Which of the following statements say the same thing? Which are true and which are false?

    (a) Every integer has an even square and is even.

    (b) If a given integer has an even square then that integer is even.

    (c) For all integers, some will have an even square.

    (d) Any integer that has an even square will be even.

    (e) If the square of some integer is even then it is even.

    (f) All integers that are even have an even square.

2.3. For each of the following statements, construct a statement of the form, "$\forall \cdots$, if $\cdots$ then $\cdots$." that says the same thing.

    (a) Any correct algorithm, correctly coded, runs correctly.

    (b) Given any two odd integers, their product is odd.

    (c) Given any two integers whose product is odd, the integers themselves are odd.

2.4. Consider the statement, "Every computer science student needs to take Java Programming." Rewrite this in two ways, corresponding to the statement forms

    (a) "$\forall x$, if $\cdots$ then $\cdots$."

    (b) "$\forall x$, $\cdots$."

2.5. Consider the statement, "Some questions are easy." Rewrite this statement in two ways, corresponding to the statement forms

    (a) "$\exists \cdots x$ such that $\cdots$."

    (b) "$\exists x$ such that $\cdots$ and $\cdots$."

2.6. An number in $\mathbb{R} - \mathbb{Q}$ is called *irrational*. Consider the statement, "The product of any irrational number and any rational number is irrational." Is the following proposed negation of this statement the correct negation? If not, what is the correct negation? "There exists an irrational number x and an irrational number y such that the product xy is rational." Which is true, the original statement or its negation?

2.7. Consider the statement, "For all computer programs P, if P is correctly programmed then P compiles without warning messages." What is the negation of this statement? Which is true, the original statement or its negation?

2.8. Consider the statement, "For all real numbers x and y, if $x^2 = y^2$ then $x = y$." Is the following proposed negation of this statement the correct negation? If not, what is the correct negation? "If $x \neq y$ then $x^2 \neq y^2$." Which is true, the original statement or its negation?

2.9. Consider the statement, "For all primes $p \in \mathbb{P}$, either p is odd or p is 2." What is the negation of this statement? Which is true, the original statement or its negation?

2.10. Consider the statement, "For all animals x, if x is a tiger then x has stripes and x has claws." What is the negation of this statement? Which is true, the original statement or its negation?

2.11. Start with the statement, "$\exists\, x \in \mathbb{R}$, $\forall$ negative $y \in \mathbb{R}$, $x > y$."

    (a) Form a statement by reversing the existential and universal quantifiers. Which statement is true?

    (b) Form the negation of the original statement. Is it true?

2.12. Start with the statement, "For all computer programs P, if P is correct then P compiles without error messages." Form the contrapositive, converse, and inverse of this statement.

2.13. Start with the statement, "$\forall\, n \in \mathbb{N}$, if n^2 is even then n is even." Form the contrapositive, converse, and inverse of this statement. Which statements are true?

2.14. Start with the statement, "$\forall\, n \in \mathbb{N}$, if n is prime then n is odd or $n = 2$." Form the contrapositive, converse, and inverse of this statement. Which statements are true?

2.15. Consider the statement "A large income is a necessary condition for happiness."

    (a) Let P be the set of people. For $x \in P$, let $L(x)$ indicate that x has a large income and $H(x)$ that x is happy. Rewrite the given statement using the notation of logic rather than the English language.

    (b) Write the statement in ordinary English, without using "necessary" or "sufficient."

    (c) Write the negation of the statement in logic notation. Move the negation inside the statement as far as possible.

    (d) Write this negation in ordinary English, without using "necessary" or "sufficient."

2.16. Which of the following statements are true, which are false ($\exists!$ means "there exists exactly one").

    (a) $\exists!\, x \in \mathbb{Z} \ni 1/x \in \mathbb{Z}$.

    (b) $\forall\, x \in \mathbb{R}\, \exists!\, y \in \mathbb{R} \ni x + y = 0$.

2.17. Let S be a predicate with domain D. Write the statement, "$\exists!\, x \in D \ni S(x)$" using $\forall$ and $\exists$ instead of $\exists!$.

2.18. In each case, is the given statement true or false? Explain.

    (a) $\forall\, m \in \mathbb{N}$, $\exists\, n \geq m$, n even, $\exists\, p, q \in \mathbb{P}$, $n = p + q$.

Logic

(b) $\forall m \in \mathbb{N}, \exists n \geq m$, n odd, $\exists p, q \in \mathbb{P}, n = p + q$.

2.19. Let $P(x)$ and $Q(x)$ be predicates with domain D. For each pair of statement forms, state which are equivalent and explain your answer.

(a) $\forall x \in D, (P(x) \wedge Q(x))$ compared with $(\forall x \in D, P(x)) \wedge (\forall x \in D, Q(x))$

(b) $\exists x \in D, (P(x) \wedge Q(x))$ compared with $(\exists x \in D, P(x)) \wedge (\exists x \in D, Q(x))$

(c) $\forall x \in D, (P(x) \vee Q(x))$ compared with $(\forall x \in D, P(x)) \vee (\forall x \in D, Q(x))$

(d) $\exists x \in D, (P(x) \vee Q(x))$ compared with $(\exists x \in D, P(x)) \vee (\exists x \in D, Q(x))$

2.20. Suppose $n > 1$ is an integer. Prove:

$$\text{If } n \text{ is composite then } 2^n - 1 \text{ is composite.}$$

Thus, if you are going to search for primes of the form $2^n - 1$, you can limit your search to n a prime.

2.21. Suppose $2^n - 1$ is a prime number (that is, a Mersenne prime). Prove that $N = 2^{n-1}(2^n - 1)$ is an even perfect number. (The converse is true but harder to prove.)

Lo-22

[48]

Multiple Choice Questions for Review

In each case there is one correct answer (given at the end of the problem set). Try to work the problem first without looking at the answer. Understand both why the correct answer is correct and why the other answers are wrong.

1. Consider the statement form $p \Rightarrow q$ where $p =$ "If Tom is Jane's father then Jane is Bill's niece" and $q =$ "Bill is Tom's brother." Which of the following statements is equivalent to this statement?

   (a) If Bill is Tom's Brother, then Tom is Jane's father and Jane is not Bill's niece.

   (b) If Bill is not Tom's Brother, then Tom is Jane's father and Jane is not Bill's niece.

   (c) If Bill is not Tom's Brother, then Tom is Jane's father or Jane is Bill's niece.

   (d) If Bill is Tom's Brother, then Tom is Jane's father and Jane is Bill's niece.

   (e) If Bill is not Tom's Brother, then Tom is not Jane's father and Jane is Bill's niece.

2. Consider the statement, "If n is divisible by 30 then n is divisible by 2 and by 3 and by 5." Which of the following statements is equivalent to this statement?

   (a) If n is not divisible by 30 then n is divisible by 2 or divisible by 3 or divisible by 5.

   (b) If n is not divisible by 30 then n is not divisible by 2 or not divisible by 3 or not divisible by 5.

   (c) If n is divisible by 2 and divisible by 3 and divisible by 5 then n is divisible by 30.

   (d) If n is not divisible by 2 or not divisible by 3 or not divisible by 5 then n is not divisible by 30.

   (e) If n is divisible by 2 or divisible by 3 or divisible by 5 then n is divisible by 30.

3. Which of the following statements is the contrapositive of the statement, "You win the game if you know the rules but are not overconfident."

   (a) If you lose the game then you don't know the rules or you are overconfident.

   (b) A sufficient condition that you win the game is that you know the rules or you are not overconfident.

   (c) If you don't know the rules or are overconfident you lose the game.

   (d) If you know the rules and are overconfident then you win the game.

   (e) A necessary condition that you know the rules or you are not overconfident is that you win the game.

4. The statement form $(p \Leftrightarrow r) \Rightarrow (q \Leftrightarrow r)$ is equivalent to

   (a) $[(\sim p \vee r) \wedge (p \vee \sim r)] \vee \sim[(\sim q \vee r) \wedge (q \vee \sim r)]$

   (b) $\sim[(\sim p \vee r) \wedge (p \vee \sim r)] \wedge [(\sim q \vee r) \wedge (q \vee \sim r)]$

   (c) $[(\sim p \vee r) \wedge (p \vee \sim r)] \wedge [(\sim q \vee r) \wedge (q \vee \sim r)]$

   (d) $[(\sim p \vee r) \wedge (p \vee \sim r)] \vee [(\sim q \vee r) \wedge (q \vee \sim r)]$

Logic

(e) $\sim[(\sim p \vee r) \wedge (p \vee \sim r)] \vee [(\sim q \vee r) \wedge (q \vee \sim r)]$

5. Consider the statement, "Given that people who are in need of refuge and consolation are apt to do odd things, it is clear that people who are apt to do odd things are in need of refuge and consolation." This statement, of the form $(P \Rightarrow Q) \Rightarrow (Q \Rightarrow P)$, is logically equivalent to

   (a) People who are in need of refuge and consolation are not apt to do odd things.

   (b) People are apt to do odd things if and only if they are in need of refuge and consolation.

   (c) People who are apt to do odd things are in need of refuge and consolation.

   (d) People who are in need of refuge and consolation are apt to do odd things.

   (e) People who aren't apt to do odd things are not in need of refuge and consolation.

6. A sufficient condition that a triangle T be a right triangle is that $a^2 + b^2 = c^2$. An equivalent statement is

   (a) If T is a right triangle then $a^2 + b^2 = c^2$.

   (b) If $a^2 + b^2 = c^2$ then T is a right triangle.

   (c) If $a^2 + b^2 \neq c^2$ then T is not a right triangle.

   (d) T is a right triangle only if $a^2 + b^2 = c^2$.

   (e) T is a right triangle unless $a^2 + b^2 = c^2$.

7. Which of the following statements is **NOT** equivalent to the statement, "There exists either a computer scientist or a mathematician who knows both discrete math and Java."

   (a) There exists a person who is a computer scientist and who knows both discrete math and Java or there exists a person who is a mathematician and who knows both discrete math and Java.

   (b) There exists a person who is a computer scientist or there exists a person who is a mathematician who knows discrete math or who knows Java.

   (c) There exists a person who is a computer scientist and who knows both discrete math and Java or there exists a mathematician who knows both discrete math and Java.

   (d) There exists a computer scientist who knows both discrete math and Java or there exists a person who is a mathematician who knows both discrete math and Java.

   (e) There exists a person who is a computer scientist or a mathematician who knows both discrete math and Java.

8. Which of the following is the negation of the statement, "For all odd primes $p < q$ there exists positive non-primes $r < s$ such that $p^2 + q^2 = r^2 + s^2$."

   (a) For all odd primes $p < q$ there exists positive non-primes $r < s$ such that $p^2 + q^2 \neq r^2 + s^2$.

   (b) There exists odd primes $p < q$ such that for all positive non-primes $r < s$, $p^2 + q^2 = r^2 + s^2$.

(c) There exists odd primes $p < q$ such that for all positive non-primes $r < s$, $p^2 + q^2 \neq r^2 + s^2$.

(d) For all odd primes $p < q$ and for all positive non-primes $r < s$, $p^2 + q^2 \neq r^2 + s^2$.

(e) There exists odd primes $p < q$ and there exists positive non-primes $r < s$ such that $p^2 + q^2 \neq r^2 + s^2$

9. Consider the following assertion: "The two statements
   (1) $\exists x \in D, (P(x) \wedge Q(x))$ and
   (2) $(\exists x \in D, P(x)) \wedge (\exists x \in D, Q(x))$ have the same
   truth value." Which of the following is correct?

   (a) This assertion is false. A counterexample is $D = \mathbb{N}$, $P(x) = $ "x is divisible by 6," $Q(x) = $ "x is divisible by 3."

   (b) This assertion is true. The proof follows from the distributive law for $\wedge$.

   (c) This assertion is false. A counterexample is $D = \mathbb{Z}$, $P(x) = $ "$x < 0$," $Q(x) = $ "$x \geq 0$."

   (d) This assertion is true. To see why, let $D = \mathbb{N}$, $P(x) = $ "x is divisible by 6," $Q(x) = $ "x is divisible by 3." If $x = 6$, then x is divisible by both 3 and 6 so both statements in the assertion have the same truth value for this x.

   (e) This assertion is false. A counterexample is $D = \mathbb{N}$, $P(x) = $ "x is a square," $Q(x) = $ "x is odd."

10. Which of the following is an unsolved conjecture?

    (a) $\exists n \in \mathbb{N}, 2^{2^n} + 1 \notin \mathbb{P}$

    (b) $\exists K \in \mathbb{N}, \forall n \geq K, $ n odd, $\exists p, q, r \in \mathbb{P}, n = p + q + r$

    (c) $(\exists x, y, z, n \in \mathbb{N}^+, x^n + y^n = z^n) \Leftrightarrow (n = 1, 2)$

    (d) $\forall m \in \mathbb{N}, \exists n \geq m, $ n even, $\exists p, q \in \mathbb{P}, n = p + q$

    (e) $\forall m \in \mathbb{N}, \exists n \geq m, n \in \mathbb{P}$ and $n + 2 \in \mathbb{P}$

11. Which of the following is a solved conjecture?

    (a) $\forall m \in \mathbb{N}, \exists n \geq m, $ n odd, $\exists p, q \in \mathbb{P}, n = p + q$

    (b) $\forall m \in \mathbb{N}, \exists n \geq m, n \in \mathbb{P}$ and $n + 2 \in \mathbb{P}$

    (c) $\forall m \in \mathbb{N}, \exists n \geq m, 2^{2^n} + 1 \in \mathbb{P}$

    (d) $\forall k \in \mathbb{N}, \exists p \in P, p \geq k, 2^p - 1 \in \mathbb{P}$

    (e) $\forall n \geq 4, $ n even, $\exists p, q \in \mathbb{P}, n = p + q$

Answers: 1 (b), 2 (d), 3 (a), 4 (e), 5 (c), 6 (b), 7 (b), 8 (c), 9 (c), 10 (e), 11 (a).

Number Theory and Cryptography

Section 1: Basic Facts About Numbers

In this section, we shall take a look at some of the most basic properties of $\mathbb{Z}$, the set of integers. We look at properties related to parity (even, odd), prime factorization, irrationality of square roots, and modular arithmetic.

First we recall some standard notation for sets of various basic types of numbers.

- $\mathbb{R}$ denotes the real numbers,

- $\mathbb{Z}$ denotes the integers,

- $\mathbb{Q}$ denotes the rational numbers (ratios of integers),

- $\mathbb{N}$ denotes the nonnegative integers (the "natural numbers"),

- $\mathbb{N}^+$ denotes the nonzero natural numbers (the positive integers),

- $\mathbb{N}_2^+$ denotes the set of natural numbers greater than or equal to 2.

Note that $\mathbb{R} - \mathbb{Q}$ is the set of irrational numbers.

Example 1 (Odd and even integers)   A basic subdivision of $\mathbb{Z}$ is into the odd integers and the even integers. An element of $\mathbb{Z}$ is even if it is "of the form $2t$," where $t \in \mathbb{Z}$. An element of $\mathbb{Z}$ is odd if it is not even. The odd integers are all of the form $2t + 1$, where $t \in \mathbb{Z}$. (This should be proved, but we will not do so.) The phrase "of the form $2t$" can be written precisely as

$$\forall n \in \mathbb{Z}, \ (n \text{ is even}) \text{ if and only if } (\exists t \in \mathbb{Z} \text{ such that } n = 2t).$$

The most elementary mathematical facts about odd and even integers concern the *closure properties*.[1] Here is the closure property for multiplication:

The integers m and n are both odd if and only if mn is odd.

(Equivalently, by negating both sides of "if and only if," at least one the integers m or n is even if and only if mn is even.) To show the "only if" part, suppose that if m and n are both odd, say $m = 2j+1$ and $m = 2k+1$. Then $mn = 4jk+2j+2k+1 = 2(2jk+j+k)+1$ is of the form $2t + 1$ where $t = 2jk + j + k$. Thus, mn is odd. To show the "if" part, we use the inverse. Suppose that at least one of m or n is even. Without loss of generality, we may suppose that m is even, say $m = 2j$. Then $mn = 2jn$ is of the form $2t$ where $t = jn$. Thus, mn is even. A similar statement for addition is that, for integers m and n, $m + n$ is odd if and only if one of them is odd and the other is even.

---

[1]   A function on $S \times S$ has the closure property on S if its image is contained in S. Here S is the odd integers and the function is multiplication.

From the closure property for multiplication of odd integers, you can prove by induction that for any $k \geq 1$, and any integer m, m^k is odd if and only if m is odd. Logically equivalent is that m^k is even if and only if m is even. The fact that m^k is odd if m is odd can also be proved using the binomial theorem, which you should have seen in high school:

$$(x+y)^k = \sum_{i=1}^{k} \binom{k}{i} x^i y^{k-i}.$$

Since m is odd, $m = 2j+1$ for some integer j. Let $x = 2j$ and $y = 1$. Written another way,

$$m^k = (2j+1)^k = 1 + (2j)^1 \binom{k}{1} + (2j)^2 \binom{k}{2} + \cdots + (2j)^k \binom{k}{k}.$$

In this form m^k is obviously 1 plus an even integer and hence odd. $\blacksquare$

Prime Numbers and Factorization

Most mathematicians would agree that the most important concept in number theory is the notion of a prime.

Definition 1 (Prime and composite numbers) *A natural number n is prime if $n \geq 2$ and the only divisors of n are n and 1. We denote the set of prime numbers by $\mathbb{P}$. An integer $n \geq 2$ that is not prime is composite.*

The number 2 is the smallest prime and the only even prime. The other primes less than 20 are 3, 5, 7, 11, 13, 17, 19.

Example 2 (Prime factorization of any integer $n \geq 2$) Consider the integer 226512. It ends in 2 so it is divisible by 2. (We say that "n is divisible by m," indicated by the notation $m \mid n$, if $n = qm$ for some integer q.) In fact, $226512/2 = 113256$. We can divide by 2 again, $113256/2 = 56628$; and again, $56628/2 = 28314$; and again, $38314/2 = 14157$. That's it. We can't divide by 2 anymore, so we have $226512 = 2^4 \times 14157$. But, it is easy to check that 14157 is divisible by 3 to get 4719 which is again divisible by 3 to get 1573. That's it for dividing by 3, so we have $226512 = 2^4 \times 3^2 \times 1573$. Continuing in this manner, we end up with $226512 = 2^4 \times 3^2 \times 11^2 \times 13$. We have written 226512 as a product of primes. Also, the notation $m \nmid n$ means that n is not divisible by m.

Can every integer greater than 1 be written as a product of primes? What about a single prime p? It is convenient to adopt the terminology that a single prime p is a product of one prime, itself.[2]

---

[2] We could go even further and say that 1 is also can be written as an empty product. In fact, mathematicians do this: They say that an empty sum is 0 and an empty product is 1. You may think this strange, but you've already seen it with exponents: The notation a^n stands for the product of n copies of a. Thus a^0 is the product of no copies of a, and you learned that we define $a^0 = 1$ when you studied exponents. This is done so that the rule $a^{n+m} = a^n a^m$ will work when $n = 0$.

In Unit IS (Induction, Sequences and Series) we use induction to prove the assertion $A(n)$ for every integer $n \geq 2$ where

$$A(n) = \text{"}n \text{ is a product of primes."}$$

You might find it helpful to read the first two pages of Unit IS at this time. We start (base case) with $n = 2$, which is a prime and hence a product of primes. The induction hypothesis is the following:

"Suppose that for some $n > 2$, the assertion $A(k)$ is true for all k such that $2 \leq k < n$."

Assume the induction hypothesis and consider n. If n is a prime, then it is a product of primes (itself). Otherwise, $n = st$ where $1 < s < n$ and $1 < t < n$. By the induction hypothesis, s and t are each a product of primes. Hence $n = st$ is a product of primes. Thus $A(n)$ is true and the assertion is proved by induction.

If $n \geq 2$ is an integer, the notation $n = p_1^{e_1} p_2^{e_2} \cdots p_k^{e_k}$ is commonly used to designate its prime factorization, where $p_1, p_2, \ldots p_k$ are distinct primes and all $e_i > 0$. In other words, each prime factor is raised to its highest power that divides n. Thus, $226512 = 2^4 \times 3^2 \times 1573^1$. Of course, exponents with value 1 are usually omitted, thus 1573^1 would be written 1573.

It is important to note (We won't give a proof.) that prime factorization is *unique* in the following sense. Suppose one student correctly computes a prime factorization of n and gets $n = p_1^{e_1} p_2^{e_2} \cdots p_k^{e_k}$ where she has ordered the prime factors so that $p_1 < p_2 < \cdots < p_k$. Suppose that another student also correctly computes a prime factorization of n and gets $n = q_1^{f_1} q_2^{f_2} \cdots q_j^{f_j}$ with $q_1 < q_2 < \cdots < q_j$, then $k = j$, $q_i = p_i$, and $e_i = f_i$, for $i = 1, \ldots, k$. Let's call this a theorem:

Theorem 1 (Unique prime factorization) *Every integer $n \geq 2$ can be factored into a product of primes. This factorization is unique in the sense that any two such factorizations differ only in the order in which the primes are written.*

Sometimes people think it is "obvious" that prime factorization is unique. That's not true. There are sets other than the integers where prime factorization can be defined, but it may not be unique.[3] The assumption that it is unique was used in a "proof" of Fermat's Last Theorem about a century ago. Of course, the proof was false because factorization was not unique in the set being studied. Understanding the problem led to what is known as "algebraic number theory," which eventually led to a correct proof of Fermat's Last Theorem. ◻

---

[3] When $a, b \in \mathbb{Z}$, complex numbers of the form $a + b\sqrt{-5}$ are a type of "algebraic integer." The set of these "integers" is denoted by $\mathbb{Z}\left[\sqrt{-5}\right] = \left\{a + b\sqrt{-5} \mid a, b \in \mathbb{Z}\right\}$. We have

$$6 = 2 \times 3 = \left(1 + \sqrt{-5}\right)\left(1 - \sqrt{-5}\right).$$

Since $2, 3, 1 + \sqrt{-5}$ and $1 - \sqrt{-5}$ cannot be factored further in $\mathbb{Z}\left[\sqrt{-5}\right]$, they are "primes." Hence prime factorization is not unique for $\mathbb{Z}\left[\sqrt{-5}\right]$. The desire for uniqueness led to the concept of "ideals" in $\mathbb{Z}\left[\sqrt{-5}\right]$ and the development of "algebraic number theory."

Number Theory and Cryptography

Now that we know that every integer $n \geq 2$ is a product of powers of primes, we can show

Theorem 2 (Infinitely many primes)   *There are infinitely many primes.*

Proof: Suppose that there were only finitely many primes, say the k primes

$$\mathbb{P} = \{p_1, p_2, \ldots, p_k\}.$$

Consider the integer $n = (p_1 p_2 \cdots p_k) + 1$ gotten by taking the product of all of the primes in $\mathbb{P}$ and adding one. Clearly, $n \notin \mathbb{P}$ (it's too big). That means n is a product of primes. Let p be one of the prime factors of n. Hence n/p is an integer. For $p_i \in \mathbb{P}$, dividing n by p_i leaves a remainder of 1 and so n/p_i is not an integer. Since n/p is an integer and n/p_i is not, we cannot have $p = p_i$. Hence $p \notin \mathbb{P}$. Contradiction! Thus there cannot be finitely many primes. $\square$

Prime factorization can be used to prove things that apparently do not depend on primes. Our next example illustrates this.

Example 3 (For all $n \in \mathbb{N}$, $\sqrt{n}$ is either an integer or irrational)   The integer 36 is nice because $\sqrt{36} = 6$ and 6 is an integer. Thus 36 is called a *perfect square*. A perfect square is an integer whose square root is also an integer. Suppose $\sqrt{n}$ is not an integer. How "bad" is it? For example, maybe, though not an integer, $\sqrt{n}$ is rational; that is, $\sqrt{n} = a/b$ for some integers a and b. Sadly, that can't happen. We prove this by contradiction

Suppose $\sqrt{n} = a/b$ where $b \geq 2$ and we have cancelled common factors from the numerator and denominator. Since $\sqrt{n} = a/b$, we have $nb^2 = a^2$. Let p be a prime factor of b (p exists since $b \geq 2$). Since prime factorization is unique, p is a prime factor of $nb^2 = a^2$. On the other hand, since p is a prime factor of b, it is not a prime factor of a since we have cancelled common factors to get a and b. So far, we have shown that p is a prime factor of a^2 but not a prime factor of a. In the next paragraph, we show that this is a contradiction.

For any integer x, if the prime factorization of x is $x = p_1^{e_1} p_2^{e_2} \cdots p_k^{e_k}$ then the prime factorization of x^2 is $x^2 = p_1^{2e_1} p_2^{2e_2} \cdots p_k^{2e_k}$. In other words, any integer x has exactly the same prime divisors as its square, x^2. Apply this with $x = a$. We have proved

Theorem 3 (Irrational square roots)   *For all $n \in \mathbb{N}$, $\sqrt{n}$ is either an integer or irrational.*

We can use this to get a lot of irrational numbers. Suppose that $k^2 < n < (k+1)^2$ for some $k \in \mathbb{N}$. Taking square roots, we have $k < \sqrt{n} < k+1$. Thus $\sqrt{n}$ cannot be an integer and so it must be irrational. In particular $\sqrt{2}, \sqrt{3}, \sqrt{5}, \sqrt{6}, \sqrt{7}, \sqrt{8}$ are all irrational.[4]

---

[4] Some classical Greeks were bothered by this. They thought there should be a basic unit of length such that all the lines in a geometrical construction were integer multiples of that length, but they could prove that this was impossible: By the Pythagorean Theorem, the diagonal of a unit square has length $\sqrt{2}$, which they knew was irrational. If the side of the square were b basic units long and the diagonal were a, then $\sqrt{2} = a/b$.

There are some basic properties of irrational and rational numbers lurking beneath the surface here. *If the product xy of two numbers is irrational, one of the numbers must be irrational.* Equivalently (the contrapositive), if x and y are both rational, say $x = a/b$ and $y = c/d$, then $xy = ac/bd$ is rational. Likewise, *if the sum $x + y$ of two numbers is irrational, one of the numbers must be irrational* (prove this).

Some students think these statements mean that the product of two nonzero irrational numbers is irrational and the sum of two irrational numbers is irrational, both statements are false: $\sqrt{2} \times \sqrt{2} = 2$ and $(-\sqrt{2}) + \sqrt{2} = 0$. It is true, however, that *if $x \neq 0$ is rational and y is irrational, then the product xy is irrational.* To prove this statement, use the contrapositive. If $xy = a/b$ then $y = a/bx$. Since $x \neq 0$ is rational, say $x = c/d$, this implies that $y = ad/cb$ is rational. $\Box$

Example 4 (The rational numbers are countable) We want to show that we can create a list $a_1, a_2, a_3, \ldots$ such that every rational number appears on the list. We do this as follows:

Step 1. Start the list with $0, 1/1, -1/1$ and set $k = 3$.

Step 2. Append to the list all rational numbers in reduced form where the sum of the numerator and denominator (ignoring signs) is k. Begin with the largest numerators and proceed to the smallest, listing positive numbers and then negative ones. (Thus, for $k = 3$ we append $2/1, 1/2, -1/2, -2/1$ and for $k = 4$ we append $3/1, 1/3, -1/3, -3/1$.)

Step 3. Increase k by one and go to Step 2.

The list begins

$$
\begin{array}{llll}
 & a_1 = 0, & a_2 = 1/1, & a_3 = -1/1, \\
k = 3: & a_4 = 2/1, & a_5 = 1/2, & a_6 = -1/2, \quad a_7 = -2/1, \\
k = 4: & a_8 = 3/1, & a_9 = 1/3, & a_{10} = -1/3, \quad a_{11} = -3/1, \\
k = 5: & a_{12} = 4/1, & a_{13} = 3/2, & a_{14} = 2/3, \quad a_{15} = 1/4, \\
 & a_{15} = -1/4, & a_{16} = -2/3, & a_{17} = -3/2, \quad a_{18} - 4/1,
\end{array}
$$

Note that each rational number occurs exactly once in the list. In some sense, the number of rational numbers is the same as the number of positive integers since we have one rational number for each positive integer (the subscript of a)!

Because we can form such a list, we say that the set of rational numbers is *countable*. More simply, people say that the rationals are countable. $\Box$

Example 5 (The real numbers are *not* countable) We must show that it is impossible to form a list of the real numbers. How can we do this? We must show that, *no matter what list of real numbers we have*, there is some real number that is not on the list.

Suppose we have a list $a_1, a_2, \ldots$ of real numbers. Let d_k be the kth digit after the decimal point in a_k. For example, if $a_4 = 2.718281828\ldots$ (the number e), then $d_4 = 2$. If $d_k = 1$, let $b_k = 2$ and, if $d_k \neq 1$, let $b_k = 1$. Look at the number $r = 0.b_1 b_2 b_3 \ldots$. We claim it is not in the list. Why is this? Suppose someone claims, for example that $a_{99} = r$. By definition, d_{99} is the ninety-ninth digit of a_{99} after the decimal point. Since $b_{99} \neq d_{99}$, the numbers r and a_{99} differ in their ninety-ninth digits. Thus $r \neq a_{99}$.

Number Theory and Cryptography

Arguments of this type are called *diagonal arguments*. Why is this? A picture can help. Here $*$ stands for a digit we are not interested in and we have dropped all the digits before the decimal points.

$$a_1 =.d_1\ *\ \ *\ \ *\ \ *\ \ *\ldots$$
$$a_2 =.*\ \ d_2\ *\ \ *\ \ *\ \ *\ldots$$
$$a_3 =.*\ \ *\ \ d_3\ *\ \ *\ \ *\ldots$$
$$a_4 =.*\ \ *\ \ *\ \ d_4\ *\ \ *\ldots$$
$$a_5 =.*\ \ *\ \ *\ \ *\ \ d_5\ *\ldots$$

The digits $d_1, d_2, \ldots$ that we are changing appear in a diagonal pattern. The diagonal is not always so straightforward in a diagonal argument. $\blacksquare$

Remainders and Modular Arithmetic

We all know from elementary school that if we divide one integer x by another $d > 0$, we get a quotient q and a remainder r, where $0 \leq r < d$. In other words, $x = qd+r, 0 \leq r < d$. For example, if $x = 234$ and $d = 21$, then $q = 11$ and $r = 3$. Thus, $234 = 11 \times 21 + 3$. There are 21 possible remainders that can be gotten by dividing some randomly chosen integer by 21. These remainders belong to the set $\{0, 1, 2, \ldots, 20\}$. The set $\mathbb{Z}$ of all integers can be partitioned (divided up) into 21 subsets

$$21\mathbb{Z}, \quad 21\mathbb{Z}+1, \quad 21\mathbb{Z}+2, \ldots, 21\mathbb{Z}+20$$

according to these remainders. Note that, for a set S of numbers $aS + b = \{as + b \mid s \in S\}$ so that $21\mathbb{Z}+4 = \{\ldots, -17, 4, 25, \ldots\}$. We have just seen that 234 belongs to the subset $21\mathbb{Z}+3$. (The set $21\mathbb{Z}+3$ equals $\{3 + 21k \mid k = 0, \pm 1, \pm 2, \ldots\}$.) For general $d > 0$, instead of $d = 21$, we get

$$d\mathbb{Z}, \quad d\mathbb{Z}+1, \quad d\mathbb{Z}+2, \ldots, d\mathbb{Z}+(d-1)$$

The sets $d\mathbb{Z} + j$ are called *residue classes modulo d*.

If $x = qd+r, 0 \leq r < d$, then we denote this fact by x modulo $d = r$ or by x mod $d = r$. In this usage, "mod" is called a *binary operation*. Given any pair of integers x and $d > 0$, computing x mod d always results in some integer $r, 0 \leq r < d$.

The word "mod" is also used to convey the information that "x and x' belong to the same residue class mod d." The notation is $x = x'$ (mod d) or $x \neq x'$ (mod d) to express the facts (respectively) that "x and x' belong to the same residue class mod d," or, "x and x' do not belong to the same residue class mod d." Often you will see $\equiv$ used instead of $=$ in these expressions.

Because of the possible confusion between these two uses, we will use the C programming language notation for the binary operation. Let's summarize all this in a definition.

Definition 2 (Residue classes and "mod") Let $d \geq 2$ be an integer For $0 \leq j < d$ the set $d\mathbb{Z} + j = \{nd + j \mid n \in \mathbb{Z}\}$ is called a *residue class modulo d*. The notation "mod" is used in two ways:

- $x = x'$ (mod d)   This means that x and x' belong to the same reside class modulo d. In other words, when x and x' are divided by d they have the same remainder. We say that x and y are equal modulo d (or mod d). For reasons we will learn later, this is referred to as "using mod as an equivalence relation." The notation $x \equiv x'$ (mod d) is also used to indicate that x and y are equal modulo d. If the value of d is clear, people often write $x \equiv x'$, omitting (mod d).

- x mod $d = r$ or $x \% d = r$ This means that when x is divided by d the remainder is r where $0 \le r < d$. Used this way, "mod" is a binary operator. To avoid confusion, we will use the C programming language notation $r = x \% d$.

Since the two uses of "mod" involve different placement of "mod," you should not be confused as to which use is intended.

Example 6 (A fact about remainders) There is something important about remainders that they may not have discussed in elementary school. Suppose $x = qd + r$ and $x' = q'd + r'$. Then, subtracting and dividing by d gives

$$\frac{x - x'}{d} = \frac{(q - q')d + (r - r')}{d} = q - q' + \frac{r - r'}{d}.$$

Note that since $0 \le r < d$ and $0 \le r' < d$ we must have $0 \le |r - r'| < d$. This means that the only way that $\frac{r-r'}{d}$ can be an integer is that $|r - r'| = 0$ or $r = r'$. This seems like a trivial point, but it is very important. It means that x and x' have the same remainder when divided by d (i.e., belong to the same residue class mod d) if and only if d divides $x - x'$. For example 7666 and 7652 belong to the same residue class modulo 7 since $7666 - 7652 = 14$, which is 0 modulo 7. ∎

The notation $x = x'$ (mod d) behaves like equality in many ways. The following theorem lists three of them.

Theorem 4 (Arithmetic with mod)   The notation $x = x'$ (mod d) behaves like equality for addition, subtraction and multiplication. In other words, if $x = x'$ (mod d) and $y = y'$ (mod d) then

$$x + y = x' + y' \ (\text{mod } d), \quad x - y = x' - y' \ (\text{mod } d) \quad \text{and} \quad xy = x'y' \ (\text{mod } d).$$

We talk about *addition modulo d* or simply *modular addition*, and similarly for subtraction and multiplication. Notice that we did not say $x/y = x'/y'$ mod d. It is not true in general. For example, $2 = 8$ (mod 6) and $2 = 2$ (mod 6) but $2/2 \ne 8/2$ (mod 6).

Proof: We prove addition. By definition $x + y = x' + y'$ (mod d) means that $(x + y) - (x' + y')$ is divisible by d. But

$$\frac{(x + y) - (x' + y')}{d} = \frac{(x - x') + (y - y')}{d} = \frac{x - x'}{d} + \frac{y - y'}{d}.$$

Number Theory and Cryptography

Since $x = x'$ (mod d) and $y = y'$ (mod d), both $x - x'$ and $y - y'$ are divisible by d. Thus, $(x + y) - (x' + y')$ is divisible by d.

The proof for subtraction is nearly the same as for addition, so we omit it.

We now prove multiplication. Again, we show that $xy - x'y'$ is divisible by d:

$$\frac{xy - x'y'}{d} = \frac{x(y - y') + y'(x - x')}{d} = x\frac{y - y'}{d} + y'\frac{x - x'}{d}.$$

Since, $x = x'$ (mod d) and $y = y'$ (mod d), both $x - x'$ and $y - y'$ are divisible by d. Thus, $xy - x'y'$ is divisible by d. $\square$

Example 7 (Powers of $d\mathbb{Z} + 1$) Suppose $x \in d\mathbb{Z} + 1$. We could equally well write this as x mod $d = 1$ or $x = 1$ (mod d) or even just $x \equiv 1$ provided we know we are doing arithmetic modulo d. We claim that $x^n \equiv 1$ for all $n \in \mathbb{N}$. The proof is by induction on n.

For $n = 0$, $x^0 = 1$ and so $x^0 \equiv 1$. For $n = 1$, $x^1 = x$ and so $x^1 \equiv 1$ since we are given that $x \equiv 1$. For $n > 1$, $x^n = (x^{n-1})x$. By induction $x^{n-1} \equiv 1$. By the theorem, $x^{n-1}x \equiv 1 \times 1 = 1$. We are done.

When $d = 2$, you should be able to see that this simply states that powers of odd numbers are odd, a fact we proved in Example 1. $\square$

Example 8 (Using modular arithmetic cleverly) There are smart ways and dumb ways to use Theorem 4. It is interesting to look first at a dumb way, just to see the power of these statements. Suppose you want to find the remainder when the number $N = 113 \times (167 + 484) + 192 \times 145$ is divided by 21. That is, we wish to know N (mod 21). A friend says he is going to help. He tells you that $113 = 95180$ (mod 21), $167 = 5159244761$ (mod 21), $484 = 9073$ (mod 21), $192 = 207441$ (mod 21) and $145 = 19857871$ (mod 21). He suggests you substitute those larger numbers for the original numbers in the expression $N = 113 \times (167 + 484) + 192 \times 145$ to get

$$M = 95180 \times (5159244761 + 9073) + 207441 \times 19857871.$$

He assures you that, if you compute M and divide by 21 you will get the desired remainder r. He says he would like to borrow your car while you do the computations. After several hours work, you get $M = 495177116538231$. Dividing by 21 gives 15 as a remainder. Thus, $r = 15$, so N (mod 21) = 15. That is the right answer but it is a dumb way to do it!

Another way is to just compute

$$N = 113 \times (167 + 484) + 192 \times 145 = 101403$$

and divide that by 21 to get the remainder 15. That is not too dumb.

Another way is to note that $113 = 8$ (mod 21), $167 = 20$ (mod 21), $484 = 1$ (mod 21), $192 = 3$ (mod 21), $145 = 19$ (mod 21). Substitute those for the corresponding numbers to get $L = 8(20 + 1) + 3 * 19 = 225$. Now divide 225 by 21 to get 15 as the remainder.

A modification on the above is to note that $20 = -1$ (mod 21) and $19 = -2$ (mod 21) to get $L' = 8(-1 + 1) + 3(-2) = -6$. Dividing -6 by 21 gives a remainder of 15. Did you learn that in elementary school? The remainder r must always be positive, $0 \leq r < 21$. Thus, writing $-6 = q \times 21 + r$ gives $-6 = (-1) \times 21 + 15$. Do you see the power of these techniques? Don't be afraid to use them (wisely). Note that they apply to multiplying and adding, not dividing. For example, $484 = 1 \pmod{21}$, $22 = 1 \pmod{21}$, but $484/21 \pmod{21} \neq 1/1 \pmod{21}$. The number $484/21$ is not even an integer. $\square$

The Floor and Ceiling Functions

In computer science, many basic concepts are naturally expressed in terms of integer values (e.g., running time, input size, memory blocks) but are analyzed by functions that return real numbers. The conversion of the real numbers to integers that have direct meaning in terms of original problems sometimes involves the special functions "floor" and "ceiling."

Let $x \in \mathbb{R}$ be a real number. The *floor function* of x, denoted by $\lfloor x \rfloor$, is the largest integer less than or equal to x. It is also called the *greatest integer* function. The *ceiling function* of x, denoted by $\lceil x \rceil$, is the least integer greater than or equal to x. It is also called the *least integer* function.

Here are some examples:

$$\lceil 2.8 \rceil = 3, \quad \lceil 5 \rceil = 5, \quad \lceil -2.8 \rceil = -2,$$
$$\lfloor 2.8 \rfloor = 2, \quad \lfloor 5 \rfloor = 5, \quad \lfloor -2.8 \rfloor = -3,$$
$$\lceil 55 + 2.8 \rceil = 55 + \lceil 2.8 \rceil = 55 + 3 = 58,$$
$$\lfloor -5.6 \rfloor = -6 = -\lceil -(-5.6) \rceil,$$

Geometrically, the idea is simple. The floor of x moves you to the next integer less than or equal to x on the number line. The ceiling moves you to the next integer greater than or equal to x. For computation, notice that

$$\forall\, n \in \mathbb{Z},\ \forall\, x \subset \mathbb{R},\ \lfloor n + x \rfloor = n + \lfloor x \rfloor.$$
$$\forall\, n \in \mathbb{Z},\ \forall\, x \in \mathbb{R},\ \lceil n + x \rceil = n + \lceil x \rceil.$$

This is easily shown and we omit the proof. Note also that

$$\lfloor x \rfloor = -\lceil -x \rceil \quad \text{and} \quad \lceil x \rceil = -\lfloor -x \rfloor.$$

For example, $\lfloor 2.1 \rfloor = -\lceil -2.1 \rceil$.

For proofs and exercises, it is often helpful to know that any real number can be written as the sum of an integer n and a fraction f, $-1 < f < +1$. Thus, $4.9 = 4 + 0.9$, $-3.6 = -3 - 0.6 = -4 + 0.4$. If $x = n + f$, then, since $\lfloor x \rfloor = n + \lfloor f \rfloor$ and $\lceil x \rceil = n + \lceil f \rceil$. you only have to think about the fractional part in your computations. For example,

$$\lfloor 4.9 \rfloor = 4 + \lfloor 0.9 \rfloor = 4 + 0 = 4,$$
$$\lceil -3.6 \rceil = -4 + \lceil 0.4 \rceil = -4 + 1 = -3.$$

If you prefer, $\lceil -3.6 \rceil = -3 + \lceil -.6 \rceil = -3 + 0 = -3$.

Exercises for Section 1

1.1. Prove the statement if true, otherwise find a counterexample.

    (a) For all natural numbers x and y, $x + y$ is odd if one of x and y even and the other is odd.

    (b) For all natural numbers x and y, if $x + y$ is odd then one of x and y even and the other is odd.

1.2. Prove the statement if true, otherwise find a counterexample.

    (a) The difference of any two odd integers is odd.

    (b) If the sum of two integers is even, one of them must be even.

1.3. Prove the statement if true, otherwise find a counterexample.

    (a) The product of two integers is even if and only if at least one of them is even.

    (b) The product of two integers is odd if and only if at least one of them is odd.

1.4. Prove the statement if true, otherwise find a counterexample.

    (a) For any integers m and n, $m^3 - n^3$ is even if and only if $m - n$ is even.

    (b) For any integers m and n, $m^3 - n^3$ is odd if and only if $m - n$ is odd.

1.5. Prove the statement if true, otherwise find a counterexample.

    (a) For all integers $n > 2$, $n^3 - 8$ is composite.

    (b) For all integers n, $(-1)^n = -1$ if and only if n is odd.

1.6. Prove the statement if true, otherwise find a counterexample.

    (a) $\forall\, n \in \mathbb{Z}$, $n^2 + n + 5$ is odd.

    (b) $\forall\, n \in \mathbb{Z}$, $(6(n^2 + n + 1) - (5n^2 - 3)$ is a perfect square).

    (c) $\exists\, M > 0$, $\forall\, n > M$, $(n^2 - n + 11$ is prime).

    (d) There is a unique prime p of the form $n^2 + 2n - 3$.

1.7. Prove the statement if true, otherwise find a counterexample.

    (a) For all integers $n > 0$, either n is a perfect square or, $n = x + y$ where x and y are perfect squares or, $n = x + y + z$ where x, y, and z perfect squares.

    (b) The product of four consecutive positive integers is never a perfect square.

1.8. Prove the statement if true, otherwise find a counterexample.

Section 1: Basic Facts About Numbers

(a) For all distinct positive integers m and n, either $m^{1/2} + n^{1/2}$ and $m^{1/2} - n^{1/2}$ are both rational or both irrational.
Hint: Consider $\left(m^{1/2} + n^{1/2}\right)\left(m^{1/2} - n^{1/2}\right)$.

(b) For all distinct positive integers, if either $m^{1/2} + n^{1/2}$ or $m^{1/2} - n^{1/2}$ are rational then both m and n are perfect squares.

(c) For all distinct positive integers m and n, both m and n are perfect squares if and only if $m + 2m^{1/2}n^{1/2} + n$ is a perfect square.

(d) Which of (a), (b) and (c) are true if $m \neq n$ is changed to $m = n$?

1.9. Prove that an integer $n > 1$ is composite if and only if p divides n for some prime $p \leq n^{1/2}$.

1.10. Write the following rational numbers as the ratio a/b of two integers a and $b > 0$.

(a) 3.1415

(b) 0.30303030...

(c) 6.32152152152152...

1.11. Let $x \in \mathbb{R}$ satisfy the equation $\frac{ax+b}{cx+d} = 1$ where a, b, c, and d are rational. Is x rational? Explain.

1.12. In each case, if the statement is true, prove it, if false, give a counterexample.

(a) The sum of three consecutive integers is zero (mod 3).

(b) The product of two even integers is zero (mod 4).

(c) An integer is divisible by 16 only if it is divisible by 8.

(d) For all odd integers n, $3n + 3$ is divisible by 6.

1.13. In each case, if the statement is true, prove it, if false, give a counterexample.

(a) $\forall\, a, b, c \in \mathbb{Z}$, if $a \mid b$ then $a \mid bc$.

(b) $\forall\, a, b, c \in \mathbb{Z}$, if $a \mid b$ and $b \mid c$, then $a \mid c$

(c) $\forall\, a, b, c \in \mathbb{Z}$, if $a \mid c$ then $ab \mid c$.

1.14. In each case, if the statement is true, prove it, if false, give a counterexample.

(a) $\forall\, a, b, c \in \mathbb{Z}$, if $a \mid (b + c)$ then $a \mid b$ and $a \mid c$.

(b) $\forall\, a, b, c \in \mathbb{Z}$, if $a \mid bc$ then $a \mid b$ or $a \mid c$.

(c) $\forall\, a, b \in \mathbb{Z}$, if $a \mid b$ then $a^2 \mid b^2$.

(d) $\forall\, a, b \in \mathbb{Z}$, if $a \mid 6b$ then $a \mid 6$ or $a \mid b$.

1.15. In each case, factor the given number into a product of powers of distinct primes.

NT-11

(a) 1404.   (b) 9702.   (c) 89250.

1.16. Let $n = p_1^{e_1} \cdots p_k^{e_k}$ be the factorization of n into powers of distinct primes. Let $m \geq 1$ be an integer.

(a) What is the factorization of n^m into powers of distinct primes?

(b) If $s > 0$ is an integer but $s^{1/m}$ is not, must $s^{1/m}$ be irrational? Explain your answer.

1.17. In each case, factor the given number into a product of powers of distinct primes. Recall that $n! = n(n-1)(n-2) \cdots 1$ is the product of the first n integers.

(a) 20!. How many terminal zeros in this number?

(b) $(20!)^2$. How many terminal zeros in this number?

(c) $(20!)^3$. How many terminal zeros in this number?

1.18. Prove that if x is a nonzero natural number then $3 \mid x$ if and only if 3 divides the sum of the decimal digits of x.

1.19. Prove or give a counterexample: The product of any four consecutive integers is equal to 0 (mod 8).

1.20. Prove that, for all integers $n > 1$, $n^2 - 3 \neq 0$ (mod 4).

1.21. Prove that, for all odd integers n, $n^4 = 1$ (mod 16).

1.22. If $m - n$ has remainder 0 when divided by d does that mean the m and n each have the same remainder when divided by d? Support your answer by giving a counterexample or a proof.

1.23. For all integers m, n, a, b, if $m \bmod d = a$ and $n \bmod d = b$ does that mean that $(m + n) \bmod d = a + b$?

1.24. (a) Prove: If $j = k$ (mod d), then $d\mathbb{Z} + j = d\mathbb{Z} + k$.

(b) Prove: If $j \neq k$ (mod d), then $(d\mathbb{Z} + j) \cap (d\mathbb{Z} + k)$ is the empty set.

1.25. If $a > 0$, $\log_a(x)$ is the unique number such that $a^{\log_a(x)} = x$.

(a) Suppose that p and q are two different primes. Prove that $\log_p(q)$ is irrational.

(b) Is the result in (a) true if p and q are allowed to be composite numbers? Justify your answer.

(c) For integers k and m, prove that $\log_a(b) = k/m$ if and only if $a^k = b^m$.

1.26. In each case, if the statement is true, prove it, if false, give a counterexample.

(a) $\forall\, x, y \in \mathbb{R}$, $(\lfloor x - y \rfloor = \lfloor x \rfloor - \lfloor y \rfloor)$.

(b) $\forall\, x \in \mathbb{R}$, $\forall\, k \in \mathbb{Z}$, $(\lfloor x - k \rfloor = \lfloor x \rfloor - k)$.

(c) $\forall\, x \in \mathbb{R}$, $k \in \mathbb{N}$, $(\lfloor x^k \rfloor = \lfloor x \rfloor^k)$.

1.27. In each case, if the statement is true, prove it, if false, give a counterexample.

(a) $\forall\, n \in \mathbb{Z}$, $k \in \mathbb{N}^+$, $(\lfloor \frac{n}{k} \rfloor = \frac{n-r}{k})$ where $r = n \% k$).

(b) $\forall\, x \in \mathbb{R}$, $\forall\, a, b \in \mathbb{N}^+$, $(\lfloor ax + b \rfloor = a\lfloor x \rfloor + b)$.

1.28. Prove each of the following statements or give a counterexample.

(a) $\forall\, x \in \mathbb{R} - \mathbb{Z}$, $(\lfloor x \rfloor + \lfloor -x \rfloor = -1)$.

(b) $\forall\, x \in \mathbb{R} - \mathbb{Z}$, $(\lceil x \rceil + \lceil -x \rceil = +1)$.

Section 2: Cryptography and Secrecy

Cryptography is concerned with secret messages. Cryptanalysis is the name for the general area of breaking secret codes so the messages can be read. This general topic represents a vast body of knowledge. We begin by introducing the basic ideas and problems. Then we take time out to study some number theory functions that are useful for cryptography on the internet. Finally, we look at two protocols that are currently used — Diffie-Hellman and RSA.

Basic Ideas

Suppose that Alice wishes to send a message to Bob in such a way that anyone else receiving her message will not be able to understand it. She can communicate in code. There are three pieces of data involved:

- The *plaintext*, which is what Alice wants to tell Bob.

- The *ciphertext*, which is the message Alice actually sends Bob.

- The *key*, which tells how to convert plaintext to ciphertext and vice versa. Since the key is known to Alice and Bob, it is sometimes called the *shared key*.

The rules for converting can be thought of as functions. If $\mathcal{P}$ is the set of all possible plaintext messages and $\mathcal{C}$ is the set of all possible ciphertext messages, then the key K determines a function $f_K : \mathcal{P} \to \mathcal{C}$ that Alice uses to *encipher* the message. Bob uses the inverse function f_K^{-1} to *decipher* the message. Notice that, in order to decipher, f_K^{-1} must exist. Thus f_K must be an injection. The next example illustrates a simple scheme for doing this.

Example 9 (A simple code) Instead of Alice and Bob, we have two factories A and B that are going to exchange goods. There are 64 different items (coded $0, 1, 2, \ldots 63$) to be shipped and four methods of shipping (regular mail represented by the code 00; priority mail, code 01; air mail, code 10; and next day air, code 11). A shipment request looks something like 10101001. The two least significant bits, 01 in this case specify the method of shipping and the other six bits the item in base 2 (101010 or item 42 in this case).

The factories want to keep the orders they are requesting from each other secret from their competitors. To keep things secret, the factories agree on a simple encipherment procedure. They agree on a fixed eight bit binary string that they share as a secret. Here is the secret string that they happen to choose: $K = 11000111$. This is the *shared key*, also called the *secret key* or, simply, the key.

Factory A wants to place order $r = 10101001$ with factory B. To do this, the folks at A add r to K bit-by-bit using addition mod 2. That is, $0 + 0 = 0$, $0 + 1 = 1 + 0 = 1$, $1 + 1 = 0$. Here is what happens:

$$
\begin{array}{ll}
10101001 & \text{plaintext} \\
11000111 & \text{key } K \\
\hline
01101110 & \text{ciphertext}
\end{array}
$$

The first line is the message, the second line is the key, and the third line is the mod 2 bit-by-bit sum of the message and the key. We have just computed $f_K(10101001)$. Actually, this is done in the computer. When someone wants to place an order, they type in 10101001. The computer does the addition and sends the result to factory B.

When factory B's computer receives the ciphertext, it adds the shared key to the ciphertext as follows:

$$
\begin{array}{ll}
01101110 & \text{ciphertext} \\
11000111 & \text{key } K \\
\hline
10101001 & \text{plaintext}
\end{array}
$$

This reverses the process and reveals the correct order from factory A. Pretty nifty — the function and its inverse are the same, i.e. $f_K^{-1} = f_K$. $\square$

In the previous example, $f_K^{-1} = f_K$. This makes programming easier since the software for deciphering is the same as the software for enciphering. As a result, many systems are designed to have $f_K^{-1} = f_K$.

There is a problem with our simple system (other than the fact that it's too simple): We can only send an 8-bit message.

- What if we want to send English instead of bits? This is no problem since computers store *everything* as bits. For example, text is stored using ASCII.

- What if we want to send longer messages? Well, we could break it into pieces that are 8-bits long and add the key to each 8-bit piece. For reasons we won't go into, using the same key K for each 8-bit piece is bad. Therefore there should be some rule for changing K. A simple rule is to replace the K for the current piece with $3K \bmod 2^8$ for the next piece.

Example 10 (Industrial espionage)   Let's return to our factories that have been happily communicating secretly with each other.

Suppose Joe, who does industrial espionage for a competitor is able to intercept the ciphertext as it passes over the internet. He wants to know what orders are being placed; that is, he wants to find the plaintext. (He knows how to interpret the plaintext since lots of people at factories A and B know what it means.)

Joe manages to get an employee to place a fake order, say 11110000.

$$
\begin{array}{ll}
11110000 & \text{plaintext} \\
11000111 & \text{key } K \\
\hline
00110111 & \text{ciphertext}
\end{array}
$$

Bob intercepts the ciphertext and adds it to the plaintext as follows:

$$
\begin{array}{ll}
00110111 & \text{ciphertext} \\
11110000 & \text{plaintext} \\
\hline
11000111 & \text{key } K
\end{array}
$$

Now Joe has the key. Clever guy!

Except that the key and messages are much longer and the function f_K is not so simple, this sort of stuff goes on in the real world all of the time. For example, K might be anywhere from 64 to 128 bits, so there are anywhere from 2^{64} to 2^{128} possibilities for K.

You might ask why Joe didn't just get an employee to tell him key. The key is in the computer program. Only a few people, if any, know what it is. Well then, how did Joe know that f_K was plaintext plus key? In the real world, people use standard encryption algorithms (i.e., standard functions) that are public knowledge. When your computer browser is in secure mode, it is using a standard algorithm that Joe knows about. □

How can a company prevent Joe from getting their secrets this way? When we're thinking about this, we should imagine that the key is longer (64 to 128 bits) and that the plaintext is much longer. Here are some possibilities.

- Make it harder for Joe to get K.

    o We could improve employee loyalty. This may be difficult. A more reliable solution would be preferred.

    o We could invent an encryption system so that, even with plaintext and ciphertext, it is hard for Joe to compute K. Later, we'll discuss a way to do this.

- Change K frequently.

    o Sending out a new K may be feasible with two factories. It's much harder if there are a hundred — there are logistic and security problems. Why can't we simply encrypt the new K and send it out? Because, if Joe has the old K, he can read the message and get the new one.

    o When two computers want to communicate, have them decide on a K for that communication. This sounds impossible since Joe could eavesdrop. Later, we'll discuss a way to do this.

- Make Joe's knowledge of K useless.

    o We could invent an encryption system so that, even with K and ciphertext it is hard for Joe to compute plaintext without some additional (secret) information. Later, we'll discuss a way to do this.

The gcd, lcm and ϕ Functions

We now discuss some number theory functions that are important in cryptography. After we understand them, we'll use them in the Diffie-Hellman and RSA protocols.

Definition 3 (Greatest common divisor and least common multiple) *If k, n and n/k are integers, we write $k \mid n$ (read "k divides n") and we call k a divisor of n and we call n a multiple of k. The greatest common divisor of m and n is the largest (positive) k such that k is a divisor of m and k is a divisor of n. It is denoted by $\gcd(m, n)$. The least common multiple of m and n is the smallest positive integer k such that k is a multiple of m and k is a multiple of n. It is denoted by $\operatorname{lcm}(m, n)$.*

For example, if $m = 6$, its positive divisors are 1, 2, 3 and 6. Its positive multiples are 6, 12, 18, ... The greatest common divisor of 6 and 9 is 3, written $\gcd(6, 9) = 3$. Similarly, $\operatorname{lcm}(6, 9) = 18$.

The $\gcd(120, 26) = 2$. It is also the case that $5 \times 120 - 23 \times 26 = 2$. In other words, there are integers $a = 5$ and $b = -23$ such that $am + bn = \gcd(m, n)$ where $m = 120$ and $n = 26$. This is a fact that is true for any m and n. That is, we claim

Theorem 5 (The gcd as a linear combination) *The greatest common divisor of m and n is a linear combination, with integral coefficients, of m and n.*

Corollary (All common divisors) *An integer k divides m and n if and only if it divides $\gcd(m, n)$.*

Proof: We can see why this must be true without knowing how to compute the coefficients a and b. The set $S = \{Am + Bn \mid A, B \in \mathbb{Z}, Am + Bn > 0\}$ is a nonempty set of positive integers (since $|m| \in S$) and therefore has a least element (by common sense at this point). Let $am + bn = L$ be this least element. Note that $L \mid m$. If not, we would have $m = qL + r$, $0 < r < L$. Thus,

$$r = m - qL = m - q(am + bn) = (1 - qa)m - (qb)n \in S.$$

This would contradict the minimality of L since $0 < r < L$. Similarly, $L \mid n$. Thus, L is a common divisor of m and n. Any integer x that is a common divisor of m and n divides any element $Am + Bn$ of S and thus $x \mid L$. Thus, $L = \gcd(m, n)$ is the greatest common divisor of m and n. This proves that $am + bn = \gcd(m, n)$.

In the last couple of sentences of the previous paragraph, we concluded that, if x divides both m n, then $x \mid \gcd(m, n)$. Conversely, suppose $x \mid \gcd(m, n)$. This means that x divides both m and n. This proves the corollary. $\square$

Example 11 (Some properties of gcd and lcm) Let $n > 0$ and $m > 0$ be positive integers and let $n = p_1^{e_1} p_2^{e_2} \cdots p_k^{e_k}$ and $m = p_1^{f_1} p_2^{f_2} \cdots p_k^{f_k}$ be factorizations of m and n into primes where some of the exponents f_i or e_i may be zero (in order to make k and the list of p_i the same for both factorizations). For example, $n = 6500 = 2^2 \times 5^3 \times 13$ and $m = 24696 = 2^3 \times 3^2 \times 7^3$ would, using this convention, be written as $n = 2^2 \times 3^0 \times 5^3 \times 7^0 \times 13^1$ and $m = 2^3 \times 3^2 \times 5^0 \times 7^3 \times 13^0$. The following theorem is the general result of which this example is a special case. We will not prove it. You should think carefully about the example and make up some of your own until you see why the theorem is true.

Theorem 6 (Computing gcd and lcm) If $n = p_1^{e_1} p_2^{e_2} \cdots p_k^{e_k}$ and $m = p_1^{f_1} p_2^{f_2} \cdots p_k^{f_k}$, then

$$\gcd(m,n) = p_1^{\min(e_1,f_1)} p_2^{\min(e_2,f_2)} \cdots p_k^{\min(e_k,f_k)}$$

and

$$\text{lcm}(m,n) = p_1^{\max(e_1,f_1)} p_2^{\max(e_2,f_2)} \cdots p_k^{\max(e_k,f_k)}.$$

Applying this to

$$6500 = 2^2 \times 3^0 \times 5^3 \times 7^0 \times 13^1 \quad \text{and} \quad 24696 = 2^3 \times 3^2 \times 5^0 \times 7^3 \times 13^0$$

gives

$$\gcd(6500, 24696) = 2^2 \times 3^0 \times 5^0 \times 7^0 \times 13^0 = 4$$

and

$$\text{lcm}(6500, 24696) = 2^3 \times 3^2 \times 5^3 \times 7^3 \times 13^1 = 40131000.$$

This is really pretty easy!

The theorem has various consequences.

- Every divisor $d = p_1^{d_1} p_2^{d_2} \cdots p_k^{d_k}$ of m and n has $d_i \leq e_i$ and $d_i \leq f_i$. Thus $d_i \leq \min(e_i, f_i)$ and so d is also a divisor of $\gcd(m, n)$. That is, every common divisor of m and n is a divisor of $\gcd(m, n)$. (We also proved this in the process of proving Theorem 5.) Conversely, every divisor of $\gcd(m, n)$ is a common divisor of m and n.

- Similarly, every common multiple of m and n is a multiple of $\text{lcm}(m, n)$. Conversely, every multiple of $\text{lcm}(m, n)$ is a common multiple of m and n.

- $\gcd(m, n)\text{lcm}(m, n) = mn$ because $\min(e_i, f_i) + \max(e_i, f_i) = e_i + f_i$ and so the p_i term in $\gcd(m, n)\text{lcm}(m, n)$ is

$$p_i^{\min(e_i,f_i)} p_i^{\max(e_i,f_i)} = p_i^{\min(e_i,f_i)+\max(e_i,f_i)} = p_i^{e_i+f_i} = p_i^{e_i} p_i^{f_i}.$$

- If d is a common divisor of m and n, then $\gcd(m/d, n/d) = \gcd(m, n)/d$. In particular, when $d = \gcd(m, n)$, we have $\gcd(m/d, n/d) = 1$. We omit the proof. $\square$

The one thing you have to do to use the previous method for computing greatest common divisors and least common multiples is to factor n and m into primes. That can be difficult for big numbers. This method for computing gcd and lcm is more of theoretical or conceptual interest than of practical interest. Commonly available software for your computer will compute the gcd and the lcm quickly and efficiently for most integers that you may be interested in, without having to factor the integers. In the next example, we discuss the method that the software uses.

Number Theory and Cryptography

Example 12 (The Euclidean algorithm) Suppose we want to compute $\gcd(330, 156)$. Here's a "magical" procedure for doing it.

- We form a sequence that starts 330, 156.

- To get the next term in the sequence, divide 156 into 330 and keep the remainder: 330, 156, 18.

- To get the next term in the sequence, divide 18 into 156 and keep the remainder: 330, 156, 18, 12.

- To get the next term in the sequence, divide 12 into 18 and keep the remainder: 330, 156, 18, 12, 6.

- To get the next term in the sequence, divide 6 into 12 and keep the remainder: 330, 156, 18, 12, 6, 0.

Since we've reached zero, we stop and the term just before it (namely six) is the greatest common divisor. We could have started with 156, 330. Then we would have 156, 330, 156, 18, 12, 6, 0.

We need to formulate this in general and we need to prove that it works; that is, it isn't magic.

Here's the general procedure. Given two numbers $m > 0$ and $n > 0$, let $X_1 = m$ and $X_2 = n$. Define X_{k+1} to be the remainder when X_{k-1} is divided by X_k. Since X_{k+1} is a remainder, $X_{k+1} < X_k$. Thus we have $X_2 > X_3 > \cdots$. This eventually must reach zero, say $X_{t+1} = 0$. Then $\gcd(m, n) = X_t$. This is known as the *Euclidean algorithm*.

Why does it work? We claim that $\gcd(X_{k+1}, X_k) = \gcd(X_k, X_{k-1})$ for $k = 2, 3, \ldots, t$. Before proving this, let's see why it tells us that the algorithm works. We have

$$\gcd(m, n) = \gcd(X_1, X_2) = \gcd(X_2, X_3) = \cdots = \gcd(X_t, X_{t+1}) = \gcd(X_t, 0) = X_t,$$

where $\gcd(X_t, 0) = X_t$ since all numbers divide zero.

Now for the proof of the claim. Since X_{k+1} is the remainder after dividing X_{k-1} by X_k, it follows that $X_{k+1} = X_{k-1} - qX_k$ where q is the quotient when we divide X_{k-1} by X_k. Our claim states that

$$\gcd(X_{k-1} - qX_k, \ X_k) = \gcd(X_k, X_{k-1}).$$

More generally, we claim that $\gcd(a, \ b - ca) = \gcd(a, b)$ for any integers a, b, c. Suppose $d \mid a$ and $d \mid b$, then $a = Ad$ and $b = Bd$ for some integers A and B. Then

$$b - ca = Bd - cAd = (B - cA)d \quad \text{and so} \quad d \mid (b - ca).$$

Suppose $d \mid a$ and $d \mid (b - ca)$, then $a = Ad$ and $b - ca = Cd$ for some integers A and C. Then

$$b = (b - ca) + ca = Cd + cAd = (C + cA)d \quad \text{and so} \quad d \mid b.$$

We've now shown that d is a common divisor of a and b if and only if it is a common divisor of a and $b - ca$. This completes the proof. $\square$

Example 13 (The Euclidean algorithm and Theorem 5) In Theorem 5 we showed that there are $a, b \in \mathbb{Z}$ so that $\gcd(m,n) = am + bn$, but we had no idea how to compute a and b. The Euclidean algorithm, with a slight modification, allows us to compute the a and b. Suppose we start with $m = X_1$ and $n = X_2$ and apply the Euclidean algorithm to get $X_t = d = \gcd(m,n)$:

$$X_1 > X_2 > X_3 > \cdots > X_t > X_{t+1} = 0.$$

Let $Q_2, Q_3, \ldots, Q_{t-1}$ be the list of quotients associated with the nonzero remainders in this list. Thus, $X_{i-1} = Q_i X_i + X_{i+1}$ for $i = 2, \ldots, t - 1$. Note that $X_{t-2} = Q_{t-1} X_{t-1} + X_t$ so $\gcd(m,n) = X_t = X_{t-2} - Q_{t-1} X_{t-1}$. If $t = 3$ we would have $am + bn = \gcd(m,n)$ with $a = 1$, $b = -Q_{t-1}$, and our work would be done!

If $t > 3$, we can continue in the same way. We still have $X_t = X_{t-2} - Q_{t-1} X_{t-1}$. We also have $X_{t-1} = X_{t-3} - Q_{t-2} X_{t-2}$. If we substitute the second equation into the first, we get $X_t = \gcd(m,n)$ as a linear combination with integral coefficients of X_{t-3} and X_{t-2}. If $t = 4$, we are done. Otherwise, using $X_{t-2} = X_{t-4} - Q_{t-3} X_{t-3}$, we get X_t as a linear combination of X_{t-3} and X_{t-4}. Note that we are working our way towards getting $X_t = \gcd(m,n)$ as a linear combination with integral coefficients of X_1 and X_2. At this point we abandon the general discussion and move to an example.

Consider $X_1 = 60$ and $X_2 = 13$. Here is the list of nonzero remainders produced by the Euclidean algorithm:
$$60 > 13 > 8 > 5 > 3 > 2 > 1.$$

Thus, $t = 7$ and $\gcd(60, 13) = 1$. We kept track of the quotients: 4, 1, 1, 1, 1. To make it easier to see the connection between quotients and remainders we can write them in this way

60	>	13	>	8	>	5	>	3	>	2	>	1
		4		1		1		1		1		

where we see that $60 = 4 \times 13 + 8$, $13 = 1 \times 8 + 5$, $\ldots$, $3 = 1 \times 2 + 1$. Now we start working backwards. $1 = 3 - 1 \times 2$, $2 = 5 - 1 \times 3$, so $1 = 2 \times 3 - 1 \times 5$. Next we have $3 = 8 - 1 \times 5$, so

$$1 = 2(8 - 5) - 1 \times 5 = 2 \times 8 - 3 \times 5.$$

Next, $5 = 13 - 1 \times 8$, so $1 = 2 \times 8 - 3(13 - 8) = 5 \times 8 - 3 \times 13$. Finally, $8 = 60 - 4 \times 13$, so $1 = 5 \times 60 - 23 \times 13$. This is the final answer: $1 = \gcd(m,n) = am + bn$ where $m = 60$, $n = 13$, $a = 5$, and $b = -13$. You should make up some examples on your own and carry out this computation. □

The positive integers $k = 1, 5, 7, 11$ are less than 12 and have no common factors with 12 (i.e., are *relatively prime* to 12). Another way to say this is $\gcd(k, 12) = 1$. The four numbers, 1, 5, 7, and 11 are the only numbers k with $\gcd(k, 12) = 1$ and $1 \le k \le 12$. For this reason, we say $\phi(12) = 4$. More generally:

Definition 4 (The Euler ϕ function) *We define a function $\phi(n)$, with domain the positive integers, to be the number of integers k, $1 \le k \le n$, such that $\gcd(k, n) = 1$. This function is called the Euler ϕ function.*

Number Theory and Cryptography

Example 14 (Properties of the Euler ϕ function)   We have noted that $\phi(12) = 4$. Since $\gcd(1,1) = 1$, we have $\phi(1) = 1$. For any prime p, we have $\phi(p) = p - 1$ because $\gcd(k,p) = 1$ for $k = 1, 2, \ldots, p - 1$.

Suppose $n = pq$ is the prime factorization of n and $p \neq q$. We can list the positive integers less than n that are *not* relatively prime to n. There are two classes of such numbers. The q multiples of p: $p, 2p, 3p, \ldots, qp$ and the p multiples of q: $q, 2q, 3q, \ldots, pq$. Except for $qp = pq$, these two lists have no numbers in common (why?). Thus, the total number of positive integers less than or equal to n that *are not* relatively prime to n is $q + p - 1$. Thus, the number of number less than or equal to $n = pq$ that *are* relatively prime to n is $pq - (p + q - 1) = (p - 1)(q - 1)$.

The set of numbers less than n that *are* relatively prime to n has a name. It is called the *group of units of n* and the numbers in that set are called *units*. The reason for this name is beyond the scope of our course, but does not involve difficult concepts. The Euler ϕ function and the group of units come into computer science in connection with computer security. It is the basis for a certain type of encryption known as *RSA* (discussed below) and is used in a common encryption protocol called PGP (Pretty Good Privacy). The key property that makes the group of units useful in this context is that $a^{\phi(n)} = 1 \pmod{n}$ whenever a is a unit. We won't prove this fact, but let's look at an example. Suppose $n = 12$. We know that $\phi(12) = 4$ and that the units are $\{1, 5, 7, 11\}$. Clearly $1^{\phi(12)} = 1$ $\pmod{12}$. What about the other units? We have $5^2 = 25 = 1 \pmod{12}$. Thus $5^4 = 1^2 = 1$ $\pmod{12}$. We could do the same calculations for 7 and 11. Here's another way. Since $7 = -5 \pmod{12}$, $7^4 = (-1)^4 5^4 = 5^4 = 1 \pmod{12}$. Likewise, $11 = -1 \pmod{12}$ and so $11^4 = (-1)^4 = 1 \pmod{12}$. You may have noticed that $a^2 = 1 \pmod{12}$ for all units a. There's no guarantee that $\phi(n)$ is the least power for which $a^{\phi(n)} = 1 \pmod{n}$ for all units a.

If $n = pq$ then, since $\phi(n) = (p - 1)(q - 1)$, this property becomes

$$m^{(p-1)(q-1)} = 1 \pmod{pq} \quad \text{when} \quad \gcd(m, pq) = 1.$$

This fact will be important in our discussion of the RSA protocol.   □

Cryptography on the Internet

Suppose two people — Alice and Bob — wish to communicate secretly, but anyone can eavesdrop on there conversation. How can they do this? We already saw in Example 9 how they could do this, and we saw how some problems could arise because of espionage. There's another problem we haven't mentioned. What if Alice and Bob don't have a secret key K that they both know?

Cryptography on the internet addresses this. It uses "public-information algorithms": No prior secret communication between Alice and Bob is needed — it's all done publicly. There are two approaches in use.

- Somehow Alice and Bob can develop a secret key even though someone is eavesdropping on their conversation. In this process, Alice and Bob usually play similar roles and so this is known as *symmetric encryption*.

• Alice can make known to the world data that allows people to encrypt messages to send to her but makes it hard for people other than Alice to decrypt them. Bob can do the same. Since this information (the key) is publicly known, this approach is called *public key cryptography*.

These approaches depend on what are called *trapdoor functions*. A trapdoor function is an invertible function g such that, given $g(x)$ it is hard to compute x. Such functions are also called *one-way functions*, but this is a bit misleading since it suggests that g is not invertible. We will discuss protocols that use two different trapdoor functions.

Example 15 (Discrete logs and better encryption) There are many ways to design a system such that, knowing the plaintext and ciphertext, it is still hard to recover the key. The method we describe here is not actually used, but it lays some of the groundwork for our next example.

If you use your calculator, you can easily compute $11^7 = 19487171$. If you know that 19487171 is of the form 11^x, for some x, you can equally well use your calculator to get x. From high school, you should remember that $x = \log_{11}(19487171)$. Probably, you would do that calculation using the LOG or LN button on your calculator as follows: $LOG(19487171)/LOG(11) = 7$. In any case, it is pretty easy. But, a seemingly innocent modification makes this sort of calculation very difficult in many cases.

If we compute $11^t \% 163$ for $t = 0, 2, \ldots, 161$, we get each of the numbers $1, 2, \ldots, 162$ exactly once — but they are in a mixed up order. Instead of 11^7, let's compute $11^7 \% 163$. The answer is 32. Now its not so easy to solve the equation $32 = 11^x \% 163$ for x even though we know there is a unique x between 0 and 161. For small numbers like these, it can be done by trying all $0 \le x < 162$. But, for big numbers with hundreds of digits, it seems to be all but impossible by any presently available methods. This problem of recovering an exponent from an exponentiated expression after it has been reduced modulo some number is called the *discrete logarithm problem* and the exponent is called the *discrete logarithm*.

Here is how we might use discrete logarithms to make it very hard for Joe's espionage when Alice and Bob have a secret key K. We choose a large modulus p that never changes. When someone wants to send a message P, the computer chooses a "base" b at random and computes $b^K \% p$. Call the result of this computation L.

The computer uses L to encrypt P by whatever method is being used for encryption. Thus, the computer obtains $f_L(P) = C$. It sends b and C. The computer at the other end computes $b^K \% p$ to obtain L and uses it to decrypt the message.

What can the spy Joe do? Suppose the encryption method is the one used in Example 10: We simply write L as a binary number and add it bitwise to the message P. Since the modulus p is fixed, we'll assume Joe knows what it is. As before, Joe gets his friend to send a message, so he has P, C and b for this particular message — call them P_1, C_1 and b_1. From P_1 and C_1, Joe recovers L_1. Later, someone else sends a message P_2. The computer chooses a random b_2, computes $b_2^K \% p = L_2$ and C_2. By eavesdropping Joe gets b_2 and C_2.

• To decrypt the message, Joe needs to find L_2 so that he can add it bitwise to C_2.

• To get L_2 he needs K because $L_2 = b_2^K \pmod{p}$ and he knows b_2.

[73]   NT-21

Number Theory and Cryptography

- To get K he needs to solve the discrete log problem because he has b_1 and L_1 and $b_1^K = L_1 \pmod{p}$.

This is too hard, so Joe gives up.

There was nothing special about adding L bitwise to P. Whatever method was used, Joe would still want to recover K and so would need to carry out the steps in the previous paragraph. □

Suppose the values of b and p are known and fixed. The function g, defined by $g(n) = b^n \% p$, is thought to be a trapdoor function. Finding n from $g(n)$ is referred to as computing the discrete log of b^n. As remarked in the previous example, computing the discrete log is believed to be very difficult. Thus g is believed to be a trapdoor function.

Suppose Alice and Bob want to communicate over the internet in secrecy, but have no shared key K. They must somehow construct K even though Joe can read their communications.

Example 16 (Diffie-Hellman: a symmetric key-exchange protocol) Here is how two computers can use the difficulty of the discrete log problem to generate a key K that they will share. Everyone agrees on a modulus p that is built into a program all computers can use. They also agree on a base b. Thus everyone, including the spy Joe, knows p and b. For purposes of illustration, we take $p = 163$ and $b = 11$. The values actually used on the internet are *much* bigger. We call the two computers that want to communicate A and B.

Computer A chooses, in secret, a random number s with $1 < s < p - 1$. Let us say 13 is chosen by A. Then A computes $b^s \% p = S$ and sends S to computer B. In our example, $S = 19$ since $11^{13} \% 163 = 19$. Meanwhile, B carries out the same process, choosing t and computing T, which it sends to A. Let us say B chooses $t = 23$. Thus B computes[5] $T = 11^{23} \% 163 = 116$.

Where are we now? Both computers and the spy Joe know that $S = 19$ and $T = 116$. Only computer A knows that $s = 13$ and only computer B knows that $t = 23$. In general, the public information is b, p, S and T; however, s and t are *not* public information since they were never sent over the internet.

What do the computers do now? Computer A uses its secret number s and computes $T^s \% p = K$. In our case, $116^{13} \% 163 = 154$, so $K = 154$. Likewise, B computes $S^t \% p = K$, which is $19^{23} \% 163 = 154$ in our case. That's amazing — A and B have the same number! Why is this? With all calculations modulo p, we have

$$T^s = (b^t)^s = b^{ts} = (b^s)^t = S^t \pmod{p}.$$

Where does this leave Joe? The obvious way for him to get key is to find either s or t since he already knows S and T. To find s, he needs to solve the discrete log problem $b^s = S \pmod{p}$. Likewise for T. Maybe there is a clever way for Joe to get K easily from b, p, S and T. At the present time, nobody knows of any such method, so Joe is stuck.

---

[5] The following computations and others like it can be done by using software packages such as GNU-MP, Maple® and Mathematica®. If you have to do it on a pocket calculator, it's best to do it in steps taking advantage of the properties of modular arithmetic.

The method of key exchange just discussed is called the Diffie-Hellman algorithm. It was discovered in 1976 and was the first public-information algorithm invented — invented *in public* that is! Apparently, the same algorithm, as well as other, later-to-be-discovered algorithms (such as RSA — Rivest, Shamir, Adleman, published by them in 1978), were discovered by British cryptanalysts working in secret in the Communications-Electronics Security Group in Britain during the early 1970's. Working in that group, Malcolm Williamson discovered the "Diffie-Hellman" algorithm in 1974. □

Our next example is based on the difficulty of factoring. In this case, g is the function from pairs of primes $p < q$ to their product; that is, $g(p,q) = pq$. This is believed to be a trapdoor function when both p and q are large. To put this another way, all known methods of factoring take a long time. The protocol in this example is due to Rivest, Shamir and Adleman and so is called the RSA protocol.

*Example 17 (The RSA protocol)   This encryption system is based on the choice of some integer N that is a product of two primes. Suppose we take $N = 77$. We see easily that $N = pq$ where $p = 7$, $q = 11$. In real applications of this protocol p and q are primes with hundreds of digits, so given $N = pq$, it is very hard (or so it seems with present techniques) to factor N to get p and q. This is where the security of this method resides. Let's pretend that Alice makes known to the public her integer 77, and that Bob wants to send her a message. Suppose the spy Joe can't figure out how to factor 77. (In RSA this is true because much larger primes are used and multiplication is believed to be a trapdoor function.)

Alice is going to make known some more information. She picks two numbers e and d such that $ed = 1 \pmod{60}$. Why 60? Because $60 = (p-1)(q-1) = \phi(77)$. Suppose Alice picks $e = 13$ and $d = 37$. In this case $ed = 13 \times 37 = 481$. Check it out: $481 = 1 \pmod{60}$. She makes known to the public $e = 13$ and keeps $d = 37$ secret. Since Joe can't factor 77, he can't get the values $p = 7$ and $q = 11$. Hence Joe can't get the number $(p-1)(q-1) = 60$, and so he can't figure out that $d = 37$, given the publicly displayed number $e = 13$.

By the way, we didn't say how Alice chose the pair $e = 13$ and $d = 37$. Well, she just picked the e because it "seemed like a nice number." So that's her choice, as long as $\gcd(e, 60) = 1$. Clearly $\gcd(13, 60) = 1$, so she did all right there. To pick the $d = 37$ she used the method in Example 13 applied to $m = 13$, $n = 60$. You should reconstruct her calculations.

So now we have all that Alice is willing to tell the world: $N = 77$ and $e = 13$. In other words N and e are Alice's public information. The factorization $N = pq$ and the value of d are *not* public information because they were not sent over the internet.

In this encryption scheme, the only messages that Bob can send are numbers M, $1 \le M < 77 = N$, where $\gcd(M, N) = 1$. There are $\phi(77) = 60$ such messages. For example, Bob may decide to send the message $M = 5$. To send his message, he looks at Alice's public information (77 and 13) and sends $M^e \% 77 = 5^{13} \% 77$. You can easily check on your calculator that $5^{13} = 26 \pmod{77}$. In general, $M^e \% N$ is sent by Bob. Call it C.

So now Alice receives the message 26. Here is what she does to decrypt the message. She computes $26^{37} \% 77$ and gets 5. Recall that 37 was her secret number paired with 13.

This is the RSA protocol.

Suppose Joe intercepts C by eavesdropping. (In this case, the value was 26.) What can he do? If he knew $d = 37$, his life would be simple since he could do what Alice has done to decrypt the message. As far as is known, he'd have to be able to factor N in order to compute d — too hard! Could he do something else? Nobody knows of anything Joe could do that would not be hard.

Some of you might think that Joe had to solve the discrete log problem rather than the factoring problem since he saw $M^e \% N$. In the discrete log problem for $M^e \bmod N$, we know M and want to find e. Joe's problem is just the reverse — he knows e and wants to find M. This is believed to be a hard problem and is believed to be equivalent to factoring.

Why does Alice's decryption method work? In general, she is sent C, which is $M^e \% N$, and computes $C^d = (M^e)^d = M^{ed} \pmod{N}$. Recall, that $ed = 1 \pmod{\phi(N)}$. Thus $ed = 1 + k\phi(N)$ for some integer k. Hence

$$M^{ed} = M^{1+k\phi(N)} = M \times \left(M^{\phi(N)} \right)^k.$$

Since $\gcd(M, N) = 1$, M is a unit (see Example 14) and so, by the property at the end of Example 14, $M^{\phi(N)} = 1 \bmod N$. Thus

$$M^{ed} = M \times (1)^k = M \pmod{N}.$$

Since $1 \le M < N$, we have recovered M exactly, not just "mod N."

The RSA protocol is not a good scheme for amateurs to set up for themselves. The primes p and q must be hundreds of digits. The numbers e and d must be chosen with care. We must be sure that $\gcd(M, N) = 1$. Since $\phi(N)/N = (1 - \frac{1}{p})(1 - \frac{1}{q})$, it is close to 1 for large p and q. So the probability of selecting a bad message M (one that is not relatively prime to N) is quite small. $\square$

Exercises for Section 2

2.1. Use the Euclidean algorithm to find all common divisors of

(a) 1001 and 544   (b) 3510 and 652

2.2. Find all common divisors of 252 and 180 using the Euclidean algorithm.

2.3. How many common divisors are there of 59400 and 16200?

2.4. Using the Euclidean algorithm, find A and B such that $Am + Bn = \gcd(m.n)$ where $m = 252$ and $n = 180$.

2.5. Using the Euclidean algorithm, find A and B such that $Am+Bn = \gcd(m.n)$ where $m = 59400$ and $n = 16200$.

2.6. Using the Euclidean algorithm, find A and B such that $Am+Bn = \gcd(m.n)$ where $m = 163$ and $n = 86$.

2.7. Prove that $\gcd(a,b)$ divides $\operatorname{lcm}(a,b)$.

2.8. In each case find $\operatorname{lcm}(120,108)$ (a) by prime factorization and (b) by the Euclidean algorithm.

2.9. Suppose a and b are positive integers. Prove directly from the definition of the least common multiple that $a \mid b$ if and only if $\operatorname{lcm}(a,b) = b$.

2.10. Following Example 16, suppose $p = 163$, $b = 11$. Computer A still chooses 13, but B chooses 15 instead of 23. What is the common key that results?

2.11. Suppose that, in Example 16, one of the computers chooses 1. Explain how the spy Joe can detect that and get their shared key.

*2.12. Suppose that N is a prime in the RSA protocol of Example 17. How can the spy Joe find the message M if he has e, N and the encrypted message $M^e \% N = C$?

*2.13. Using the same numbers as in Example 17, decrypt the message 2.

*2.14. Consider the RSA protocol (Example 17). Suppose that $N = 5 \times 13$ and $e = 7$. What is d?

*2.15. Consider the RSA protocol (Example 17). Explain why d and e must both be chosen to be odd.

Number Theory and Cryptography

Multiple Choice Questions for Review

In each case there is one correct answer (given at the end of the problem set). Try to work the problem first without looking at the answer. Understand both why the correct answer is correct and why the other answers are wrong.

1. "If $k > 1$ then $2^k - 1$ is not a perfect square." Which of the following is a correct proof?

   (a) If $2^k - 1 = n^2$ then $2^{k-1} - 1 = (n-1)^2$ and $\frac{n^2+1}{(n-1)^2+1} = \frac{2^k}{2^{k-1}} = 2$. But this latter ratio is 2 if and only if $n = 1$ or $n = 3$. Thus, $2^k - 1 = n^2$ leads to a contradiction.

   (b) If $2^k - 1 = n^2$ then $2^k = n^2 + 1$. Since 2 divides n^2, 2 does not divide $n^2 + 1$. This is a contradiction since obviously 2 divides 2^k.

   (c) $2^k - 1$ is odd and an odd number which is a perfect square can't differ from a power of two by one.

   (d) $2^k - 1$ is odd and an odd number can never be a perfect square.

   (e) If $2^k - 1 = n^2$ then n is odd. If $n = 2j + 1$ then $2^k - 1 = (2j+1)^2 = 4j^2 + 4j + 1$ which implies that 2^k, $k > 1$ is divisible by 2 but not by 4. This is a contradiction.

2. The repeating decimal number $3.14159265265265\ldots$ written as a ratio of two integers a/b is

   (a) $313845111/99990000$

   (b) $313844841/99900000$

   (c) $313845006/99990000$

   (d) $313845106/99900000$

   (e) $313845123/99000000$

3. Which of the following statements is true:

   (a) A number is rational if and only if its square is rational.

   (b) An integer n is odd if and only if $n^2 + 2n$ is odd.

   (c) A number is irrational if and only if its square is irrational.

   (d) A number n is odd if and only if $n(n + 1)$ is even

   (e) At least one of two numbers x and y is irrational if and only if the product xy is irrational.

4. Which of the following statements is true:

   (a) A number k divides the sum of three consecutive integers n, $n + 1$, and $n + 2$ if and only if it divides the middle integer $n + 1$.

   (b) An integer n is divisible by 6 if and only if it is divisible by 3.

   (c) For all integers a, b, and c, $a \mid bc$ if and only if $a \mid b$ and $a \mid c$.

   (d) For all integers a, b, and c, $a \mid (b + c)$ if and only if $a \mid b$ and $a \mid c$.

Review Questions

(e) If r and s are integers, then $r \mid s$ if and only if $r^2 \mid s^2$.

5. For all $N \geq 0$, if $N = k(k+1)(k+2)$ is the product of three consecutive non-negative integers then for some integer $s > k$, N is divisible by a number of the form

(a) $s^2 - 1$

(b) $s^2 - 2$

(c) s^2

(d) $s^2 + 1$

(e) $s^2 + 2$

6. To one percent accuracy, the number of integers n in the list $0^4, 1^4, 2^4, \ldots, 1000^4$ such that $n \% 16 = 1$ is

(a) 20 percent

(b) 50 percent

(c) 30 percent

(d) 35 percent

(e) 25 percent

7. Which of the following statements is TRUE:

(a) For all odd integers n, $\lceil n/2 \rceil = \frac{n+1}{2}$.

(b) For all real numbers x and y, $\lceil x + y \rceil = \lceil x \rceil + \lceil y \rceil$.

(c) For all real numbers x, $\lceil x^2 \rceil = (\lceil x \rceil)^2$.

(d) For all real numbers x and y, $\lfloor x + y \rfloor = \lfloor x \rfloor + \lfloor y \rfloor$.

(e) For all real numbers x and y, $\lfloor xy \rfloor = \lfloor x \rfloor \lfloor y \rfloor$.

8. Which of the following statements is logically equivalent to the statement, "If a and $b \neq 0$ are rational numbers and $r \neq 0$ is an irrational number, then $a + br$ is irrational."

(a) If a and $b \neq 0$ are rational and $r \neq 0$ is real, then $a + br$ is rational only if r is irrational.

(b) If a and $b \neq 0$ are rational and $r \neq 0$ is real, then $a + br$ is irrational only if r is irrational.

(c) If a and $b \neq 0$ are rational and $r \neq 0$ is real, then r is rational only if $a + br$ is rational.

(d) If a and $b \neq 0$ are rational and $r \neq 0$ is real, then $a + br$ is rational only if r is rational.

(e) If a and $b \neq 0$ are rational and $r \neq 0$ is real, then $a + br$ is irrational only if r is rational.

9. The number of primes of the form $|n^2 - 6n + 5|$ where n is an integer is

(a) 0   (b) 1   (c) 2   (d) 3   (e) 4

[79]

NT-27

Number Theory and Cryptography

10. The Euclidean Algorithm is used to produce a sequence $X_1 > X_2 > \cdots > X_{k-1} > X_k = 0$ of positive integers where each X_t, $2 < t \leq k$, is the remainder gotten by dividing X_{t-2} by X_{t-1}. If $X_{k-1} = 45$ then the set of all (positive) common divisors of X_1 and X_2 is

    (a) $\{1, 3, 5\}$

    (b) $\{1, 3, 5, 9, 15, \}$

    (c) $\{1, 9, 15, 45\}$

    (d) $\{1, 3, 5, 15\}$

    (e) $\{1, 3, 5, 9, 15, 45\}$

11. Let L be the least common multiple of 175 and 105. Among all of the common divisors $x > 1$ of 175 and 105, let D be the smallest. Which is correct of the following:

    (a) $D = 5$ and $L = 1050$

    (b) $D = 5$ and $L = 35$

    (c) $D = 7$ and $L = 525$

    (d) $D = 5$ and $L = 525$

    (e) $D = 7$ and $L = 1050$

12. The Euclidean Algorithm is used to produce a sequence $X_1 > X_2 > X_3 > X_4 > X_5 = 0$ of positive integers where $X_t = q_{t+1} X_{t+1} + X_{t+2}, t = 1, 2, 3$. The quotients are $q_2 = 3$, $q_3 = 2$, and $q_4 = 2$. Which of the following is correct?

    (a) $\gcd(X_1, X_2) = -2X_1 + 6X_2$

    (b) $\gcd(X_1, X_2) = -2X_1 - 6X_2$

    (c) $\gcd(X_1, X_2) = -2X_1 - 7X_2$

    (d) $\gcd(X_1, X_2) = 2X_1 + 7X_2$

    (e) $\gcd(X_1, X_2) = -2X_1 + 7X_2$

Answers: **1** (e), **2** (d), **3** (b), **4** (e), **5** (a), **6** (b), **7** (a), **8** (d), **9** (c), **10** (e), **11** (d), **12** (e).

Sets and Functions

Section 1: Sets

The basic concepts of sets and functions are topics covered in high school math courses and are thus familiar to most university students. We take the intuitive point of view that sets are unordered collections of objects. We first recall some standard terminology and notation associated with sets. When we speak about sets, we usually have a "universal set" U in mind, to which the various sets of our discourse belong.

Definition 1 (Set notation)   *A set is an unordered collection of distinct objects. We use the notation $x \in S$ to mean "x is an element of S" and $x \notin S$ to mean "x is not an element of S." Given two subsets (subcollections) of U, X and Y, we say "X is a subset of Y," written $X \subseteq Y$, if $x \in X$ implies that $x \in Y$. Alternatively, we may say that "Y is a superset of X." $X \subseteq Y$ and $Y \supseteq X$ mean the same thing. We say that two subsets X and Y of U are equal if $X \subseteq Y$ and $Y \subseteq X$. We use braces to designate sets when we wish to specify or describe them in terms of their elements: $A = \{a, b, c\}$, $B = \{2, 4, 6, \ldots\}$. A set with k elements is called a k-set or set with cardinality k. The cardinality of a set A is denoted by $|A|$.*

Since a set is an unordered collection of distinct objects, the following all describe the same 3-element set

$$\{a, b, c\} = \{b, a, c\} = \{c, b, a\} = \{a, b, b, c, b\}.$$

The first three are simply listing the elements in a different order. The last happens to mention some elements more than once. But, since a set consists of distinct objects, the elements of the set are still just a, b, c. Another way to think of this is:

> Two sets A and B are equal if and only if every element of A is an element of B and every element of B is an element of A.

Thus, with $A = \{a, b, c\}$ and $B = \{a, b, b, c, b\}$, we can see that everything in A is in B and everything in B is in A. You might think "When we write a set, the elements are in the order written, so why do you say a set is not ordered?" When we write something down we're stuck — we have to list them in some order. You can think of a set differently: Write each element on a separate slip of paper and put the slips in a paper bag. No matter how you shake the bag, it's still the same set.

If we are given that A is a set and no other information about A, then there is no ordering to the elements of A. Thus, we cannot speak of "the second element of the set A" unless we have specified an ordering of the elements of A. If we wish to regard A as ordered in some way, then we specify this fact explicitly: "The elements of A are ordered a, b, c," or "$A = (a, b, c)$." The latter notation replaces the braces with parentheses and designates that A is ordered, left to right, as indicated. We call this an *ordered set*. An ordered set is also called a *linear order*. Various other names are also used: *list, vector, string, word*

Sets and Functions

— all with **no repeated elements**.[1] Of course, you've seen repeated elements in vectors, for example the point in the plane at the coordinates (1,1). That's fine, it's just not an ordered set. If there are k elements in the ordered set, it is referred to as a k-*list*, k-*vector*, etc., or as a list, vector, etc., of *length* k — all with no repeated elements because they are ordered **sets**.

Sometimes we cannot list the elements of a set explicitly. What do we do if we want to describe the set of all real numbers greater than 1 without writing it out in words? We write

$$\{x \mid x \in \mathbb{R}, \ x > 1\} \quad \text{or} \quad \{x \mid x > 1\} \quad \text{or} \quad \{x : x > 1\}.$$

These are read "the set of all x such that ..." In the first example we mentioned that x was a real number ($x \in \mathbb{R}$). In the other two we didn't because we assumed the reader knew from context that we were talking about real numbers.

For the most part, we shall be dealing with finite sets. Let U be a set and let A and B be subsets of U. The sets

$$A \cap B = \{x \mid x \in A \text{ and } x \in B\}$$

and

$$A \cup B = \{x \mid x \in A \text{ or } x \in B\}$$

are the *intersection* and *union* of A and B. The set $A \setminus B$ or $A - B$ is the *set difference* of A and B (i.e., the set $\{x \mid x \in A, x \notin B\}$). The set $U \setminus A$ (also A^c, A' or $\sim A$) is the *complement* of A (relative to U). Note that $A - B = \{x \mid x \in A, x \notin B\} = A \cap B^c$. The empty set, denoted by $\emptyset$, equals U^c. Also note that, for any set $A \subseteq U$, $A \cup A^c = U$ and $A \cap A^c = \emptyset$.

The set $A \oplus B = (A \setminus B) \cup (B \setminus A)$ is the *symmetric difference* of A and B. We use $A \times B = \{(x, y) \mid x \in A, y \in B\}$ to denote the *product* or *Cartesian product* of A and B. If we want to consider the product of k sets, $A_1, \ldots, A_k$, this is denoted by $\times_{i=1}^k A_i$. If we want to consider the product of a set A with itself k times, we write $\times^k A$.

Set Properties and Proofs

The algebraic rules for operating with sets are also familiar to most beginning university students. Here is such a list of the basic rules. In each case the standard name of the rule is given first, followed by the rule as applied first to $\cap$ and then to $\cup$.

Theorem 1 (Algebraic rules for sets) *The universal set U is not mentioned explicitly but is implicit when we use the notation $\sim X = U - X$ for the complement of X. An*

---

[1] Why is it okay to specify a set $S = \{a, b, c, a\}$ where the element a has been repeated, but it is not okay to have repeated elements in an ordering of S? When we say $S = \{a, b, c, a\}$, we know that S contains just the three elements a, b and c. If we were to talk about the ordered set (a, b, c, a) it would not make sense because it would say that the element a is in two places at once: the first position and the last position.

alternative notation is $X^c = \sim X$.

Associative:	$(P \cap Q) \cap R = P \cap (Q \cap R)$	$(P \cup Q) \cup R = P \cup (Q \cup R)$
Distributive:	$P \cap (Q \cup R) = (P \cap Q) \cup (P \cap R)$	$P \cup (Q \cap R) = (P \cup Q) \cap (P \cup R)$
Idempotent:	$P \cap P = P$	$P \cup P = P$
Double Negation:	$\sim\sim P = P$	
DeMorgan:	$\sim(P \cap Q) = \sim P \cup \sim Q$	$\sim(P \cup Q) = \sim P \cap \sim Q$
Absorption:	$P \cup (P \cap Q) = P$	$P \cap (P \cup Q) = P$
Commutative:	$P \cap Q = Q \cap P$	$P \cup Q = Q \cup P$

These rules are "algebraic" rules for working with ∩, ∪, and ∼. You should memorize them as you use them. They are used just like rules in ordinary algebra: whenever you see an expression on one side of the equal sign, you can replace it by the expression on the other side.

When we wrote "$P \cap Q \cap R$" you may have wondered if we meant "$(P \cap Q) \cap R$" or "$P \cap (Q \cap R)$." The associative law says it doesn't matter. That is why you will see the notation $P \cap Q \cap R$ or $P \cup Q \cup R$ without anyone getting excited about it. On the other hand $P \cap (Q \cup R)$ and $(P \cap Q) \cup R$ may not be equal, so we need parentheses here.

The best way to "prove" the rules or to understand their validity is through the geometric device of a *Venn diagram*.

Example 1 (Venn diagrams and proofs of set equations) Here is a Venn diagram for three sets, P, Q, and R, with universal set U:

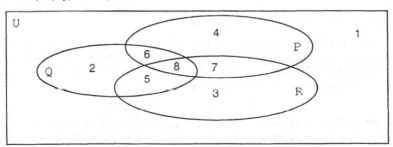

The three oval regions labeled P, Q, and R represent the sets of those names. The rectangular region represents the universal set U. There are eight subregions, labeled 1 through 8 in the picture. Region 8 represents the subset $P \cap Q \cap R$; region 1 represents $U - (P \cup Q \cup R)$; region 2 represents the elements of $Q - (P \cup R)$; and so on.

Let's use the above Venn diagram to verify that the distributive rule, $P \cup (Q \cap R) = (P \cup Q) \cap (P \cup R)$, is valid. The idea is to replace the sets P, Q, and R by their corresponding sets of regions from the Venn diagram. Thus, Q is replaced by $\{2,5,6,8\}$, P is replaced by $\{4,6,7,8\}$, and R is replaced by $\{3,5,7,8\}$. Even though the sets P, Q, and R are arbitrary, perhaps even infinite, the distributive rule reduces to verifying the same rule for these simplified sets:

$$\{4,6,7,8\} \cup (\{2,5,6,8\} \cap \{3,5,7,8\}) = (\{4,6,7,8\} \cup \{2,5,6,8\}) \cap (\{4,6,7,8\} \cup \{3,5,7,8\}).$$

SF-3

Sets and Functions

This identity is trivial to check directly: Both sides reduce to the set $\{4, 5, 6, 7, 8\}$.

This "Venn diagram" approach reduces a set identity that involves potentially infinitely many elements to subsets of a set of eight elements. It is fine for proofs and especially good for checking out "set identities" to see quickly if they are true or not. For example, is it true that $Q - (P \cap R) = Q - (P \cap Q \cap R)$? Checking the Venn diagram shows that both sides correspond to the set of regions $\{2, 5, 6\}$. The identity is true. You will get a chance to practice this technique in the exercises. $\square$

There are, of course, other ways to verify set identities. One way is called the *element method*:

Example 2 (The element method for proofs of set equations) To use that method, you simply translate the identity $X = Y$ into basic statements about what conditions a single element must satisfy to be (first) in the set on the left and then (second) in the set on the right. Thus, to show that $X = Y$, you assert that if $x \in X$ then blah, blah, blah (a bunch of words that make sense) implies that $x \in Y$. This shows that $X \subseteq Y$. Then, you reverse the argument and assert that if $y \in Y$ then blah, blah, blah (a bunch of words that make sense) implies that $y \in X$. This shows that $Y \subseteq X$. Thus $X = Y$.

Here is an example. Show, by the element method that, for all subsets P, Q, and R of U, $(P - Q) \cap (R - Q) = (P \cap R) - Q$.

(1) If $x \in (P - Q) \cap (R - Q)$ then (here comes the blah, blah, blah) x is in P but not in Q AND x is in R but not in Q.

(2) Thus x is in P **and** R, but x is not in Q.

(3) Thus x is in $(P \cap R) - Q$. This shows that $(P - Q) \cap (R - Q) \subseteq (P \cap R) - Q$. We leave it to you to use the element method to show the reverse, $(P - Q) \cap (R - Q) \supseteq (P \cap R) - Q$, and hence that $(P - Q) \cap (R - Q) = (P \cap R) - Q$. You should start your argument by saying, "Suppose $x \in (P \cap R) - Q$." $\square$

A different sort of element approach looks at each element of the universal set U and asks which sets contain it. The reult can be put in tabular form. When this is done, each row of the table corresponds to a region in the Venn diagram. The next example illustrates this *tabular method*.

Example 3 (The tabular method for proofs of set equations) We redo the identity of the previous example: $(P - Q) \cap (R - Q) = (P \cap R) - Q$. To do this we construct a table whose columns are labeled by various sets and whose entries answer the question "Is x in the set?" The first three columns in the following table are set up to allow all possible answers to the three questions "Is x in P?" "Is x in Q?" "Is x in R?" "Left" and "Right" refer to $(P - Q) \cap (R - Q)$ and $(P \cap R) - Q$, the two sides of the equation we want to prove. "Venn" refers to the region in the Venn diagram of Example 1. Normally that column would not be in the table, but we've inserted it so that you can see how each row corresponds to a Venn diagram region.

P	Q	R	$P-Q$	$R-Q$	Left	$P \cap R$	Right	Venn
No	No	No	No	No	No	No	No	1
No	No	Yes	No	Yes	No	No	No	3
No	Yes	No	No	No	No	No	No	2
No	Yes	Yes	No	No	No	No	No	5
Yes	No	No	Yes	No	No	No	No	4
Yes	No	Yes	Yes	Yes	Yes	Yes	Yes	7
Yes	Yes	No	No	No	No	No	No	6
Yes	Yes	Yes	No	No	No	Yes	No	8

Since the answers are identical in the columns labeled "Left" and "Right," the identity is proved.

We can prove more from the table. For example,

$$\text{If } P \subseteq Q \cup R, \text{ then } P - Q = (P \cap R) - Q.$$

How does the table prove this? Because of the condition, the row that begins "Yes No No" is impossible. Therefore, we throw out that row and compare columns "$P - Q$" and "Right." $\square$

Another way to prove set identities is to use the basic algebraic identities of Theorem 1. This is called the *algebraic method*.

Example 4 (An algebraic proof) It is probably a good idea for you to label the steps with the appropriate rule (e.g., DeMorgan's rule, associative rule, distributive rule, etc.) the first few times you do such a proof. Therefore, we'll do that in this example. Mathematicians, however, would rarely bother to do it. A proof is accepted if others who know the basic rules of set theory can read it, understand it, and believe it is true.

Let's prove that $Q - (P \cap R) = Q - (P \cap Q \cap R)$. Here it is

$$
\begin{aligned}
Q - (P \cap R) &= Q \cap (P \cap R)^c & & \text{since } A - B = A \cap B^c \\
&= Q \cap (P^c \cup R^c) & & \text{DeMorgan's rule} \\
&= (Q \cap P^c) \cup (Q \cap R^c) & & \text{distributive rule} \\
&= (Q \cap P^c) \cup \emptyset \cup (Q \cap R^c) & & \text{since } A \cup \emptyset = A \\
&= (Q \cap P^c) \cup (Q \cap Q^c) \cup (Q \cap R^c) & & \text{since } Q \cap Q^c = \emptyset \\
&= Q \cap (P^c \cup Q^c \cup R^c) & & \text{distributive rule} \\
&= Q \cap (P \cap Q \cap R)^c & & \text{DeMorgan's rule} \\
&= Q - (P \cap Q \cap R) & & \text{since } A - B = A \cap B^c
\end{aligned}
$$

Some steps in this proof are baffling. For example, why did we introduce $\emptyset$ in the fourth line? We knew where we were going because we worked from both "ends" of the proof. In other words, we came up with a proof that moved both ends toward the middle and then

[85]

rearranged the steps so that we could go from one end to the other. Unfortunately, proofs are often presented this way.

Here's another way to write the proof of $Q - (P \cap R) = Q - (P \cap Q \cap R)$ that shows more clearly how we got the proof. Note first that this identity is equivalent to showing that

$$Q \cap (P \cap R)^c = Q \cap (P \cap Q \cap R)^c$$

since $A - B = A \cap B^c$. This is equivalent, by DeMorgan's rules, to showing that

$$Q \cap (P^c \cup R^c) = Q \cap (P^c \cup Q^c \cup R^c).$$

But

$$Q \cap (P^c \cup Q^c \cup R^c) = Q \cap \left((P^c \cup R^c) \cup Q^c\right) = \left(Q \cap (P^c \cup R^c)\right) \cup \left(Q \cap Q^c\right) = Q \cap (P^c \cup R^c).$$

This latter identity follows from the fact that $Q \cap Q^c = \emptyset$ and $X \cup \emptyset = X$ for any set X. This completes the proof.

How should you write an algebraic proof? You can use whichever method you prefer. The first approach can be read mechanically because of the way it's laid out. However, if you use the first approach, you may sometimes need to use the second method for yourself first. $\square$

Ordering Sets

In computer programming, you will store and compute with sets of all sorts (sets of number, letters, geometric figures, addresses to arrays, pointers to structures, etc.). In almost all cases, you will work with these sets as lists (also called "linear orders") of some type where order does matter. The order matters in terms of the efficiency of your computations, not in terms of the rules of set theory.

In many cases, the linear ordering of the elements of a set is inherited from the universal set U. For example, the sets $A = \{1, 2, 3\}$ and $B = \{a, b, c\}$ inherit a natural linear ordering from the integers and the alphabet, respectively. But what about $C = \{?, >, <, \}$? There is no standard convention for C. You could use the ASCII code order $(<, >, ?)$, but if you do, some explanation should be given.

Example 5 (Lexicographic order) If you have decided on linear orders (i.e., listings) for a set X and a set Y, there is a commonly used and natural linear ordering for $X \times Y$ called *lexicographic order*. Suppose we list X and Y in some manner:

$$(x_1, x_2, \ldots, x_n) \quad \text{and} \quad (y_1, y_2, \ldots, y_m).$$

Given pairs $(a, b) \in X \times Y$ and $(c, d) \in X \times Y$, we say that (a, b) is *lexicographically less than or equal* to (c, d) if

(1) a is before c in the linear order on X or

(2) $a = c$ and b is equal to or before d in the linear order on Y.

For example, the lexicographic order for $\{1, 2, 7\} \times \{a, b\}$ is

$$\big\{(1, a),\ (1, b),\ (2, a),\ (2, b),\ (7, a),\ (7, b)\big\}.$$

Lexicographic order is called *lex order* for short.

Once you have ordered $X \times Y$ lexicographically, you can order $(X \times Y) \times Z$ by the same two rules (1) and (2) above, provided an order is specified on Z. You can use lex order on $X \times Y$ and the given linear order on Z. Likewise, you can apply (1) and (2) to $X \times (Y \times Z)$ using the given linear order on X and lex order on $Y \times Z$. The sets $(X \times Y) \times Z$ and $X \times (Y \times Z)$ are different sets – elements in the former have the form $((x, y), z)$ and those in the latter have the form $(x, (y, z))$. Imagine listing $(X \times Y) \times Z$ in lex order and stripping off the inner parentheses so that $((x, y), z)$ becomes (x, y, z). Now do the same with $X \times (Y \times Z)$. The two lists will contain the same elements in the same order. It's easy to see why the elements are the same, but it's not so easy to see why the orders are the same. Let's prove the orders are the same.

That means we have to prove

$$((x_1, y_1), z_1) \text{ precedes } ((x_2, y_2), z_2) \text{ if and only if } (x_1, (y_1, z_1)) \text{ precedes } (x_2, (y_2, z_2)).$$

It will make things easier if we write "$((x_1, y_1), z_1) < ((x_2, y_2), z_2)$" for "$((x_1, y_1), z_1)$ precedes $((x_2, y_2), z_2)$." By definition the definition of lex order, $((x_1, y_1), z_1) < ((x_2, y_2), z_2)$ means that $(x_1, y_1) < (x_2, y_2)$ or $(x_1, y_1) = (x_2, y_2)$ and $z_1 < z_2$. By the definition of lex order, the first case means that either $x_1 < x_2$ or $x_1 = x_2$ and $y_1 < y_2$. Note that $(x_1, y_1) = (x_2, y_2)$ means $x_1 = x_2$ and $y_1 = y_2$. Putting all this together, we have shown that $((x_1, y_1), z_1) < ((x_2, y_2), z_2)$ means that either

(1) $x_1 < x_2$ or

(2) $x_1 = x_2$ and $y_1 < y_2$ or

(3) $x_1 = x_2$ and $y_1 = y_2$ and $z_1 < z_2$.

In other words, look for the first position where (x_1, y_1, z_1) and (x_2, y_2, z_2) disagree and use that position to determine the order. We leave it to you to show that the same conditions describe $(x_1, (y_1, z_1)) < (x_2, (y_2, z_2))$. This completes the proof.

Since $(X \times Y) \times Z$ and $X \times (Y \times Z)$ have the same order when inner parentheses are dropped, one usually does that. We just write $X \times Y \times Z$ and call the elements (x, y, z).

Sets and Functions

When you think about what we've just done, you should be able to see that, if $S_1, S_2, \ldots, S_n$ are sets, we can write $S_1 \times S_2 \times \cdots \times S_n$, leaving out parentheses. The product $S_1 \times S_2 \times \cdots \times S_n$ is also written $\times_{i=1}^{n} S_i$.

Suppose $(s_1, s_2, \ldots, s_n)$ and $(t_1, t_2, \ldots, t_n)$ are in $S_1 \times S_2 \times \cdots \times S_n$. Which comes first? You should be able to see that we determine which precedes which as follows: By going left to right, find the first position where they disagree. Say position k is where this disagreement occurs. Use the order of s_k and t_k to determine the order of $(s_1, s_2, \ldots, s_n)$ and $(t_1, t_2, \ldots, t_n)$. This is the same order you use when you look things up in a dictionary. $\square$

Example 6 (Dictionary order on words or strings) The order of words in the dictionary is called "dictionary order." Lex order appears to be the same as dictionary order, but there is a problem with this. We've only defined lex order for n-tuples where n has some fixed value, but words in the dictionary have different length. Let's look at this more carefully.

Let S be a finite set. Let $S^k = \times^k S$ be the product of S with itself k times. The set S^0 is special and consists of one string ϵ called the *empty string*. Let $S^* = \cup_{k=0}^{\infty} S^k$. In words, S^* consists of all strings (words, vectors) of all possible lengths (including length zero) over S. Assume S is linearly ordered. We now define an order relation on S^* called *lexicographic order* or *dictionary order*, denoted by $\leq_L$, on S^*.

Let $(a_1, a_2, \ldots, a_m)$ and $(b_1, b_2, \ldots, b_n)$ be two elements of S^* with $m, n > 0$. We say that
$$(a_1, a_2, \ldots, a_m) \leq_L (b_1, b_2, \ldots, b_n)$$
if either of the following two conditions hold:

(D1) $m \leq n$ and $a_i = b_i$ for $i = 1, \ldots, m$.

(D2) For some $k < \min(m, n)$, $a_i = b_i$, $i = 1, \ldots, k$, $a_{k+1} \neq b_{k+1}$, and a_{k+1} is before b_{k+1} in the linear order on S.

Since $m, n > 0$, we have not discussed the empty string. Thus we need:

(D3) The empty string $\epsilon \leq_L x$ for any string x.

We have just defined dictionary order and also called it lex order. Is this the same as our previous definition of lex order? Yes because the two definitions of lex order agree when the strings have the same length.

We shall study this ordering on words carefully when we study order relations in general. For now we just give an example. Let $S = \{x, y\}$ with the ordering on S the alphabetic order. If $u = (x, x, y)$ and $v = (x, x, y, x)$, then $u \leq_L v$ by (D1). If $s = (x, y, x)$ and $t = (x, x, x, y)$, then $t \leq_L s$ by (D2). More examples will be given in the exercises. The standard English dictionary is an example where this linear order is applied to a subset of all words on the standard English alphabet (the words that have meaning in English).

A variation on this dictionary order is to order all words first by length and then by lex order. Thus, $u = (y, y, y)$ comes before $v = (x, x, x, x)$ because u has length three (three components) and v has length one. This order on S^* is called *length-first lex order* or *short lex order*. $\square$

Subsets of Sets

We use the notation $\mathcal{P}(A)$ to denote the set of all subsets of A and $\mathcal{P}_k(A)$ the set of all subsets of A of size (or cardinality) k. We call $\mathcal{P}(A)$ "the set of all subsets of A" or simply the *power set* of A. Let $C(n,k) = |\mathcal{P}_k(A)|$ denote the number of different k-subsets that can be formed from an n-set. The notation $\binom{n}{k}$ is also frequently used. These are called *binomial coefficients* and are read "n choose k." We now prove

Theorem 2 (Binomial coefficient formula) *The value of the binomial coefficient is*

$$\binom{n}{k} = C(n,k) = \frac{n(n-1)\cdots(n-k+1)}{k!} = \frac{n!}{k!\,(n-k)!},$$

where $0! = 1$ and, for $j > 0$, $j!$ is the product of the first j integers. We read $j!$ as "j factorial".

Proof: Let A be a set of size n. The elements of $\mathcal{P}_k(A)$ are sets and are thus unordered. Generally speaking, unordered things are harder to count than ordered ones. Suppose, instead of a set of size k chosen from A, you wanted to construct an ordered list L of k elements from A (L is called a "k-list"). We could construct L in two stages.

- First choose an element of $S \in \mathcal{P}_k(A)$ (a subset of A with k elements). This can be done in $C(n,k)$ ways since $C(n,k) = |\mathcal{P}_k(A)|$.

- Next order S to obtain L. This ordering can be done in $k! = k(k-1)\cdots 1$ ways. Why? You have k choices for the element of S to appear first in the list L, $k-1$ choices for the next element, $k-2$ choices for the next element, etc.

From this two-stage process, we see that there are $C(n,k)\,k!$ ordered k-lists with no repeats. (The factor $C(n,k)$ is the number of ways to carry out the first stage and the factor $k!$ is the number of ways to carry out the second stage.)

Theorem 3 (Number of ordered lists) *The number of ordered k-lists L that can be made from and n-set A is*

- n^k *if repeats are allowed and*

- $n(n-1)\cdots(n-k+1) = n!/(n-k)!$ *if repeats are not allowed. One also uses the notation $(n)_k$ for these values. This is called the "falling factorial" and is read "n falling k".*

Why? With repeats allowed, there are n choices of elements in A for the first entry in the k-list L, n choices for the second entry, etc. If repeats are not allowed, there are n choices of elements in A for the first entry in the k-list L, $n-1$ choices for the second entry, etc.

Sets and Functions

Since we've counted the same thing (k-lists made from A) in two different ways, the two answers must be equal; that is, $C(n,k)\, k! = n!/(n-k)!$. Dividing by $k!$, we have the theorem. □

In high school, you learned about "Pascal's Triangle" for computing binomial coefficients. We review this idea in the next example.

Example 7 (Binomial recursion) Let $X = \{x_1, \ldots, x_n\}$. We'll think of $C(n,k)$ as counting k-subsets of X. Imagine that we are going to construct a subset S of X with k elements. Either the element x_n is in our subset S or it is not. The cases where it is in the subset S are all formed by taking the various $(k-1)$-subsets of $X - \{x_n\}$ and adding x_n to them. By the definition of binomial coefficients, there are $\binom{n-1}{k-1}$ such subsets. The cases where it is not in the subset S are all formed by taking the various k-subsets of $X - \{x_n\}$. By the definition of binomial coefficients, there are $\binom{n-1}{k}$ such subsets. What we've done is describe how to build all k-subsets of X from certain subsets of $X - \{x_n\}$. Since this gives each subset exactly once,

$$\binom{n}{k} = \binom{n-1}{k-1} + \binom{n-1}{k},$$

which can be written $C(n,k) = C(n-1,k-1) + C(n-1,k)$. This equation is called a *recursion* because it tells how to compute the function $C(n,k)$ from values of the function with smaller arguments. Here are the starting values together with the basic recursion:

$$C(1,0) = C(1,1) = 1,$$
$$C(1,k) = 0 \quad \text{for } k \neq 0, 1 \quad \text{and}$$
$$C(n,k) = C(n-1,k-1) + C(n-1,k) \quad \text{for } n > 1.$$

Below we have made a table of values for $C(n,k)$.

$n \backslash^k$	0	1	2	3	4	5	6
0	1						
1	1	1					
2	1	2	1				
3	1	3	3	1			
4	1	4	6	4	1		
5	1	5	10	10	5	1	
6	1	6	15	20	15	6	1

$C(n,k)$

This tabular representation of $C(n,k)$ is called "Pascal's Triangle." □

Definition 2 (Characteristic function) Let U be the universal set and let $A \subseteq U$. The *characteristic function* of A, denoted χ_A is defined for each $x \in U$ by

$$\chi_A(x) = \begin{cases} 1, & \text{if } x \in A, \\ 0, & \text{if } x \notin A. \end{cases}$$

Thus the domain of χ_A is U and the range of χ_A is $\{0,1\}$.[2]

---

[2] If you are not familiar with "domain" and "range", see the definition at the beginning

Example 8 (Subsets as (0,1)-vectors) If A has n elements, listed $(a_1, a_2, \ldots, a_n)$, then you can specify any subset $X \subset A$ by a sequence $(\epsilon_1, \epsilon_2, \ldots, \epsilon_n)$ where $\epsilon_k = 0$ if the element $a_k \notin X$ and $\epsilon_k = 1$ if the element $a_k \in X$. The vector $(\epsilon_1, \epsilon_2, \ldots, \epsilon_n)$ is just the characteristic function of X since $\epsilon_k = \chi_X(a_k)$.

How many different subsets of A are there? We'll show that there are 2^n choices for $(\epsilon_1, \epsilon_2, \ldots, \epsilon_n)$ and thus $|\mathcal{P}(A)| = 2^n$. Why 2^n? There are clearly two choices for ϵ_1 and two choices for ϵ_2 and so forth. Thus there are $2 \times 2 \times \cdots = 2^n$ choices for $(\epsilon_1, \epsilon_2, \ldots, \epsilon_n)$. $\square$

Example 9 (Sets with sets as elements) Sets can have sets as elements. In the first exercise of this section, you will be asked such questions as "Is $\{1, 2\} \in \{\{1, 2\}, \{3, 4\}\}$?" or "Is $1 \in \{\{1\}, \{2\}, \{3\}\}$?" Easy stuff if you understand the definitions: You can see that the set $\{1, 2\}$ is indeed an element of the set $\{\{1, 2\}, \{3, 4\}\}$ because this latter set has just two elements, each of them a set of size two, one of which is $\{1, 2\}$. You can also see that every element of $\{\{1\}, \{2\}, \{3\}\}$ is a set and that the number 1 is nowhere to be found as an element of this set.

You have already seen $\mathcal{P}(A)$, which is a set whose elements are sets, namely the subsets of A.

Another important class of sets with sets as elements are the *set partitions*. Some of the elementary aspects of set partitions fit into our present discussion. More advanced aspects of them will be discussed in Section 2. Here is a preview. Let $A = \{1, 2, \ldots, 15\}$. Consider the following set whose elements are themselves subsets of A.

$$\alpha = \{\{1\}, \{2\}, \{9\}, \{3, 5\}, \{4, 7\}, \{6, 8, 10, 15\}, \{11, 12, 13, 14\}\}.$$

This set is a subset of the power set $\mathcal{P}(A)$. But, it is a very special type of subset, called a *set partition* of A because it satisfies the three conditions:

(1) every element of α is nonempty,

(2) the union of the elements of α is A, and

(3) if you pick sets $X \in \alpha$ and $Y \in \alpha$, either $X = Y$ or $X \cap Y = \emptyset$.

Any collection of subsets of a set A satisfying (1), (2), and (3) is a set partition of A or simply a partition of A. Each element of α (which is, of course, a subset of A) is called a *block* of the partition α.

How many partitions are there of a set A? This is a tricky number to compute and there is no simple formula like $C(n, k) = \frac{n!}{k!(n-k)!}$ for it. We will discuss it in the Section 2. The number of partitions of a set of size n is denoted by B_n. These numbers are called *Bell numbers* after Eric Temple Bell. The first few Bell numbers are $B_1 = 1$, $B_2 = 2$, $B_3 = 5$, $B_4 = 15$, $B_5 = 52$.

We can *refine* the partition α by splitting blocks into smaller blocks. For example, we might split the block $\{6, 8, 10, 15\}$ into two blocks, say $\{6, 15\}$ and $\{8, 10\}$, and also split the block $\{11, 12, 13, 14\}$ into three blocks, say $\{13\}$, $\{14\}$, and $\{11, 12\}$. The resulting partition is called a *refinement* of α and equals

$$\{\{1\}, \{2\}, \{9\}, \{3, 5\}, \{4, 7\}, \{6, 15\}, \{8, 10\}, \{13\}, \{14\}, \{11, 12\}\}.$$

of the next section.

Sets and Functions

Note that a refinement of a partition is another partition of the same set. We also consider a partition α to be a refinement of itself. We shall gain a deeper understanding of the notion of refinement when we study order relations. $\square$

Exercises for Section 1

1.1. Answer the following about the $\in$ and $\subseteq$ operators.

(a) Is $\{1,2\} \in \{\{1,2\},\{3,4\}\}$?

(b) Is $\{2\} \in \{1,2,3,4\}$?

(c) Is $\{3\} \in \{1,\{2\},\{3\}\}$

(d) Is $\{1,2\} \subseteq \{1,\ 2,\ \{1,2\},\ \{3,4\}\}$?

(e) Is $1 \in \{\{1\},\ \{2\},\ \{3\}\}$?

(f) Is $\{1,2,1\} \subseteq \{1,2\}$?

1.2. For each of the following, draw a Venn diagram.

(a) $A \subseteq B,\ C \subseteq B,\ A \cap C = \emptyset$

(b) $A \supseteq C,\ B \cap C = \emptyset$.

1.3. Let $A = \{w,x,y,z\}$ and $B = \{a,b\}$. Take the linear orders on A and B to be alphabetic order. List the elements in each of the following sets in lexicographic order.

(a) $A \times B$

(b) $B \times A$

(c) $A \times A$

(d) $B \times B$

1.4. Let $A = \{1,2,3\}$, $B = \{u,v\}$, and $C = \{m,n\}$. Take the linear order on A to be numeric and the linear orders on B and C to be alphabetic. List the elements in each of the following sets in lexicographic order.

(a) $A \times (B \times C)$ (use lex order on $B \times C$).

(b) $(A \times B) \times C$ (use lex order on $A \times B$).

(c) $A \times B \times C$.

1.5. Let $\Sigma = (x,y)$ be an alphabet. List each of the following sets of strings over this alphabet in the order indicated.

(a) All palindromes (strings that read the same forward and backward) of length less than or equal to 4. List them in dictionary order.

(b) All strings (words) that begin with x and have length less than four. List them in both dictionary and length-first lex order.

(c) List all strings of length four in lex order.

1.6. Each of the following statements about subsets of a set U is FALSE. Draw a Venn diagram to represent the situation being described. In each case case, show that the assertion is false by specializing the sets.

(a) For all A, B, and C, if $A \not\subseteq B$ and $B \not\subseteq C$ then $A \not\subseteq C$.

(b) For all sets A, B, and C, $(A \cup B) \cap C = A \cup (B \cap C)$.

(c) For all sets A, B, and C, $(A - B) \cap (C - B) = A - (B \cup C)$.

(d) For all A, B, and C, if $A \cap C \subseteq B \cap C$ and $A \cup C \subseteq B \cup C$ then $A = B$.

(e) For all A, B, and C, if $A \cup C = B \cup C$ then $A = B$.

(f) For all sets A, B, and C, $(A - B) - C = A - (B - C)$.

1.7. Prove each statement directly from the definitions.

(a) If A, B, and C are subsets of U, then $A \subseteq B$ and $A \subseteq C$ implies that $A \subseteq B \cap C$.

(b) If A, B, and C are subsets of U, then $A \subseteq B$ and $A \subseteq C$ implies that $A \subseteq B \cup C$.

1.8. Prove, using the definition of set equality, that for all sets A, B, and C, $(A - B) \cap (C - B) = (A \cap C) - B$.

1.9. Prove each statement by the method indicated.

(a) Prove using element arguments that if U is the universal set and A and B subsets of U, then $A \subseteq B$ implies that $U - A \supseteq U - B$ (alternative notation: $A \subseteq B$ implies $A^c \supseteq B^c$, or $A' \supseteq B'$)

(b) Prove, using element arguments and the definition of set inclusion, that for all A, B, and C, if $A \subseteq B$ then $A \cap C \subseteq B \cap C$.

(c) Prove, using (a), (b), and DeMorgan's law, that for all A, B, and C, if $A \subseteq B$ then $A \cup C \subseteq B \cup C$.

1.10. Prove each statement by the "element method."

(a) If A, B, and C are subsets of U, then $A \times (B \cup C) = (A \times B) \cup (A \times C)$.

(b) If A, B, and C are subsets of U, then $A \times (B \cap C) = (A \times B) \cap (A \times C)$.

1.11. Prove each of the following identities from the basic algebraic rules for sets. You may want to use the fact that $D - E = D \cap E^c$.

[93]

Sets and Functions

   (a) If A, B, and C are subsets of U, then $(A - B) - C = A - (B \cup C)$.

   (b) If A, B, and C are subsets of U, then $(A - B) - C = (A - C) - B$.

   (c) If A and B are subsets of U, then $(A - B) \cup (B - A) = (A \cup B) - (A \cap B)$.

1.12. Prove or give a counterexample. Use a Venn diagram argument for the proof. For the counterexample, use a Venn diagram or use set specialization.

   (a) If A, B, and C are subsets of U, then $(A - C) \cap (B - C) \cap (A - B) = \emptyset$.

   (b) If A and B are subsets of U and if $A \subseteq B$, then $A \cap (U - B) = \emptyset$.

   (c) If A, B, and C are subsets of U, and if $A \subseteq B$, then $A \cap (U - (B \cap C)) = \emptyset$.

   (d) If A, B, and C are subsets of U, and if $(B \cap C) \subseteq A$, then $(A - B) \cap (A - C) = \emptyset$.

   (e) If A and B are subsets of U and if $A \cap B = \emptyset$, then $A \times B = \emptyset$.

1.13. Recall that the symmetric difference of sets A and B is $A \oplus B = (A - B) \cup (B - A)$. It is evident from the definition that $A \oplus B = B \oplus A$, the commutative law. Let U be the universal set. Prove each of the following properties either using a Venn diagram argument or algebraically or directly from the definition.

   (a) $A \oplus (B \oplus C) = (A \oplus B) \oplus C$ (associative law for $\oplus$).

   (b) $A \oplus \emptyset = A$.

   (c) $A \oplus A^c = U$.

   (d) $A \oplus A = \emptyset$.

   (e) If $A \oplus C = B \oplus C$ then $A = B$.

1.14. Let A, B, and C be subsets of U. Prove or disprove using Venn diagrams.

   (a) $A - B$ and $B - C$ are disjoint.

   (b) $A - B$ and $C - B$ are disjoint.

   (c) $A - (B \cup C)$ and $B - (A \cup C)$ are disjoint.

   (d) $A - (B \cap C)$ and $B - (A \cap C)$ are disjoint.

1.15. Which of the following are partitions of $\{1, 2, \ldots, 8\}$? Explain your answers.

   (a) $\{\{1, 3, 5\}, \{1, 2, 6\}, \{4, 7, 8\}\}$

   (b) $\{\{1, 3, 5\}, \{2, 6, 7\}, \{4, 8\}\}$

   (c) $\{\{1, 3, 5\}, \{2, 6\}, \{2, 6\}, \{4, 7, 8\}\}$

   (d) $\{\{1, 5\}, \{2, 6\}, \{4, 8\}\}$

1.16. How many refinements are there of the partition $\{\{1, 3, 5\}, \{2, 6\}, \{4, 7, 8, 9\}\}$? Explain.

1.17. Suppose S and T are sets with $S \cap T = \emptyset$. Suppose σ is a partition of S and τ is a partition of T.

(a) Prove that $\sigma \cup \tau$ is a partition of $S \cup T$.

(b) If σ has n_σ refinements and τ has n_τ refinements, how many refinements does $\sigma \cup \tau$ have? Explain.

1.18. Use the characteristic function format to list the power set of the following sets. That is, describe each element of the power set as a vector of zeroes and ones.

(a) $\{1, 2, 3\}$

(b) $X \times Y$ where $X = \{a, b\}$ and $Y = \{x, y\}$.

1.19. Find the following power sets:

(a) $\mathcal{P}(\emptyset)$

(b) $\mathcal{P}(\mathcal{P}(\emptyset))$

(c) $\mathcal{P}(\mathcal{P}(\mathcal{P}(\emptyset)))$

1.20. Compare the following pairs of sets. Can they be equal? Is one a subset of the other? Can they have the same size (number of elements)?

(a) $\mathcal{P}(A \cup B)$ and $\mathcal{P}(A) \cup \mathcal{P}(B)$

(b) $\mathcal{P}(A \cap B)$ and $\mathcal{P}(A) \cap \mathcal{P}(B)$

(c) $\mathcal{P}(A \times B)$ and $\mathcal{P}(A) \times \mathcal{P}(B)$

1.21. Let $S = \{1, 2, \ldots, n\}$. Let S_1 be the set of all subsets of S that contain 1. Let T_1 denote the set of all subsets of S that don't contain 1. Prove $|T_1| = |S_1| = 2^{(n-1)}$.

Section 2: Functions

Functions, such as linear functions, polynomial functions, trigonometric functions, exponential functions, and logarithmic functions are familiar to all students who have had mathematics in high school. For discrete mathematics, we need to understand functions at a basic set theoretic level. We begin with a familiar definition.

Definition 3 (Function) *If A and B are sets, a function from A to B is a rule that tells us how to find a unique $b \in B$ for each $a \in A$. We write $f(a) = b$ and say that f maps a to b. We also say the value of f at a is b.*

We write $f : A \to B$ to indicate that f is a function from A to B. We call the set A the domain of f and the set B the range or, equivalently, codomain of f.

To specify a function completely you must give its domain, range and rule.

If $X \subseteq A$, then $f(X) = \{f(x) \mid x \in X\}$. In particular $f(\emptyset) = \emptyset$ and $f(A)$ is called the image of f.

Sets and Functions

Some people define "range" to be the values that the function *actually* takes on. Most people call that the *image*.

In high school, you dealt with functions whose ranges were $\mathbb{R}$ and whose domains were contained in $\mathbb{R}$; for example, $f(x) = 1/(x^2 - 1)$ is a function from $\mathbb{R} - \{-1, 1\}$ to $\mathbb{R}$. If you have had some calculus, you also studied functions of functions! The derivative is a function whose domain is all differentiable functions and whose range is all functions. If we wanted to use functional notation we could write $D(f)$ to indicate the function that the derivative associates with f.

The set of all functions from A to B is written B^A. One reason for this notation, as we shall see below, is that $|B^A| = |B|^{|A|}$. Thus $f : A \to B$ and $f \in B^A$ say the same thing.

To avoid the cumbersome notation $\{1, 2, 3, \ldots, n\}$, we will often use $\underline{n}$ instead.

Example 10 (Functions as relations) There is a fundamental set-theoretic way of defining functions. Let A and B be sets. A *relation from A to B* is a subset of $A \times B$. For example, if $A = \underline{3} = \{1, 2, 3\}$ and $B = \underline{4}$, then $R = \{(1, 4), (1, 2), (3, 3), (2, 3)\}$ is a relation from A to B. To specify a relation, you must define three sets: A, B and R.

If the relation R satisfies the condition $\forall x \in A \; \exists! \, y \in B, \, (x, y) \in R$, then the relation R is called a *functional relation*. We used some shorthand notation here that is worth remembering:

$\forall$	means	"for all"
$\exists$	means	"for some" or "there exists"
$\exists!$	means	"for exactly one"

If you think about Definition 3, you will realize that a "functional relation" is just one possible way of giving all of the information required to specify a function.

Given any relation $R \subseteq A \times B$, the inverse relation R^{-1} from B to A is $\{(y, x) : (x, y) \in R\}$. For $R = \{(1, 4), (1, 2), (3, 3), (2, 3)\}$, $A = \underline{3}$ and $B = \underline{4}$, the inverse relation is $R^{-1} = \{(4, 1), (2, 1), (3, 3), (3, 2)\}$. Note that neither R nor R^{-1} is a functional relation in this example. You should make sure that you understand why this statement is true. (Hint: R fails the "$\exists!$" test and R^{-1} fails the "$\forall$" part of the definition of a functional relation.) Note also that if R and R^{-1} are functional then $|A| = |B|$. In algebra or calculus, when you draw a graph of a real-valued function $f : \mathbf{D} \to \mathbb{R}$ (such as $f(x) = x^3$, $f(x) = x/(1-x)$, etc.), you are attempting a pictorial representation of the set $\{(x, f(x)) : x \in \mathbf{D} \subseteq \mathbb{R}\}$, which is the subset of $\mathbf{D} \times \mathbb{R}$. This subset is a "functional relation from $\mathbf{D}$ to $\mathbb{R}$."

In our notation, we would write $(a, b) \in R$ to indicate that the pair (a, b) is in the relation R from A to B. People also use the notation $a \, R \, b$ to indicate this. For example, the "less than" relation $\{(a, b) \mid a < b\}$ is written $a < b$. $\square$

In many cases in discrete mathematics, we are concerned with functions whose domain is finite. Special notation is used for specifying such functions.

Definition 4 (One-line notation) *Let A be a finite ordered set with elements ordered $(a_1, a_2, \ldots, a_{|A|})$. Let B be any set. A function $f : A \to B$ can be written in one-line notation as $f = (f(a_1), f(a_2), \ldots, f(a_{|A|}))$. Thus the values of the function are written as* *list, which is also called a vector or a string. In other words the function f assigns to a_k the kth element of the list $(f(a_1), f(a_2), \ldots, f(a_{|A|}))$ for each value of k from 1 to $|A|$.*

It follows from the definition that we can think of function as an element of $B^{|A|} = B \times B \times \cdots \times B$, where there are $|A|$ copies of B. This is another reason for the notation B^A for all functions from A to B. Do you see why we don't use $B^{|A|}$ instead? No, it's not because B^A is easier to write. It's because $B^{|A|}$ does not specify the domain A. Instead, only its size $|A|$ is given.

Example 11 (Using the notation) To get a feeling for the notation used to specify a function, it may be helpful to imagine that you have an envelope or box that contains a function. In other words, this envelope contains all the information needed to completely describe the function. Think about what you're going to see when you open the envelope.

You might see

$$P = \{a, b, c\}, \qquad g : P \to \underline{4}, \qquad g(a) = 3, \quad g(b) = 1 \quad \text{and} \quad g(c) = 4.$$

This tells you that the name of the function is g, the domain of g is P, which is $\{a, b, c\}$, and the range of g is $\underline{4} = \{1, 2, 3, 4\}$. It also tells you the values in $\underline{4}$ that g assigns to each of the values in its domain. Someone else may have put

$$g \in \underline{4}^{\{a,b,c\}}, \qquad \text{ordering: } a, b, c, \qquad g = (3, 1, 4).$$

in the envelope instead. This describes the same function. It doesn't give a name for the domain, but that's okay since all we need to know is what's in the domain. On the other hand, it gives an order on the domain so that the function can be given in one-line form. Since the domain is ordered a, b, c and since $g = (3, 1, 4)$, by the definition of one-line notation $g(a) = 3$, $g(b) = 1$ and $g(c) = 4$. Can you describe other possible envelopes for the same function?

What if the envelope contained only $g = (3, 1, 4)$? If you think you have been given the one-line notation for g, you are mistaken. You *must* know the *ordered* domain of g before you can interpret $g = (3, 1, 4)$. Here we don't even know the domain as a set (or the range). The domain might by $\{a, b, c\}$, or $\{<, >, ?\}$, or any other 3-set.

What if the envelope contained

$$\text{the domain of } g \text{ is } \{a, b, c\}, \qquad \text{ordering: } a, b, c, \qquad g = (3, 1, 4)?$$

We haven't specified the range of g, but is it necessary since we know the values of the function? Our definition included the requirement that the range be specified, so this is not a complete definition. Some definitions of a function do not require that the range be specified. For such definitions, this would be a complete specification of the function g. □

Sets and Functions

Example 12 (Counting functions) Think about specifying $f : A \to B$ in one-line notation: $(f(a_1), f(a_2), \ldots, f(a_{|A|}))$. There are $|B|$ ways to choose $f(a_1)$, $|B|$ ways to choose $f(a_2)$, etc., and finally $|B|$ ways to choose $f(a_{|A|})$. This means that the cardinality of the set of all functions $f : A \to B$ is $|B|^{|A|}$. In other words, $|B^A| = |B|^{|A|}$.

We can represent a subset S of A by a unique function $f : A \to \underline{2}$ where

$$f(x) = \begin{cases} 1, & \text{if } x \notin S, \\ 2, & \text{if } x \in S. \end{cases}$$

This proves that there are $2^{|S|}$ such subsets. We proved this result in Example 9. You should verify that this is essentially the same proof that was given there.

We can represent a list of k elements of a set S with repetition allowed by a unique function $f : \underline{k} \to S$. In this representation, the list corresponds to the function written in one-line notation. (Recall that the ordering on $\underline{k}$ is the numerical ordering.) This proves that there are exactly $|S|^k$ such lists. $\square$

Definition 5 (Types of functions) Let $f : A \to B$ be a function. If for every $b \in B$ there is an $a \in A$ such that $f(a) = b$, then f is called a *surjection* (or an *onto function*). Another way to describe a surjection is to say that it takes on each value in its range at least once.

If $f(x) = f(y)$ implies $x = y$, then f is called an *injection* (or a *one-to-one function*). Another way to describe an injection is to say that it takes on each value in its range at most once.

If f is both an injection and a surjection, it is a called a *bijection*. The bijections of A^A are called the *permutations* of A. The set of permutations on a set A is denoted in various ways. Two notations are $\mathrm{PER}(A)$ and $S(A)$.

If $f : A \to B$ is a bijection, we may talk about the *inverse bijection* of f, written f^{-1}, which reverses what f does. Thus $f^{-1} : B \to A$ and $f^{-1}(b)$ is that unique $a \in A$ such that $f(a) = b$.

Note that $f(f^{-1}(b)) = b$ and $f^{-1}(f(a)) = a$. Do not confuse f^{-1} with $1/f$. For example, if $f : \mathbb{R} \to \mathbb{R}$ is given by $f(x) = x^3 + 1$, then $1/f(x) = 1/(x^3 + 1)$ and $f^{-1}(x) = (x - 1)^{1/3}$.

Example 13 (Surjections, injections and bijections as lists) Lists provide another fundamental way to think about the various types of functions we've just defined. We'll illustrate this with some examples.

Let $A = \underline{4}$, $B = \{a, b, c, d, e\}$ and $f = (d, c, d, a)$ describe the function f in one-line notation. Since the list (d, c, d, a) contains d twice, f is not an injection. The function $(b, d, c, e) \in B^A$ is an injection since there are no repeats in the list of values taken on by the function. The 4-lists without repeats that can be formed from B correspond to the injections from $\underline{4}$ to B. In general, the injections in $S^{\underline{k}}$ correspond to k-lists without repeats whose elements are taken from S.

With the same f as in the previous paragraph, note that the value b is not taken on by f. Thus f is not a surjection. (We could have said e is not taken on, instead.)

Now let $A = \underline{4}$, $B = \{x, y, z\}$ and $g = (x, y, x, z)$. Since every element of B appears at least once in the list of values taken on, g is a surjection.

Finally, let $A = B = \underline{4}$ and $h = (3, 1, 4, 2)$. The function is both an injection and a surjection. Hence, it is a bijection. Since the domain and range are the same and f is a bijection, it is a permutation of $A = \underline{4}$. The list $(3, 1, 4, 2)$ is a rearrangement (a permutation) of the ordered listing $(1, 2, 3, 4)$ of A. That's why we call h a permutation. The inverse of h is $(2, 4, 1, 3)$. $\square$

Example 14 (Encryption) Suppose we want to send data (a text message, a JPEG file, etc.) to someone and want to be sure no one else can read the data. Then we use *encryption*. We can describe encryption as a function $f : D \to R$ where D is the set of possible messages. Of course, there are a huge number of *possible* messages, so what do we do? We can break the message into pieces. For example, we could break an ordinary text message into pieces with one character (with the space as a character) per piece. Then apply a function f to each piece. Here's a simple example: If x is a letter, $f(x)$ is the next letter in the alphabet with $Z \to A$ and $f(\text{space}) = \text{space}$. Then we would encrypt "HELLO THERE" as "IFMMP UIFSF." This is too simple for encryption.

What can we do? Let S be the set of symbols that we are using (A to Z and space in the previous paragraph). We could choose a more complicated function $f : S \to S$ than our simple function. What properties should it have?

- It must have an inverse so that we can decrypt.

- The encryption and decryption must be quick on a computer.

- It must be hard for someone else to figure out f^{-1}.

Since $f : S \to S$ and it has an inverse, f must be a bijection (in fact, a permutation of S). How can we make f hard to figure out? That is a problem in the design of encryption systems. One key ingredient is to make S large. For example, in systems like PGP (Pretty Good Privacy) and DES (Data Encryption Standard) S consists of all n-long vectors of zeroes and ones, typically with $n = 64$. In this case $|S| = 2^{64} \approx 10^{19}$, which is quite large. $\square$

Example 15 (Hashing) *Hashing* is a procedure for mapping a large space into a smaller one. For example, a hash function h may have as its domain all sequences of zeroes and ones of all possible lengths. It's range might be all n-long sequences of zeroes and ones for some n. There are some publicly available hash functions h that seem to be good.

Why would we want such a function? Suppose we want to be sure no one changes a document that is stored in a computer. We could apply h to the document and then save $h(\text{document})$. By giving $h(\text{document})$ to people, they could later check to see if the document had been changed — if the function h is well chosen it would be hard to change the document without changing the value of h, even if you know how to compute h. Suppose you email a document to a friend, but you're concerned that someone may intercept the email and change the document. You can call up your friend and tell him $h(\text{document})$ so that he can check it.

Another use for a hash function is storing data. Suppose we have an n-long array in which we want to store information about students at the university. We want a hash

Sets and Functions

function that maps student ID numbers into $\{1, 2, \ldots n\}$. Then $h(\text{ID})$ tells us which array position to use. Of course two student ID numbers may hash to the same value (array position). There are methods for dealing with such conflicts. $\square$

Example 16 (Two-line notation) Since one-line notation is a simple, brief way to specify functions, it is used frequently. If the domain is not a set of numbers, the notation is poor because we must first pause and order the domain. There are other ways to write functions which overcome this problem. For example, we could write $f(a) = 4$, $f(b) = 3$, $f(c) = 4$ and $f(d) = 1$. This could be shortened up somewhat to $a \to 4$, $b \to 3$, $c \to 4$ and $d \to 1$. By turning each of these sideways, we can shorten it even more: $\begin{pmatrix} a & b & c & d \\ 4 & 3 & 4 & 1 \end{pmatrix}$. For obvious reasons, this is called *two-line notation*. Since x always appears directly over $f(x)$, there is no need to order the domain; in fact, we need not even specify the domain separately since it is given by the top line. If the function is a bijection, its inverse is obtained by interchanging the top and bottom lines.

The arrows we introduced in the last paragraph can be used to help visualize different properties of functions. Imagine that you've listed the elements of the domain A in one column and the elements of the range B in another column to the right of the domain. Draw an arrow from a to b if $f(a) = b$. Thus the heads of arrows are on elements of B and the tails are on elements of A. Since f is a function, no two arrows have the same tail. If f is an injection, no two arrows have the same head. If f is a surjection, every element of B is on the head of some arrow. You should be able to describe the situation when f is a bijection. $\square$

Example 17 (Compositions of functions) Suppose that f and g are two functions such that the values f takes on are contained in the domain of g. We can write this as $f : A \to B$ and $g : C \to D$ where $f(a) \in C$ for all $a \in A$. We define the *composition* of g and f, written $gf : A \to D$ by $(gf)(x) = g(f(x))$ for all $x \in A$. The notation $g \circ f$ is also used to denote composition. Suppose that f and g are given in two-line notation by

$$f = \begin{pmatrix} p & q & r & s \\ P & R & T & U \end{pmatrix} \qquad g = \begin{pmatrix} P & Q & R & S & T & U & V \\ 1 & 3 & 5 & 2 & 4 & 6 & 7 \end{pmatrix}.$$

Then $gf = \begin{pmatrix} p & q & r & s \\ 1 & 5 & 4 & 6 \end{pmatrix}$.

Suppose $f : A \to B$, $g : B \to C$, and $h : C \to D$. We can form the compositions $g \circ f$ and $h \circ g$; however, we cannot form the composition $h \circ f$ unless C contains $f(x)$ for all $x \in A$. We can also form the compositions of all three functions, namely $h \circ (g \circ f)$ and $(h \circ g) \circ f$. These two compositions are equal — that's the "associative law" for composition of functions. How is it proved? Here's an algebraic proof that uses nothing more than the definition of $\circ$ at each step: For all $x \in A$

$$\bigl(h \circ (g \circ f)\bigr)(x) = h\bigl((g \circ f)(x)\bigr) = h(g(f(x))) = (h \circ g)(f(x)) = \bigl((h \circ g) \circ f\bigr)(x).$$

Let A be a set. Suppose that $f, g \in \mathcal{S}(A)$.; that is, f and g are permutations of a set A. Recall that a permutation is a bijection from a set to itself and so it makes sense to talk about f^{-1} and fg. We claim that fg and f^{-1} are also permutations of A. This is easy to see if you write the permutations in two-line form and note that the second line is a rearrangement of the first if and only if the function is a permutation.

Again suppose that $f \in S(A)$. Instead of $f \circ f$ or ff we write f^2. Note that $f^2(x)$ is not $(f(x))^2$. (In fact, if multiplication is not defined in A, $(f(x))^2$ has no meaning.) We could compose three copies of f. The result is written f^3. In general, we can compose k copies of f to obtain f^k. A cautious reader may be concerned that $f \circ (f \circ f)$ may not be the same as $(f \circ f) \circ f$. By the associative law for $\circ$, they're equal. In fact, $f^{k+m} = f^k \circ f^m$ for all nonnegative integers k and m, where f^0 is defined by $f^0(x) = x$ for all x in the domain. This is true even if k or m or both are negative. $\Box$

Example 18 (Composing permutations) Let's carry out some calculations for practice. Let f and g be the permutations

$$f = \begin{pmatrix} 1 & 2 & 3 & 4 & 5 \\ 2 & 1 & 4 & 5 & 3 \end{pmatrix} \qquad g = \begin{pmatrix} 1 & 2 & 3 & 4 & 5 \\ 2 & 3 & 4 & 5 & 1 \end{pmatrix}.$$

To compute fg, we must calculate $fg(x)$ for all x. This can be done fairly easily from the two-line form: For example, $(fg)(1)$ can be found by noting that the image of 1 under g is 2 and the image of 2 under f is 1. Thus $(fg)(1) = 1$. You should be able to verify that

$$fg = \begin{pmatrix} 1 & 2 & 3 & 4 & 5 \\ 1 & 4 & 5 & 3 & 2 \end{pmatrix} \qquad gf = \begin{pmatrix} 1 & 2 & 3 & 4 & 5 \\ 3 & 2 & 5 & 1 & 4 \end{pmatrix} \neq fg.$$

Thus, $f \circ g = g \circ f$ (commutative law) is *not a law* for permutations.

It is easy to get the inverse, simply interchange the two lines. Thus

$$f^{-1} = \begin{pmatrix} 2 & 1 & 4 & 5 & 3 \\ 1 & 2 & 3 & 4 & 5 \end{pmatrix} \quad \text{which is the same as } f^{-1} = \begin{pmatrix} 1 & 2 & 3 & 4 & 5 \\ 2 & 1 & 5 & 3 & 4 \end{pmatrix},$$

since the order of the columns in two-line form does not matter.

Let's compute some powers:

$$f^2 = \begin{pmatrix} 1 & 2 & 3 & 4 & 5 \\ 1 & 2 & 5 & 3 & 4 \end{pmatrix} \quad f^3 = \begin{pmatrix} 1 & 2 & 3 & 4 & 5 \\ 2 & 1 & 3 & 4 & 5 \end{pmatrix} \quad g^5 = f^6 = \begin{pmatrix} 1 & 2 & 3 & 4 & 5 \\ 1 & 2 & 3 & 4 & 5 \end{pmatrix}.$$

We computed f^6 using $f^6 = f^3 \circ f^3$. That was a bit tedious. Now imagine if you wanted to compute f^{100}. Cycle notation is an easy way to do that. $\Box$

Let f be a permutation of the set A and let $n = |A|$. If $x \in A$, we can look at the sequence $x, f(x), f(f(x)), \ldots, f^k(x), \ldots$, which is often written as $x \to f(x) \to f(f(x)) \to \cdots \to f^k(x) \to \ldots$. Using the fact that $f^0(x) = x$, we can write the sequence as $f^0(x) \to f^1(x) \to f^2(x) \to \cdots$. Since the range of f has n elements, this sequence will contain a repeated element in the first $n+1$ entries. Suppose that $f^s(x)$ is the first sequence entry that is ever repeated and that $f^p(x)$ is the first time that it is repeated.

We claim that $s = 0$. If $s > 0$, apply f^{-1} to both sides of this equality to obtain $f^{s-1}(x) = f^{p-1}(x)$, contradicting the fact that s was chosen as small as possible. Thus, in fact, $s = 0$.

It follows that the sequence cycles through a pattern of length p forever since $f^{p+1}(x) = f(f^p(x)) = f(x)$, $f^{p+2}(x) = f^2(f^p(x)) = f^2(x)$, and so on. We call $(x, f(x), \dots, f^{p-1}(x))$ the *cycle* containing x and call p the *length of the cycle*. If a cycle has length p, we call it a *p-cycle*. Cyclic shifts of a cycle are considered the same; for example, if (1,2,6,3) is the cycle containing 1 (as well as 2, 3 and 6), then (2,6,3,1), (6,3,1,2) and (3,1,2,6) are other ways of writing the cycle.

Suppose $(x_1, x_2, \dots, x_p)$ is a cycle of f and that $y_1 \in A$ is not in that cycle. We can form the cycle containing y_1: $(y_1, y_2, \dots, y_q)$. None of the y_k is in the cycle $(x_1, \dots, x_p)$. Why? If it were, we could continue in the cycle and eventually reach y_1. Written out algebraically: If $y_k = x_j$ for some k and j, then $y_1 = f^{q-k-1}(y_k) = f^{q-k-1}(x_j)$ and the right side is in the cycle $(x_1, \dots, x_p)$. We have proved.

Theorem 4 (Cycle form of a permutation) *Let f be a permutation of the finite set A be a finite set. Every element of A belongs to a cycle of f. Two cycles are either the same or have no elements in common.*

Example 19 (Using cycle notation) Consider the permutation

$$f = \begin{pmatrix} 1 & 2 & 3 & 4 & 5 & 6 & 7 & 8 & 9 \\ 2 & 4 & 8 & 1 & 5 & 9 & 3 & 7 & 6 \end{pmatrix}.$$

Since $1 \to 2 \to 4 \to 1$, the cycle containing 1 is (1,2,4). We could equally well write it (2,4,1) or (4,1,2); however, (1,4,2) is different since it corresponds to $1 \to 4 \to 2 \to 1$. The usual convention is to list the cycle starting with its smallest element. The cycles of f are (1,2,4), (3,8,7), (5) and (6,9). We write f in cycle form as

$$f = (1, 2, 4)\,(3, 8, 7)\,(5)\,(6, 9).$$

The order in which the cycles are written doesn't matter, so we have

$$f = (5)\,(6, 9)\,(1, 2, 4)\,(3, 8, 7) \quad \text{and} \quad f = (4, 1, 2)\,(5)\,(6, 9)\,(7, 3, 8),$$

and lots of other equivalent forms. It is common practice to omit the cycles of length one and write $f = (1, 2, 4)(3, 8, 7)(6, 9)$. The inverse of f is obtained by reading the cycles backwards because $f^{-1}(x)$ is the lefthand neighbor of x in a cycle. Thus

$$f^{-1} = (4, 2, 1)(7, 8, 3)(9, 6) = (1, 4, 2)(3, 7, 8)(6, 9).$$

To compute $f(x)$, we simply take one step to the right from x in its cycle. We just saw that $f^{-1}(x)$ is computed by taking one step to the left. You may be able to guess at this point that $f^k(x)$ is computed by taking k steps to the right, with the rule that a negative step to the right is the same as a positive step to the left. This makes it easy to compute powers.

When $f = (1,2,4)(3,8,7)(5)(6,9)$, what is f^{100}? Imagine starting at 1. After 3 steps to the right we're back at 1. Do this 33 times so that after $3 \times 33 = 99$ steps to the right we're back at 1. One more step takes us to 100 steps and so $f^{100}(1) = 2$. You should be able to figure out the rest: $f^{100} = (1,2,4)(3,8,7)(5)(6)(9)$. $\square$

We next take a close look at the notions of *image* and *coimage* of a function. Again, let $f : A \to B$ be a function. The *image* of f is the set of values f actually takes on: Image$(f) = \{ f(a) \mid a \in A \}$. The definition of a surjection can be rewritten Image$(f) = B$.

For each $b \in B$, the *inverse image* of b, written $f^{-1}(b)$ is the set of those elements in A whose image is b; i.e.,

$$f^{-1}(b) = \{ a \mid a \in A \text{ and } f(a) = b \}.$$

This extends our earlier definition of f^{-1} from bijections to all functions; however, such an f^{-1} can't be thought of as a function from B to A unless f is a bijection because it will not give a unique $a \in A$ for each $b \in B$. (There is a slight abuse of notation here: If $f : A \to B$ is a bijection, our new notation is $f^{-1}(b) = \{a\}$ and our old notation is $f^{-1}(b) = a$.)

Definition 6 (Coimage)   Let $f : A \to B$ be a function. The collection of nonempty inverse images of elements of B is called the coimage of f. In set-theoretic terms,

$$\text{Coimage(f)} = \{f^{-1}(b) \mid b \in B, \ f^{-1}(b) \neq \emptyset\} = \{f^{-1}(b) \mid b \in \text{Image}(f)\}.$$

To describe the structure of coimages, we need to recall that a partition of a set is an unordered collection of nonempty subsets of B such that each element of B appears in exactly one subset. Each subset is called a block of the partition.

Theorem 5 (Structure of coimage)   Suppose $f : A \to B$. The coimage of f is the partition of A whose blocks are the maximal subsets of A on which f is constant.

Wait — before we give a proof we need to understand what we just said. Let's look at an example. If $f \in \{a, b, c\}^{\underline{5}}$ is given in one-line form as (a, c, a, a, c), then

$$\text{Coimage}(f) = \{f^{-1}(a), f^{-1}(c)\} = \{\{1, 3, 4\}, \{2, 5\}\},$$

f is a on $\{1, 3, 4\}$ and is c on $\{2, 5\}$. Now let's prove the theorem.

Proof: If $x \in A$, let $y = f(x)$. Then $x \in f^{-1}(y)$ and so the union of the nonempty inverse images contains A. Clearly it does not contain anything which is not in A. If $y_1 \neq y_2$, then we cannot have $x \in f^{-1}(y_1)$ and $x \in f^{-1}(y_2)$ because this would imply $f(x) = y_1$ and $f(x) = y_2$, a contradiction of the definition of a function. Thus Coimage(f) is a partition of A. Clearly x_1 and x_2 belong to the same block if and only if $f(x_1) = f(x_2)$. Hence a block is a maximal set on which f is constant. $\square$

Since Coimage(f) is a partition of the domain A, we need to review the basic combinatorial properties of partitions.

[103]   SF-23

Sets and Functions

Example 20 (Set partitions) The 15 partitions of $\{1,2,3,4\}$, classified by number of blocks, are

1 block: $\{1,2,3,4\}$

2 blocks: $\{\{1,2,3\},\{4\}\}$ $\{\{1,2,4\},\{3\}\}$ $\{\{1,2\},\{3,4\}\}$ $\{\{1,3,4\},\{2\}\}$
            $\{\{1,3\},\{2,4\}\}$ $\{\{1,4\},\{2,3\}\}$ $\{\{1\},\{2,3,4\}\}$

3 blocks: $\{\{1,2\},\{3\},\{4\}\}$ $\{\{1,3\},\{2\},\{4\}\}$ $\{\{1,4\},\{2\},\{3\}\}$ $\{\{1\},\{2,3\},\{4\}\}$
            $\{\{1\},\{2,4\},\{3\}\}$ $\{\{1\},\{2\},\{3,4\}\}$

4 blocks: $\{\{1\},\{2\},\{3\},\{4\}\}$

Let $S(n,k)$ be the number of partitions of an n-set having exactly k blocks. These are called *Stirling numbers of the second kind*. Do not confuse $S(n,k)$ with $C(n,k) = \binom{n}{k}$. In both cases we have an n-set. For $C(n,k)$ we want to *choose a subset* containing k elements and for $S(n,k)$ we want to *partition the set* into k blocks.

What is the value of $S(n,k)$? Let's try to get a recursion. How can we build partitions of $\{1,2,\ldots,n\}$ with k blocks out of smaller cases? If we take partitions of $\{1,2,\ldots,n-1\}$ with $k-1$ blocks, we can simply add the block $\{n\}$. If we take partitions of $\{1,2,\ldots,n-1\}$ with k blocks, we can add the element n to one of the k blocks. You should convince yourself that all k block partitions of $\{1,2,\ldots,n\}$ arise in exactly one way when we do this. This gives us a recursion for $S(n,k)$. Putting n in a block by itself contributes $S(n-1,k-1)$. Putting n in a block with other elements contributes $S(n-1,k) \times k$ Thus,

$$S(n,k) = S(n-1,k-1) + k\,S(n-1,k).$$

Below is the tabular form for $S(n,k)$ analogous to the similar tabular form for $C(n,k)$.

n\\k	1	2	3	4	5	6	7
1	1						
2	1	1					
3	1	3	1				
4	1	7	6	1			
5	1	15	25	10	1		
6	1	31	90	65	15	1	
7	1	--	--	--	--	--	1

$S(n,k)$

Notice that the starting conditions for this table are that $S(n,1) = 1$ for all $n \geq 1$ and $S(n,n) = 1$ for all $n \geq 1$. The values for $n = 7$ are omitted from the table. You should fill them in to test your understanding of this computational process. For each n, the total number of partitions of a set of size n is equal to the sum $S(n,1) + S(n,2) + \cdots + S(n,n)$. These numbers, gotten by summing the entries in the rows of the above table, are the Bell numbers, $B(n)$, that we discussed in Section 1. $\square$

Example 21 (Counting functions by image size) Suppose A and B are sets. Let $|A| = m$ and $|B| = n$. Suppose $k \le m$ and $k \le n$. A basic question about functions $f : A \to B$ is the following:

$$\text{Let } S = \{f \mid f : A \to B, \ |\text{Image}(f)| = k\}. \text{ Find } |S|.$$

In other words, there are exactly k blocks in the coimage of f. This question clearly involves the Stirling numbers. In fact, the answer is $|S| = \binom{n}{k} S(m, k) \, k!$. The idea is to choose the image of the function in $\binom{n}{k}$ ways, then choose the coimage of the function in $S(m, k)$ ways and then put them together in $k!$ ways. You will get a chance to fill in the details in the last two exercises below. Here is an example. Suppose we take $|A| = 4$, $|B| = 5$, and $k = 3$. We get $|S| = \binom{5}{3} S(4, 3) \, 3! = 10 \times 6 \times 6 = 360$.

Let's look at some special cases.

- If $k = |B|$, then we are counting surjections. Why? We are given $|\text{Image}(f)| = |B|$. Thus every element in B must be in $\text{Image}(f)$.

- If $k = |A|$, then we are counting injections. Why? Suppose f is not an injection, say $f(a) = f(b)$ for some $a \ne b$. Then f can take on at most $|A| - 1$ different values. But $k = |A|$ says that $|\text{Image}(f)| = |A|$, a contradiction. $\square$

Exercises for Section 2

2.1. Let R be the relation on $\mathbb{R} \times \mathbb{R}$, the Cartesian plane, defined by xRy if $y = x^2$. Sketch a picture that represents the set R in $\mathbb{R} \times \mathbb{R}$.

2.2. Let R be a relation from the power set $\mathcal{P}(X)$ to itself, where $X = \{1, 2, 3, 4\}$, defined by $A R B$ if $A \cap B \ne \emptyset$.

 (a) Is $A R A$ for all $A \in \mathcal{P}(X)$?

 (b) For any $A, B \in \mathcal{P}(X)$, if $A R B$ is $B R A$?

 (c) For any $A, B \in \mathcal{P}(X)$, if $A R B$ and $B R C$, is $A R C$?

2.3. In each case, draw the "directed graph diagram" of the given relation (label points in your diagram with the elements of S, put an arrow from x to y if and only if (x, y) belongs to the relation).

 (a) $S = \{(a, b), (a, c), (b, c), (d, d)\}$ on $X \times X$, $X = \{a, b, c, d\}$.

 (b) Let $X = \{2, 3, 4, 5, 6, 7, 8\}$ and define $x R y$ if $x = y \pmod 3$; that is, if $|x - y|$ is divisible by 3.

2.4. Find all relations on $\{a, b\} \times \{x, y\}$, that are not functional.

2.5. Let S be the divides relation on $\{3, 4, 5\} \times \{4, 5, 6\}$; that is, $x \, S \, y$ if y/x is an integer. List the elements of S and S^{-1}.

2.6. Let A be a set with m elements and B be a set with n elements.

Sets and Functions

    (a) How many relations are there on $A \times B$?

    (b) How many functions are there from A to B?

2.7. Define a binary relation D from $\underline{10} = \{1, 2, 3, 4, 5, 6, 7, 8, 9, 10\}$ to $\underline{10}$ as follows: For all x, y, in $\underline{10}$, $x \, D \, y$ if $x < y$ and x divides y. How many edges are there in the directed graph of this relation. Explain.

2.8. This exercise lets you check your understanding of the definitions. In each case below, some information about a function is given to you. Answer the following questions and give reasons for your answers:

- Have you been given enough information to specify the function?

- Can you tell whether or not the function is an injection? a surjection? a bijection?

- If possible, give the function in two-line form.

    (a) $f \in \underline{3}^{\{>, <, +, ?\}}$, $f = (3, 1, 2, 3)$.

    (b) $f \in \{>, <, +, ?\}^{\underline{3}}$, $f = (?, <, +)$.

    (c) $f \in \underline{4}^{\underline{3}}$, $2 \to 3$, $1 \to 4$, $3 \to 2$.

2.9. This exercise lets you check your understanding of the definitions. In each case below, some information about a function is given to you. Answer the following questions and give reasons for your answers:

- Have you been given enough information to specify the function?

- Can you tell whether or not the function is an injection? a surjection? a bijection? If so, what is it?

    (a) $f \in \underline{4}^{\underline{5}}$,   Coimage$(f) = \{\{1, 3, 5\}, \{2, 4\}\}$.

    (b) $f \in \underline{5}^{\underline{5}}$,   Coimage$(f) = \{\{1\}, \{2\}, \{3\}, \{4\}, \{5\}\}$.

    (c) $f \in \underline{4}^{\underline{5}}$, $f^{-1}(2) = \{1, 3, 5\}$, $f^{-1}(4) = \{2, 4\}$.

    (d) $f \in \underline{4}^{\underline{5}}$, $|\text{Image}(f)| = 4$.

    (e) $f \in \underline{4}^{\underline{5}}$, $|\text{Image}(f)| = 5$.

    (f) $f \in \underline{4}^{\underline{5}}$, $|\text{Coimage}(f)| = 5$.

2.10. For each of the following definitions, state whether the definition is correct or not correct. If not correct, explain why.

    (a) Definition: $f : A \to B$ is one-to-one if $\forall s, t \in A$, $f(s) = f(t)$ implies $s = t$.

    (b) Definition: $f : A \to B$ is one-to-one if $\forall s, t \in A$, $s \neq t$ implies $f(s) \neq f(t)$.

    (c) Definition: $f : A \to B$ is one-to-one if $\forall s \in A$, $\exists ! \, t \in B$ such that $f(s) = t$.

(d) Definition: $f : A \to B$ is one-to-one if $\forall t \in B$, $\exists$ at most one $s \in A$ such that $f(s) = t$.

2.11. Define $g : \mathbb{Z} \to \mathbb{Z}$ by $g(n) = 3n - 1$, where $\mathbb{Z}$ is the set of integers.

(a) Is g one-to-one?

(b) Is g onto?

(c) Suppose that $g : \mathbb{R} \to \mathbb{R}$ and $g(x) = 3x - 1$ for all real numbers x. Is g onto?

2.12. In each case prove or disprove the statement "f is one-to-one."

(a) $f(x) = \frac{x}{x^2+1}$ where $f : \mathbb{R} \to \mathbb{R}$

(b) $f(x) = \frac{2x+1}{x}$ where $f : (\mathbb{R} - \{0\}) \to \mathbb{R}$

(c) $f(x) = \frac{x-1}{x+1}$ where $f : (\mathbb{R} - \{-1\}) \to \mathbb{R}$

2.13. This exercise lets you check your understanding of cycle form. A permutation is given in one-line, two-line or cycle form. Convert it to the other two forms. Give its inverse in all three forms.

(a) (1,5,7,8) (2,3) (4) (6).

(b) $\begin{pmatrix} 1 & 2 & 3 & 4 & 5 & 6 & 7 & 8 \\ 8 & 3 & 7 & 2 & 6 & 4 & 5 & 1 \end{pmatrix}$.

(c) (5,4,3,2,1), which is in one-line form.

(d) (5,4,3,2,1), which is in cycle form. (Assume the domain is $\underline{5}$.)

2.14. Let $S = \{f \mid f : A \to B,\ f \text{ one-to-one}\}$. In each case, find $|S|$.

(a) $|A| = 3$ and $|B| = 3$

(b) $|A| = 3$ and $|B| = 5$

(c) $|A| = m$ and $|B| = n$

2.15. Let $f : X \to Y$, $g : Y \to Z$ and $g \circ f : X \to Z$.

(a) If $g \circ f$ is onto, must f and g be onto?

(b) If $g \circ f$ is one-to-one, must f and g be one-to-one?

2.16. In each case, $f : X \to Y$ is an arbitrary function. Prove or disprove:

(a) For all subsets A and B of X, $f(A \cup B) = f(A) \cup f(B)$.

(b) For all subsets A and B of X, $f(A \cap B) = f(A) \cap f(B)$.

(c) For all subsets A and B of X, $f(A - B) = f(A) - f(B)$.

(d) For all subsets C and D of Y, $f^{-1}(C \cap D) = f^{-1}(C) \cap f^{-1}(D)$.

2.17. Let $f : X \to Y$, $g : Y \to Z$ and $g \circ f : X \to Z$. Prove or disprove:

(a) For all subsets $A \subseteq X$, $f^{-1}(f(A)) = A$.

(b) For all subsets $B \subseteq Y$, $f(f^{-1}(B)) = B$.

(c) For all subsets $E \subseteq Z$, $(g \circ f)^{-1}(E)) = f^{-1}(g^{-1}(E))$.

2.18. Let $S = \{f \mid f : A \to B, \ f \text{ onto}\}$. In each case, find $|S|$. Do (a)–(d) without using the general formula in the text.

(a) $|A| = 3$ and $|B| = 2$

(b) $|A| = 3$ and $|B| = 5$

(c) $|A| = 4$ and $|B| = 2$

(d) $|A| = m$ and $|B| = n$

(e) Explain how to use the general formula in the text to solve (d).

2.19. Let $S = \{f \mid f : A \to B, \ |\text{Image}(f)| = k\}$. Suppose that $|A| = m \geq k$ and $|B| = n \geq k$. Prove the formula for $|S|$ given in the text.

Multiple Choice Questions for Review

In each case there is one correct answer (given at the end of the problem set). Try to work the problem first without looking at the answer. Understand both why the correct answer is correct and why the other answers are wrong.

1. Which of the following statements is **FALSE**?

   (a) $2 \in A \cup B$ implies that if $2 \notin A$ then $2 \in B$.

   (b) $\{2,3\} \subseteq A$ implies that $2 \in A$ and $3 \in A$.

   (c) $A \cap B \supseteq \{2,3\}$ implies that $\{2,3\} \subseteq A$ and $\{2,3\} \subseteq B$.

   (d) $A - B \supseteq \{3\}$ and $\{2\} \subseteq B$ implies that $\{2,3\} \subseteq A \cup B$.

   (e) $\{2\} \in A$ and $\{3\} \in A$ implies that $\{2,3\} \subseteq A$.

2. Let $A = \{0,1\} \times \{0,1\}$ and $B = \{a,b,c\}$. Suppose A is listed in lexicographic order based on $0 < 1$ and B is in alphabetic order. If $A \times B \times A$ is listed in lexicographic order, then the next element after $((1,0), c, (1,1))$ is

   (a) $((1,0), a, (0,0))$

   (b) $((1,1), c, (0,0))$

   (c) $((1,1), a, (0,0))$

   (d) $((1,1), a, (1,1))$

   (e) $((1,1), b, (1,1))$

3. Which of the following statements is **TRUE**?

   (a) For all sets A, B, and C, $A - (B - C) = (A - B) - C$.

   (b) For all sets A, B, and C, $(A - B) \cap (C - B) = (A \cap C) - B$.

   (c) For all sets A, B, and C, $(A - B) \cap (C - B) = A - (B \cup C)$.

   (d) For all sets A, B, and C, if $A \cap C = B \cap C$ then $A = B$.

   (e) For all sets A, B, and C, if $A \cup C = B \cup C$ then $A = B$.

4. Which of the following statements is **FALSE**?

   (a) $C - (B \cup A) = (C - B) - A$

   (b) $A - (C \cup B) = (A - B) - C$

   (c) $B - (A \cup C) = (B - C) - A$

   (d) $A - (B \cup C) = (B - C) - A$

   (e) $A - (B \cup C) = (A - C) - B$

5. Consider the true theorem, "For all sets A and B, if $A \subseteq B$ then $A \cap B^c = \emptyset$." Which of the following statements is **NOT** equivalent to this statement:

   (a) For all sets A^c and B, if $A \subseteq B$ then $A^c \cap B^c = \emptyset$.

   (b) For all sets A and B, if $A^c \subseteq B$ then $A^c \cap B^c = \emptyset$.

[109]

Sets and Functions

(c) For all sets A^c and B^c, if $A \subseteq B^c$ then $A \cap B = \emptyset$.

(d) For all sets A^c and B^c, if $A^c \subseteq B^c$ then $A^c \cap B = \emptyset$.

(e) For all sets A and B, if $A^c \supseteq B$ then $A \cap B = \emptyset$.

6. The power set $\mathcal{P}((A \times B) \cup (B \times A))$ has the same number of elements as the power set $\mathcal{P}((A \times B) \cup (A \times B))$ if and only if

(a) $A = B$

(b) $A = \emptyset$ or $B = \emptyset$

(c) $B = \emptyset$ or $A = B$

(d) $A = \emptyset$ or $B = \emptyset$ or $A = B$

(e) $A = \emptyset$ or $B = \emptyset$ or $A \cap B = \emptyset$

7. Let $\sigma = 452631$ be a permutation on $\{1, 2, 3, 4, 5, 6\}$ in one-line notation (based on the usual order on integers). Which of the following is **NOT** a correct cycle notation for σ?

(a) $(614)(532)$

(b) $(461)(352)$

(c) $(253)(146)$

(d) $(325)(614)$

(e) $(614)(253)$

8. Let $f : X \to Y$. Consider the statement, "For all subsets C and D of Y, $f^{-1}(C \cap D^c) = f^{-1}(C) \cap [f^{-1}(D)]^c$. This statement is

(a) True and equivalent to:
  For all subsets C and D of Y, $f^{-1}(C - D) = f^{-1}(C) - f^{-1}(D)$.

(b) False and equivalent to:
  For all subsets C and D of Y, $f^{-1}(C - D) = f^{-1}(C) - f^{-1}(D)$.

(c) True and equivalent to:
  For all subsets C and D of Y, $f^{-1}(C - D) = f^{-1}(C) - [f^{-1}(D)]^c$.

(d) False and equivalent to:
  For all subsets C and D of Y, $f^{-1}(C - D) = f^{-1}(C) - [f^{-1}(D)]^c$.

(e) True and equivalent to:
  For all subsets C and D of Y, $f^{-1}(C - D) = [f^{-1}(C)]^c - f^{-1}(D)$.

9. Define $f(n) = \frac{n}{2} + \frac{1-(-1)^n}{4}$ for all $n \in \mathbb{Z}$. Thus, $f : \mathbb{Z} \to \mathbb{Z}$, $\mathbb{Z}$ the set of all integers. Which is correct?

(a) f is not a function from $\mathbb{Z} \to \mathbb{Z}$ because $\frac{n}{2} \notin \mathbb{Z}$.

(b) f is a function and is onto and one-to-one.

(c) f is a function and is not onto but is one-to-one.

(d) f is a function and is not onto and not one-to-one

(e) f is a function and is onto but not one-to-one.

10. The number of partitions of $\{1, 2, 3, 4, 5\}$ into three blocks is $S(5,3) = 25$. The total number of functions $f : \{1, 2, 3, 4, 5\} \rightarrow \{1, 2, 3, 4\}$ with $|\text{Image}(f)| = 3$ is

(a) 4×6

(b) 4×25

(c) 25×6

(d) $4 \times 25 \times 6$

(e) $3 \times 25 \times 6$

11. Let $f : X \rightarrow Y$ and $g : Y \rightarrow Z$. Let $h = g \circ f : X \rightarrow Z$. Suppose g is one-to-one and onto. Which of the following is **FALSE**?

(a) If f is one-to-one then h is one-to-one and onto.

(b) If f is not onto then h is not onto.

(c) If f is not one-to-one then h is not one-to-one.

(d) If f is one-to-one then h is one-to-one.

(e) If f is onto then h is onto.

12. Which of the following statements is **FALSE**?

(a) $\{2, 3, 4\} \subseteq A$ implies that $2 \in A$ and $\{3, 4\} \subseteq A$.

(b) $\{2, 3, 4\} \in A$ and $\{2, 3\} \in B$ implies that $\{4\} \subseteq A - B$.

(c) $A \cap B \supseteq \{2, 3, 4\}$ implies that $\{2, 3, 4\} \subseteq A$ and $\{2, 3, 4\} \subseteq B$.

(d) $A - B \supseteq \{3, 4\}$ and $\{1, 2\} \subseteq B$ implies that $\{1, 2, 3, 4\} \subseteq A \cup B$.

(e) $\{2, 3\} \subseteq A \cup B$ implies that if $\{2, 3\} \cap A = \emptyset$ then $\{2, 3\} \subseteq B$.

13. Let $A = \{0, 1\} \times \{0, 1\} \times \{0, 1\}$ and $B = \{a, b, c\} \times \{a, b, c\} \times \{a, b, c\}$. Suppose A is listed in lexicographic order based on $0 < 1$ and B is listed in lexicographic order based on $a < b < c$. If $A \times B \times A$ is listed in lexicographic order, then the next element after $((0, 1, 1), (c, c, c), (1, 1, 1))$ is

(a) $((1, 0, 1), (a, a, b), (0, 0, 0))$

(b) $((1, 0, 0), (b, a, a), (0, 0, 0))$

(c) $((1, 0, 0), (a, a, a), (0, 0, 1))$

(d) $((1, 0, 0), (a, a, a), (1, 0, 0))$

(e) $((1, 0, 0), (a, a, a), (0, 0, 0))$

14. Consider the true theorem, "For all sets A, B, and C if $A \subseteq B \subseteq C$ then $C^c \subseteq B^c \subseteq A^c$." Which of the following statements is **NOT** equivalent to this statement:

(a) For all sets A^c, B^c, and C^c, if $A^c \subseteq B^c \subseteq C^c$ then $C \subseteq B \subseteq A$.

(b) For all sets A^c, B, and C^c, if $A^c \subseteq B \subseteq C^c$ then $C \subseteq B^c \subseteq A$.

(c) For all sets A, B, and C^c, if $A^c \subseteq B \subseteq C$ then $C^c \subseteq B^c \subseteq A$.

Sets and Functions

(d) For all sets A^c, B, and C^c, if $A^c \subseteq B^c \subseteq C$ then $C^c \subseteq B^c \subseteq A$.

(e) For all sets A^c, B^c, and C^c, if $A^c \subseteq B^c \subseteq C$ then $C^c \subseteq B \subseteq A$.

15. Let $\mathcal{P}(A)$ denote the power set of A. If $\mathcal{P}(A) \subseteq B$ then

(a) $2^{|A|} \leq |B|$

(b) $2^{|A|} \geq |B|$

(c) $2|A| < |B|$

(d) $|A| + 2 \leq |B|$

(e) $2^{|A|} \geq 2^{|B|}$

16. Let $f : \{1,2,3,4,5,6,7,8,9\} \to \{a,b,c,d,e\}$. In one-line notation, $f = (e,a,b,b,a,c,c,a,c)$ (use number order on the domain). Which is correct?

(a) Image$(f) = \{a,b,c,d,e\}$, Coimage$(f) = \{\{6,7,9\},\{2,5,8\},\{3,4\},\{1\}\}$

(b) Image$(f) = \{a,b,c,e\}$, Coimage$(f) = \{\{6,7,9\},\{2,5,8\},\{3,4\}\}$

(c) Image$(f) = \{a,b,c,e\}$, Coimage$(f) = \{\{6,7,9\},\{2,5,8\},\{3,4\},\{1\}\}$

(d) Image$(f) = \{a,b,c,e\}$, Coimage$(f) = \{\{6,7,9,2,5,8\},\{3,4\},\{1\}\}$

(e) Image$(f) = \{a,b,c,d,e\}$, Coimage$(f) = \{\{1\},\{3,4\},\{2,5,8\},\{6,7,9\}\}$

17. Let $\Sigma = \{x,y\}$ be an alphabet. The strings of length seven over Σ are listed in dictionary (lex) order. What is the first string after $xxxxyxx$ that is a palindrome (same read forwards and backwards)?

(a) $xxxxyxy$ (b) $xxxyxxx$ (c) $xxyxyxx$ (d) $xxyyyxx$ (e) $xyxxxyx$

18. Let $\sigma = 681235947$ and $\tau = 627184593$ be permutations on $\{1,2,3,4,5,6,7,8,9\}$ in one-line notation (based on the usual order on integers). Which of the following is a correct cycle notation for $\tau \circ \sigma$?

(a) (124957368)

(b) (142597368)

(c) (142953768)

(d) (142957368)

(e) (142957386)

Answers: 1 (e), 2 (c), 3 (b), 4 (d), 5 (a), 6 (d), 7 (b), 8 (a), 9 (e), 10 (d), 11 (a), 12 (b), 13 (e), 14 (d), 15 (a), 16 (c), 17 (b), 18 (d).

Equivalence and Order

Section 1: Equivalence

The concept of an *equivalence relation* on a set is an important descriptive tool in mathematics and computer science. It is not a new concept to us, as "equivalence relation" turns out to be just another name for "partition of a set." Our emphasis in this section will be slightly different from our previous discussions of partitions in Unit SF. In particular, we shall focus on the basic conditions that a binary relation on a set must satisfy in order to define a partition. This "local" point of view regarding partitions is very helpful in many problems. We start with the definition.

Definition 1 (Equivalence relation)   *An equivalence relation on a set S is a partition $\mathcal{K}$ of S. We say that $s, t \in S$ are equivalent if and only if they belong to the same block of the partition $\mathcal{K}$. We call a block an equivalence class of the equivalence relation.*

If the symbol $\equiv$ denotes the equivalence relation, then we write $s \equiv t$ to indicate that s and t are equivalent (in the same block) and $s \not\equiv t$ to denote that they are not equivalent.

Here's a trivial equivalence relation that you use all the time. Let S be any set and let all the blocks of the partition have one element. Two elements of S are equivalent if and only if they are the same. This rather trivial equivalence relation is, of course, denoted by "=".

Example 1 (All the equivalence relations on a set) Let $S = \{a, b, c\}$. What are the possible equivalence relations on S? Every partition of S corresponds to an equivalence relation, so listing the partitions also lists the equivalence relations. Here they are with the equivalences other than $a \equiv a$, $b \equiv b$ and $c \equiv c$, which are always present.

$$
\begin{array}{ll}
\{\{a\}, \{b\}, \{c\}\} & \text{no others} \\
\{\{a\}, \{b, c\}\} & b \equiv c, \quad c \equiv b \\
\{\{b\}, \{a, c\}\} & a \equiv c, \quad c \equiv a \\
\{\{c\}, \{a, b\}\} & a \equiv b, \quad b \equiv a \\
\{\{a, b, c\}\} & a \equiv b, \quad b \equiv a, \quad a \equiv c, \quad c \equiv a, \quad b \equiv c, \quad c \equiv b
\end{array}
$$

What about the set $\{a, b, c, d\}$? There are 15 equivalence relations. For a five element set there are 52. As you can see, the number increases rapidly. $\square$

Equivalence and Order

Example 2 (Classification by remainder) Now let the set be the integers $\mathbb{Z}$. Let's try to define an equivalence relation by saying that n and k are equivalent if and only if they have the same remainder when divided by 24. In other words, n and k differ by a multiple of 24. Is this an equivalence relation? If it is we should be able to find the blocks of the partition. There are 24 of them, which we could number $0, \ldots, 23$. Block j consists of all integers which equal j plus a multiple of 24; that is, they have a remainder of j when divided by 24. We write this block as $24\mathbb{Z} + j$. If two integers a and b are equivalent, we write $a = b \pmod{24}$ or $a \equiv b \pmod{24}$.

For this to be an equivalence relation, we must verify that the 24 sets $24\mathbb{Z} + j$, $0 \le j \le 23$, are a partition of $\mathbb{Z}$. Each of these blocks is clearly nonempty. Their union is $\mathbb{Z}$ since every integer has some remainder $0 \le j \le 23$ when divided by 24. Why are they disjoint? If $x \in 24\mathbb{Z} + j$ and $x \in 24\mathbb{Z} + k$ where $j \ne k$, then x would have two different remainders j and k when divided by 24. This is impossible.

Of course there is nothing magic about 24 — any positive integer would work. We studied this in Section 1 of Unit NT. We proved an interesting property of these equivalence there: They can be added, subtracted and multiplied. In other words, if $a \equiv A$ and $b \equiv B$, then $a + b \equiv A + B$, $a - b \equiv A - B$ and $ab \equiv AB$. $\square$

Example 3 (Using coimage to define equivalence) Suppose S and T are sets and $F : S \to T$ is a function. We can use the coimage of F to define an equivalence relation. Recall from Section 2 of Unit SF that the coimage of F is the partition of S given by

$$\mathrm{Coimage(F)} = \{F^{-1}(b) \mid b \in T, \ F^{-1}(b) \ne \emptyset\} = \{F^{-1}(b) \mid b \in \mathrm{Image}(F)\}.$$

Since the coimage is a partition of S, it defines an equivalence relation. Thus $s_1 \equiv s_2$ if and only if $F(s_1) = F(s_2)$. As you can see, the idea is to define $F(s)$ to be the property of s that we are interested in. Almost all our examples are of this kind.

In the previous example, we could take $S = \mathbb{Z}$ and define $F(n)$ to be the remainder when n is divided by 24.

The sets need not be numeric. For example, let S be the set of people and let $T = S \times S$. Define

$$F(s) = (\text{mother of } s, \text{ father of } s).$$

Then $s_1 \equiv s_2$ if and only if s_1 and s_2 are siblings (i.e., have the same parents). $\square$

Example 4 (Some equivalence classes of functions) We use the notation $\underline{k}$ to denote $\{1, 2, \ldots, k\}$, the set of the first k positive integers. Consider all functions $S = \underline{m}^{\underline{n}}$. Here are some partitions based on the fact that S is a set of functions:

- We could partition the functions f into blocks according to the sum of the integers in the Image(f). In other words, $f \equiv g$ means $\sum_{i=1}^{n} f(i) = \sum_{i=1}^{n} g(i)$.

- We could partition the functions f into blocks according to the max of the integers in Image(f). In other words, $f \equiv g$ means that the maximum values of f and g are the same.

- We could partition the functions f into blocks according to the vector $v(f)$ where the ith component $v_i(f)$ is the number of times f takes the value i; that is, $v_i(f) = |f^{-1}(i)|$.

For example, suppose $f, g, h \in \underline{3}^{\underline{4}}$ in one-line notation are $f = (3, 1, 2, 1)$, $g = (1, 1, 3, 2)$ and $h = (1, 2, 3, 2)$. Then $f \equiv g \not\equiv h$ because $v(f) = v(g) = (2, 1, 1)$ but $v(h) = (1, 2, 1)$.

Each of these defines a partition of S and hence and equivalence relation on S.

We can use the coimage idea of the previous example to describe these equivalence relations:

- $F(f) = \sum_{i=1}^{n} f(i)$,
- $F(f) = \max(f)$ and
- $F(f) = v(f)$. $\Box$

Definition 2 (Binary relation on a set)   Given a set S, a *binary relation* on S is a subset R of $S \times S$. Given a binary relation R, we will write $s \, R \, t$ if and only if $(s, t) \in R$.[1]

Example 5 (Equivalence relations as binary relations) Suppose $\equiv$ is an equivalence relation on S associated with the partition $\mathcal{K}$. Then the set $R = \{(s, t) \mid s \equiv t\} \subseteq S \times S$ is a binary relation on S associated with the equivalence relation. Thus an equivalence relation is a binary relation.

The converse need not be true. For example $x \, R \, y$ if and only if $x < y$ defines a binary relation on $\mathbb{Z}$, but it is not an equivalence relation because we never have $x < x$, but an equivalence relation requires $x \equiv x$ for all x. $\Box$

When is a binary relation an equivalence relation? The next theorem provides necessary and sufficient conditions for a binary relation to be an equivalence relation. Verifying the conditions is a sometimes a useful way to prove that some particular situation is an equivalence relation.

Theorem 1 (Reflexive, symmetric, transitive)   Let S be a set and suppose that we have a binary relation R on S. This binary relation is an equivalence relation if and only if the following three conditions hold.

(i) *(Reflexive)* For all $s \in S$ we have $s \, R \, s$.

(ii) *(Symmetric)* For all $s, t \in S$ such that $s \, R \, t$ we have $t \, R \, s$.

(iii) *(Transitive)* For all $r, s, t \in S$ such that $r \, R \, s$ and $s \, R \, t$ we have $r \, R \, t$.

Proof: We first prove that an equivalence relation satisfies (i)–(iii). Suppose that $\equiv$ is an equivalence relation. Since s belongs to whatever block it is in, we have $s \equiv s$. Since $s \equiv t$ means that s and t belong to the same block, we have $s \equiv t$ if and only if we have $t \equiv s$. Now suppose that $r \equiv s$ and $s \equiv t$. Then r and s are in the same block and s and t are in the same block. Thus r and t are in the same block and so $r \equiv t$.

---

[1] A binary relation is a special case of a relation from S to T (discussed in Unit SF), namely, $T = S$.

Equivalence and Order

We now suppose that (i)–(iii) hold and prove that we have an equivalence relation; that is, a partition of the set S. What would the blocks of the partition be? Everything equivalent to a given element should be in the same block. Thus, for each $s \in S$ let $B(s)$ be the set of all $t \in S$ such that $s\ R\ t$. We must show that the set of these sets form a partition of S; that is, $\{B(s) \mid s \in S\}$ is a partition of S.

In order to have a partition of S, we must have

(a) the $B(s)$ are nonempty and every $t \in S$ is in some $B(s)$ and

(b) for every $p, q \in S$, $B(p)$ and $B(q)$ are either equal or disjoint.

Since R is reflexive, $s \in B(s)$, proving (a). We now turn our attention to (b). Suppose $B(p) \cap B(q)$ is not empty. We must prove that $B(p) = B(q)$. Suppose $x \in B(p) \cap B(q)$ and $y \in B(p)$. We have, $p\ R\ x$, $q\ R\ x$ and $p\ R\ y$. By the symmetric law, $x\ R\ p$. Using transitivity twice:

$$\big(q\ R\ x\ \text{ and }\ x\ R\ p\big)\ \text{ implies }\ q\ R\ p,$$
$$\big(q\ R\ p\ \text{ and }\ p\ R\ y\big)\ \text{ implies }\ q\ R\ y.$$

By the definition of B, this means $y \in B(q)$. Since this is true for all $y \in B(p)$, we have proved that $B(p) \subseteq B(q)$. Similarly $B(q) \subseteq B(p)$ and so $B(p) = B(q)$. This proves (b). $\square$

Example 6 (The rational numbers) Let the set be $S = \mathbb{Z} \times \mathbb{Z}^*$, where $\mathbb{Z}^*$ is the set of all integers except 0. Write $(a, b) \equiv (c, d)$ if and only if $ad = bc$.

We now use Theorem 1 to prove that, in fact, this is an equivalence relation.

(i) To verify $(a, b) \equiv (a, b)$, we must check that $ab = ba$ (since $c = a$ and $d = b$). This is obviously true.

(ii) Suppose $(a, b) \equiv (c, d)$. This means $ad = bc$. To verify that $(c, d) \equiv (a, b)$, we must check that $cb = da$. This follows easily from $ad = bc$.

(iii) Suppose $(a, b) \equiv (c, d)$ and $(c, d) \equiv (e, f)$. We must verify that $(a, b) \equiv (e, f)$. In other words, we are given $ad = bc$ and $cf = de$, and we want to conclude that $af = be$. We have $(ad)(cf) = (bc)(de)$. We're done if we can cancel c and d from both sides of this equality. We can cancel d since $d \in \mathbb{Z}^*$. If $c \neq 0$ we can cancel it, too, and we're done. What if $c = 0$? In that case, since $d \neq 0$, it follows from $ad = bc$ that $a = 0$. Similarly, it follows from $cf = de$ that $e = 0$. Thus $af = be = 0$.

With a moment's reflection, you should see that $ad = bc$ is a way to check if the two fractions a/b and c/d are equal. We can label each equivalence class with the fraction a/b that it represents.

Just because we've defined fractions, that doesn't mean we can do arithmetic with them. We need to prove that equivalence classes can be added, subtracted, multiplied, and divided. This problem was mentioned for modular arithmetic at the end of Example 2. We won't prove it for rational numbers. $\square$

In the exercises, you will acquire more practice with equivalence relations. You should find them easier to deal with than the previous two examples.

The Pigeonhole Principle

We now look at a class of problems that relate to various types of restrictions on equivalence relations. These problems are a part of a much more general and often very difficult branch of mathematics called *extremal set theory*. The following theorem is a triviality, but its name and some of its applications are interesting. You should be able to prove the theorem.

Theorem 2 (Pigeonhole principle) *Suppose $\mathcal{K}$ is a partition of a set S and $|S| = s$. If $\mathcal{K}$ has fewer than s blocks, then some block must have at least two elements.*

Where did the name come from? Old style desks often had an array of small horizontal boxes for storing various sorts of papers — unpaid bills, letters, etc. These were called pigeonholes because they often resembled the nesting boxes in pigeon coops. Imagine slips of paper, with one element of S written on each slip. Put the slips into the boxes. At least one box must receive more than one slip if there are more slips than boxes — that's the pigeonhole principle. After the slips are in the boxes, a partition of the set of slips has been defined. (The boxes are the blocks.)

Example 7 (Applying the pigeonhole principle) Designate the months of the year by the set numbers $M = \underline{12} = \{1, 2, \ldots, 12\}$. What is the smallest integer k such that among any k people, there must be at least two people with the same first letter of their last name and same birth month? This is a typical application of the pigeonhole principle.

Recall that we can define a partition of a set P by defining a function f with domain P and letting the partition of P be the coimage of f. In this case, we let P be the set of people and define $f : P \to M \times A$, where A is the set of letters in the alphabet, as follows: $f(p) = (m, a)$, where m is the month in which p was born and a is the first letter of p's last name. To be able to apply the pigeonhole principle to obtain the conclusion asked for, we must have $|M \times A| < k$, where $k = |P|$. In other words, we must have $12 \times 26 = 312 < k$. The smallest such $k = 313$. If $k = 312$ then it is possible to have a group of people, no two of which have the same first letter of their last name and same birth month.

What we did in the last example is a common way to set things up for the pigeonhole principle. We are given a set S (people) and a property (birthdays and initials) of the elements of S. We want to know that two elements of S have equal properties. To use the coimage, define a function $f : S \to T$ where T is the set of possible properties. Since the coimage can have at most $|T|$ blocks, we need $|T| < |S|$ to use the pigeonhole principle.

Example 8 (A divisibility example) Here is a somewhat less trivial example. Let $n > 0$ be a an integer. What is the smallest integer k such that given any set $\{t_1, t_2, \ldots, t_k\}$ of distinct integers, either there must be some $i \neq j$ such that $n \mid (t_i - t_j)$, or there must be some i such that $n \mid t_i$? We must give a little thought as to how to use the coimage version of the pigeonhole principle in this situation. It seems that the correct function is $f : \{t_1, t_2, \ldots, t_k\} \rightarrow \{0, 1, \ldots, n-1\}$ where $f(x) = x \pmod{n}$. Suppose first that there is no i such that $n \mid t_i$. Then $|\text{Image}(f)| < n$ because the only values that can be in the image are $1, 2, \ldots, (n-1)$. In that case, if $k \geq n$ then we can apply the pigeonhole principle. There must be at least one pair, $i < j$, such that $f(t_i) = f(t_j)$ or, equivalently, $t_i \pmod{n}$ $= t_j \pmod{n}$. For this pair, $n \mid (t_i - t_j)$. If $k = n-1$, we can take the set $\{1, 2, \ldots, n-1\}$ for which there is neither some i such that $n \mid t_i$ nor a pair $i \neq j$ such that $n \mid (t_i - t_j)$. Thus, the smallest k is $k = n$. You should think carefully about this example as it typical of one type of analysis associated with these extremal problems. The idea is that one condition (in this case $n \mid t_i$ for some i), when negated, brings about a situation where the pigeonhole principle applies.

Results gotten from the application of the pigeonhole principle are sometimes used in clever ways to get new results. Suppose that we have a sequence $a_1, a_2, \ldots, a_n$ of positive integers. Then, by the previous paragraph applied to the "partial sums" of this sequence, there must be a partial sum, $t_i = a_1 + a_2 + \cdots + a_i$ such that either $n \mid t_i$ or there is a pair $i < j$ such that $n \mid (t_j - t_i)$. Note that $t_j - t_i = a_{i+1} + \cdots + a_j$. Thus, if you have any sequence of n positive integers, there is at least one "consecutive" sum of the form $\sum_{k=p}^{q} a_k$ that is divisible by n. For example, take $n = 8$ and take the sequence to be 11, 12, 23, 5, 7, 9, 21, 9. The consecutive sum $12 + 23 + 5 = 40$ is divisible by 8. In general, this consecutive sum that is divisible by n will, as in this example, not be unique.

Example 9 (Equal sums) Given a positive integer N, how large must t be so that, for every list $A = (a_1, \ldots, a_t)$ of t integers, there are $i \neq j$ and $k \neq m$ such that $a_i + a_j = a_k + a_m \pmod{N}$ and $\{i, j\} \neq \{k, m\}$. (The last condition avoids the trivial situation $a_i + a_j = a_i + a_j$.) For example, if $N = 5$ and $A = (1, 2, 5, 8)$, then $1 + 2 = 5 + 8 \pmod{5}$. We want to look at sums of pairs of elements of A, so our set S will be pairs of indices chosen from t and $f(i, j) = a_i + a_j \pmod{N}$ defines a function $f : S \rightarrow \{0, 1, \ldots, N-1\}$. Thus there are N pigeonholes. What is $|S|$? In other words how many pairs can be chosen from t? The answer is $\binom{t}{2} = (t^2 - t)/2$. For the pigeonhole principle we need $(t^2 - t)/2 > N$, which is equivalent to $t^2 - t - 2N > 0$. The curve $y = x^2 - x - 2N$ is a parabola opening upward. Thus $y > 0$ whenever x exceeds the larger root of the quadratic. Solving for the larger root and setting t greater than it, we obtain

$$t > \frac{1 + \sqrt{1 + 8N}}{2} \quad \text{implies that} \quad a_i + a_j = a_k + a_m$$

for two distinct pairs of distinct indices $\{i, j\} \neq \{k, m\}$. $\square$

*Example 10 (Subset sums) In the set $S = \{1, 2, 3, 4\}$ there are two different (not equal as sets) subsets $P \subset S$ and $Q \subset S$ such that the sum of the entries in P (designated by $\sum P$) equals the sum of the entries in Q (take $P = \{1, 2, 3\}$ and $Q = \{2, 4\}$). We say that the set $S = \{1, 2, 3, 4\}$ has the *two-sum property*. In the set $S = \{1, 2, 4, 8\}$ there is no such pair of subsets. This set fails to have the two-sum property. Suppose we start with a set

$U = \underline{n} = \{1, 2, \ldots, n\}$. What conditions can we put on k such that *every* subset $S \subset U$ of size k has the two-sum property; i.e., S has a pair of distinct subsets P and Q with $\sum P = \sum Q$? This question has the aura of a pigeonhole problem, but how do we describe the function and coimage?

We will look just at the case $n = 16$. Thus $U = \underline{16}$. What is the largest subset of U that does *not* have the two-sum property? Call the size of this set $T(16)$. Using the pigeonhole principle, we will show that every $S \subset U$ of size 7 has the two-sum property. You can check that $\{1, 2, 4, 8, 16\}$, a subset of size 5, does not have the two-sum property.[2] Thus $T(16) \geq 5$. Assuming the above mentioned result for sets of size 7, we now know that either $T(16) = 5$ or $T(16) = 6$. After searching for a subset of size 6 that does not have the two-sum property, you may become convinced that $T(16) = 5$. Unless you looked at all of the $\binom{16}{6} = 8,008$ subsets, this is not a proof. Using a computer program to do this, we found that each of these 8,008 subsets has the two-sum property and so $T(16) = 5$.

Such gaps (Is $T(16) = 5$ or 6?) are common in applications of the pigeonhole principle. Any careful study of this problem would have to go into such gaps between the largest counterexample we have found ($k = 5$) and the smallest k for which the pigeonhole principle works ($k = 7$). This is an annoying feature of the pigeonhole principle in many cases where it is applied.

We now show that every $S \subseteq \underline{16}$ with $|S| = 7$ has the two-sum property.

A subset $S \subseteq U$ with $|S| = k$ has 2^k subsets. If we are going to find distinct subsets P and Q of S with $\sum P = \sum Q$, then clearly neither P nor Q can be empty or equal to S. It seems that we want to apply the pigeonhole principle to the set K of all subsets of a set S, except $\emptyset$ and S. There are $2^k - 2$ such subsets. The function $f : K \to R$ will be given by $f(P) = \sum P$ for $P \in K$. What is R? It is all possible subset sums. We need to work this out.

The value of the sum of the entries over such a subset can be as small as 1 and as large as $r = (16 - (k-2)) + (16 - (k-3)) + \cdots + 16$ (the largest sum of any $k-1$ elements from U). The pigeonhole principle assures us that there will be two distinct elements X and Y in K with $f(X) = f(Y)$ if $|K| > |R|$. In other words, if $2^k - 2 > r$. Using a calculator, we see that the first k that satisfies this inequality is $k = 7$. Thus, any subset $S \subseteq \underline{16}$ of size 7 or larger has the two-sum property. $\square$

The pigeonhole principle answers the question

When must a partition of a set have a block of size at least two?

Sometimes that's not enough. For example, what if we want more than two people to have the same birthday? This is as easy to answer as the original question:

Theorem 3 (Extended pigeonhole principle) *Suppose $f : S \to R$ is a function, $|R| = r$ and $|S| = s$. The coimage of f must have a block of size $\lceil s/r \rceil$ or larger.*

---

[2] You should be able to see why this is so by thinking in terms of the binary number system: every positive integer has a *unique* representation as sums of powers of 2.

EO-7

Equivalence and Order

Proof: Suppose the Coimage(f) has t blocks. Since $|R| = r$, we have $t \leq r$. Let $\{B_1, B_2, \ldots, B_t\}$ be the blocks of Coimage(f) and suppose $|B_i| < s/r$ for $i = 1, \ldots, t$. Now

$$s = |S| = |B_1| + |B_2| + \cdots + |B_t| < (s/r) + (s/r) + \cdots + (s/r) = (s/r)t \leq (s/r)r = s,$$

a contradiction. Thus some B_i has $|B_i| \geq s/r$. Since the size of a block is an integer and s/r may not be an integer, we can assert that Coimage(f) must have at least one block of size *greater than or equal to* $\lceil s/r \rceil$. $\square$

For example, how many people must we have to be sure that at least k have the same birthday? Let s be the number of people. Since there are 366 possible birthdays, $r = 366$. By the theorem, it suffices to have $\lceil s/366 \rceil \geq k$. In other words, $s/366 > k - 1$. Thus $s > 366(k - 1)$ guarantees that at least k out of s people *must* have the same birthday.

In some instances, the most difficult part of applying the pigeonhole principle is finding the right partition (equivalence relation). The next example is of this type.

*__Example 11 (Monotone subsequences)__ A sequence is *monotonic* if it is either decreasing or increasing.

Consider the sequence 7, 5, 2, 6, 8, 1, 9. Starting at a term, say 2, we can move to the right selecting a subsequence (not necessarily consecutive) that is increasing: 2, 6, 8, 9. Or, starting at 2 we could do the same, only selecting a decreasing subsequence: 2, 1.

In a general sequence, we cannot guarantee a long increasing subsequence — the numbers in the sequence might be decreasing. Similarly, we cannot guarantee a long decreasing subsequence. However, we might hope to guarantee a long *monotonic* subsequence. For example, the longest decreasing subsequence in our example is 2, 1, but there is a monotonic subsequence of length 4.

Let's use the pigeonhole principle for the general case. Suppose we start with a sequence of length m, $a_1, a_2, \ldots, a_m$, where the terms are distinct real numbers. How long an increasing or decreasing subsequence *must* be present?

Suppose $a_t = a_{d_1}, a_{d_2}, \ldots, a_{d_k}$ is a decreasing subsequence of length k starting at a_t and that $a_t = a_{i_1}, a_{i_2}, \ldots, a_{i_l}$ is an increasing subsequence of length l starting at a_t.

- If $a_{t-1} > a_t$, then $a_{t-1}, a_{d_1}, a_{d_2}, \ldots, a_{d_k}$ is a decreasing subsequence of length $k + 1$ starting at $t - 1$.

- If $a_{t-1} < a_t$, then $a_{t-1}, a_{i_1}, a_{i_2}, \ldots, a_{i_l}$ is an increasing subsequence of length $l + 1$ starting at $t - 1$.

Thus the length of either the increasing or decreasing sequence has increased by 1.

Let's formalize this a bit. Let D_t and I_t be the lengths of the longest decreasing and longest increasing subsequences starting at a_t. We've just shown that either $D_{t-1} = D_t + 1$ or $I_{t-1} = I_t + 1$. Thus $(I_{t-1}, D_{t-1}) \neq (I_t, D_t)$. All we used was that $t - 1$ is less than t. Thus we can replace $t - 1$ by any $u < t$ in the above argument to conclude that $(I_u, D_u) \neq (I_t, D_t)$ whenever $u < t$. In other words, $f : \underline{m} \to \mathbb{Z} \times \mathbb{Z}$ given by $f(t) = (I_t, D_t)$ is an injection.

We're ready to apply the pigeonhole principle. Why? The pigeonhole principle tells us when f *cannot* be an injection because the pigeonhole principle guarantees that Coimage(f) has a block of size greater than one.

Suppose that the overall longest increasing sequence has length ι ("iota") and the overall longest decreasing sequence has length δ ("delta"). Thus the image of f is contained in $\underline{\iota} \times \underline{\delta}$ and so $|\text{Coimage}(f)| \leq \iota\delta$. Suppose $\iota\delta < m$. By the pigeonhole principle, there must be $p < q$ with $f(p) = f(q)$, a contradiction because we know f is an injection.

We have shown that, if the longest decreasing subsequence has length δ and the longest increasing subsequence has length ι, then the sequence has length at most $\iota\delta$. In other words, $m \leq \iota\delta$.

Put another way, we have shown that, if $m > \alpha\beta$ for some integers α and β, then there must be either an increasing subsequence longer than α or a decreasing subsequence longer than β. For example, if $m = 100$, there is either an increasing subsequence of length at least 12 or a decreasing subsequence of length at least 10 because $100 > 99 = 11 \times 9$. If we write $99 = 9 \times 11$, we see that there is either an increasing subsequence of length at least 10 or a decreasing subsequence of length at least 12.

How long a monotonic subsequence must a sequence of length m have? In this case, we take $\alpha = \beta = n$ because we want to make sure the subsequence is long whether it is increasing or decreasing. If $m > n^2$, there must be a monotonic subsequence of length $n+1$ or greater. Of course, if there is a monotonic subsequence of length at least $n + 1$, then there is one of length exactly $n + 1$: just throw some elements if the subsequence is too long. $\square$

Sometimes, in working problems concerning partitions, there are restrictions on block sizes. In such cases, can be useful to list the possible *type vectors*. A type vector $\vec{v}$ for a partition has v_i equal to the number of blocks containing exactly i elements. For example, we list all type vectors for partitions of $S = \underline{15}$ with maximum block size 4 and exactly 5 blocks. Since $v_i = 0$ for $i > 4$, we give just $v_1, \ldots, v_4$.

1	2	3	4
1	1	0	3
1	0	2	2
0	2	1	2
0	1	3	1
0	0	5	0

The top row represents the block sizes, ranging from 1 to maximum size of 4. The remaining rows tell us how many blocks of each size there are (i.e., these rows correspond to the type vectors). You should think carefully about why this list of type vectors is complete for the restrictions given.

In the exercises for this section you are asked to solve a problem related to the following situation: "Fifteen clients are being defended against lawsuits by a group of five lawyers. Each client is assigned exactly one lawyer and no lawyer is to represent less than one or more than four clients." To see the connection between this problem and partitions, note first that the condition that each client be assigned one lawyer specifies that the correspondence between clients and lawyers is a function, with domain size 15 (the clients) and codomain

Equivalence and Order

size 5 (the lawyers). The condition that no lawyer is to represent less than one or more than four clients specifies that the function is onto and that the maximum block size of the coimage is four. That the function is onto and the codomain has five elements says that there are exactly five blocks in the coimage. Thus, the above table lists the type vectors of all coimages for allowable correspondences between clients and lawyers.

Exercises for Section 1

1.1. Let S be the set of students in a college. Define students $x \equiv y$ to be related if they have both the same age and the same number of years completed in college. Show that $\equiv$ is an equivalence relation by defining a function a whose coimage is the set of equivalence classes of $\equiv$.

1.2. Let $\mathbb{Z}$ be the integers with d a positive integer. Define integers $x \equiv y$ to be related if $d \mid (x-y)$. Show that $\equiv$ is an equivalence relation by defining a function m whose coimage is the set of equivalence classes of $\equiv$.

1.3. Let $\mathbf{F}_n$ be the set of all statement forms in n Boolean variables. Define forms $x \equiv y$ to be related if they have the same truth table. Show that $\equiv$ is an equivalence relation by defining a function t with coimage the equivalence classes of $\equiv$.

1.4. Let $\mathbb{Z}$ be the integers with d and k positive integers. Define $x \equiv y$ to be related if $d \mid (x^k - y^k)$. Show that $\equiv$ is an equivalence relation on $\mathbb{Z}$ by defining a function m whose coimage is the set of equivalence classes of $\equiv$.

1.5. Let $\mathbb{R}$ be the real numbers. Define $x \equiv y$ to be related if $x - y \in \mathbb{Z}$. Show that $\equiv$ is an equivalence relation on $\mathbb{R}$ by defining a function u whose coimage is the set of equivalence classes of $\equiv$.

1.6. In any group of 677 people, there must be at least two who have the same first and last letters of their names. Explain.

1.7. In each case give an explanation in terms of functions and coimages.

(a) Must any set of $k > 1$ integers have at least two with the same remainder when divided by $k - 1$?

(b) Must any set of $k > 1$ integers have at least two with the same remainder when divided by k?

1.8. What is the smallest integer k such that every k-element subset of the set $S = \underline{n}$ must always contain a pair of elements whose sum is $n + 1$?

1.9. Let $n \geq 1$ be an integer. What is the smallest integer k such that every k-element subset of the set $S = \{0, 1, 2, \ldots, n\}$ must always contain an even integer? Must always contain an odd integer?

1.10. What is the smallest integer k such that any set S of k integers selected from the set $\underline{50} = \{1, 2, \ldots, 50\}$ will always have two distinct integers, $x \in S$ and $y \in S$ such that $\gcd(x, y) > 1$? ($\gcd(x, y)$ is the greatest common divisor of x and y.)

1.11. What is the smallest integer k such that any set S of k people must have at least three people who were born in the same month of the year?

1.12. Let P be a group of 30 people. Let f be the function from P to M, where $M = \underline{12}$ represents the 12 months of the year, and $f(x)$ is the birth-month of x. Among such a group, there need not be any group of four people that have the same birth-month. One way this can happen is if Coimage(f) has ten blocks of size three. Describe the structure of the coimage of all other examples. What are their type vectors?

1.13. Some cultures divide a day into "quarter days" in order to pay respect to the tidal cycle (four six hour tidal cycles in each 24 hour period). There are 1461 ATCs (Annual Tidal Cycles) per solar year. What is the smallest integer k such that among k people there are at least four born in the same ATC?

1.14. There are N students in a class. Their exam scores ranged between 27 and 94. All possible scores were achieved by at least one student except for the scores 31, 43, and 55 (none of the students got these scores). What is the smallest value of N that guarantees that at least three students achieved the same score?

1.15. There are twelve 1967 pennies, seven 1968 pennies, and four 1971 pennies in a jar. Let N_k denote the smallest number of pennies you need to select to guarantee that you have k pennies of the same date. Find N_4, N_6 and N_8.

1.16. Let $t_1, t_2, \ldots t_n$ be n integers. Show that either $n \mid t_k$ for some k or $n \mid (t_i - t_j)$ for some $i \neq j$.

1.17. Let $n > 1$ be an integer. What is the smallest value of k such that, given any k distinct integers, $t_1, t_2, \ldots t_k$, there must be two of them t_i and t_j, $i \neq j$, such that either $n \mid (t_i - t_j)$ or $n \mid (t_i + t_j)$?
Hint: We want t_i and t_j to go in the same pigeonhole if either $t_i = -t_j \bmod n$ (so that $n \mid (t_i + t_j)$) or $t_i = t_j \bmod n$ (so that $n \mid (t_i - t_j)$).

1.18. Let $n > 1$ be an integer. What is the smallest value of m such that, given any m distinct integers, $t_1 < t_2 < \cdots < t_m$, chosen from the set $S = \underline{n}$, there must be

$i < j$, such that $t_i \mid t_j$.

Hint: Remove all factors of 2 from the elements of S.

*1.19. We want to show that $m \leq \iota\delta$ in Example 11 is best possible. In other words, there exist sequences of length $\iota\delta$ with longest increasing subsequence of length ι and longest decreasing subsequence of length δ.

    (a) Construct a sequence of length ι whose longest decreasing subsequence has length $\delta = 1$ and whose longest increasing subsequence has length ι.

    (b) Construct a sequence of length 2ι whose longest decreasing subsequence has length $\delta = 2$ and whose longest increasing subsequence has length ι.

    (c) Construct a sequence of length $\delta\iota$ whose longest decreasing subsequence has length δ and whose longest increasing subsequence has length ι.

*1.20. Suppose $m = pq$ and an m-long sequence of distinct real numbers does not have a p-long decreasing subsequence. Prove that it has has a q-long increasing subsequence.

*1.21. Suppose $m > n^4$. Let $(a_1, b_1), \ldots, (a_m, b_m)$ be an m-long sequence where the a_i and b_j are distinct real numbers. The goal of this exercise is to prove that there is an $(n+1)$-long subsequence $(a_{t_1}, b_{t_1}), \ldots, (a_{t_{n+1}}, b_{t_{n+1}})$ such that the sequences $a_{t_1}, \ldots, a_{t_{n+1}}$ and $b_{t_1}, \ldots, b_{t_{n+1}}$ are both monotone.

    (a) Let $k = n^2 + 1$. Prove that the sequence $a_1, \ldots, a_m$ has a k-long monotone subsequence. Call it $a_{s_1}, \ldots, a_{s_k}$.

    (b) Prove that the subsequence $b_{s_1}, \ldots, b_{s_k}$ has an $(n+1)$-long monotone subsequence. Call it $b_{t_1}, \ldots, b_{t_{n+1}}$.

    (c) Prove that the indices $t_1, \ldots, t_{n+1}$ solve the problem.

1.22. Fifteen clients are being defended against lawsuits by a group of five lawyers. Each client is assigned exactly one lawyer and no lawyer is to represent less than one or more than four clients. Show that if two lawyers are assigned less than three clients, at least two must be assigned four clients.

Section 2: Order

In Theorem 1 we showed the connection between certain binary relations on a set S and partitions of the same set. In this section we will study binary relations that are, as before, reflexive and transitive, but, instead of being symmetric, are "antisymmetric." We begin by defining the most general idea of a relation from one set to another and, specializing from that, defining the central theme of this section, order relations.

Definition 3 (Order relation, partially ordered set, poset) *Binary relations are defined in Definition 2. If R is a binary relation on S, then $(x, y) \in R$ is also denoted by $x \mathbin{R} y$. Likewise, $(x, y) \notin R$ is denoted by $x \mathbin{\not R} y$.*

A binary relation on a set S is called an *order relation* if it satisfies the following three conditions and then it is usually written $x \preceq y$ instead of $x\,R\,y$.

(i) (*Reflexive*) For all $s \in S$ we have $s \preceq s$.

(ii) (*Antisymmetric*) For all $s, t \in S$ such that $s \neq t$, if $s \preceq t$ then $t \not\preceq s$.

(iii) (*Transitive*) For all $r, s, t \in S$ such that $r \preceq s$ and $s \preceq t$ we have $r \preceq t$.

A set S together with an order relation $\preceq$ is called a *partially ordered set* or *poset*. Formally, a poset is a pair $(S, \preceq)$. We shall, once the binary relation is defined, refer to the poset by the set S alone, not the pair.

Order relations and equivalence relations appear similar: Both are reflexive and transitive. The only difference is that one is antisymmetric and the other is symmetric. Although this may seem a small difference, it makes a *big* difference in two types of relations, as you'll see if you compare the examples of order relations in this section with the examples of equivalence relations in the previous section.

If we use the alternative notation R for the relation, then the three conditions for an order relation are written as follows.

(i) For all $s \in S$ we have $(s, s) \in R$.

(ii) For all $s, t \in S$ such that $s \neq t$, if $(s, t) \in R$ then $(t, s) \notin R$.

(iii) For all $r, s, t \in S$ such that $(r, s) \in R$ and $(s, t) \in R$ we have $(r, t) \in R$.

In the exercises for this section, you will get a chance to think about various binary relations on a set S that may or may not satisfy the conditions of being reflexive, symmetric, antisymmetric, or transitive.

Example 12 (Some partially ordered sets) You are already familiar with a number of basic sets S with order relations R (*posets* for short).

Total orders: Let $S = \mathbb{Z}$, the integers, and define $n\,R\,m$ if $n \leq m$ (usual ordering on integers). Clearly $n \leq n$ for all $n \in \mathbb{Z}$ (reflexive condition). For $n, m \in \mathbb{Z}$, if $n \neq m$ and $n \leq m$, then $m \not\leq n$ (antisymmetric condition). For all $p, q,$ and r in $\mathbb{Z}$, if $p \leq q$ and $q \leq r$, then $p \leq r$ (transitive condition). We call the relation $\leq$, the *natural ordering* of the integers. This same ordering applies to any subset of $\mathbb{Z}$. For example, take $S = \underline{n}$ (the first n positive integers) ordered by $\leq$. This S is a poset. The natural ordering on the integers has an additional property, namely, for all $n, m \in \mathbb{Z}$, either $m \leq n$ or $n \leq m$. The order relation $\leq$ is called a *total ordering* or *linear ordering* because, for any two elements x and y, either $x \leq y$ or $y \leq x$ (or both if $x = y$). The relation $\leq$ can be extended to the real numbers $\mathbb{R}$ and the rational numbers $\mathbb{Q}$.

Subset lattice: Given a set X, let $S = \mathcal{P}(X)$ be the power set of X (the set of all subsets of X). For $A, B \in \mathcal{P}(X)$ we can define $A\,R\,B$ by $A \subseteq B$. The relation $\subseteq$ is an order relation (called *set inclusion*). To check the conditions that an order relation must satisfy, note that $A \subseteq A$ for all $A \in \mathcal{P}(X)$. For all $A, B \in \mathcal{P}(X)$, if $A \neq B$ and $A \subseteq B$, then there is some $x \in B$, $x \notin A$. Thus, $B \not\subseteq A$. We leave transitivity for you to check. The poset $\mathcal{P}(X)$ with the relation $\subseteq$ is called the *lattice of subsets* of X.[3] The subset lattice has the

---

[3] "Lattice" is a technical term whose meaning we will not explain. It is a poset with certain additional properties.

Equivalence and Order

property that, if $|X| > 1$ then there are always elements $A, B \in \mathcal{P}(X)$ such that $A \not\subseteq B$ and $B \not\subseteq A$. For example, if $X = \{a, b\}$, then $A = \{a\}$ and $B = \{b\}$ are such *incomparable* subsets. If $X = \{a, b, c\}$, then all three subsets of size two are pairwise incomparable. If $|X| = n > 1$, then the set $\mathcal{P}_k(X)$ of all subsets of size k, for any $0 < k < n$, is always a nontrivial (at least two elements) collection of pairwise incomparable subsets of X. The number of elements in $\mathcal{P}_k(X)$ is the binomial coefficient $C(n, k)$.

Divides relation: Another familiar poset is gotten by taking a collection of positive integers, say $S = \underline{n}$, and defining the relation $i \, R \, j$ to be $i \mid j$ (the *divides* relation). Clearly, for all $i \in S$, $i \mid i$. For all $i, j \in S$, if $i \neq j$ and $i \mid j$, then j does not divide i (antisymmetry). If $i \mid j$ and $j \mid k$, then $i \mid k$ (transitivity). $\square$

We recall some definitions that were mentioned in the previous example.

Definition 4 (Incomparable elements, linear order, total order, chain) *Let $(S, \preceq)$ be a poset. If $x, y \in S$ and neither $x \preceq y$ nor $y \preceq x$ are true, we call x and y incomparable. If either $x \preceq y$ or $y \preceq x$, we say that x and y are comparable.*

*If **every** two elements of S are comparable, $(S, \preceq)$ is called a linear order. It is also called a total order or a chain.*

Since some students have the most trouble with so called "trivial" situations, we take a look at those in the next example.

Example 13 (Trivial examples of binary relations) Suppose first that $S = \emptyset$. Since S is empty, so is $S \times S$. Hence the only binary relation on S is $R = \emptyset$. Is R reflexive? symmetric? transitive? Yes. One way to see this is to note that the conditions talk about all $s \in S$ (and possibly r and t). Since there is nothing in S, there is nothing to check. Another way to see this is to look at how you show a condition is *not* satisfied. For example, the way you show something is **not** reflexive is to find an $x \in S$ such that $(x, x) \notin R$. Since S is empty, it is impossible to find such an $x \in S$. Thus, R is reflexive. Similarly, it is also symmetric, antisymmetric and transitive.

Suppose S is not empty, but R is empty. Then, if we choose any $x \in S$, $(x, x) \notin R$. Thus, R is not reflexive. What about symmetric? To show that R is *not* symmetric, we need to find $x, y \in S$ such that $(x, y) \in R$, but ... (stop right here). We can't do this because R is empty. Thus R is symmetric. For the same basic reason, R is also antisymmetric and transitive.

Most cases are of the form $R \subseteq S \times S$ with R (and thus S) not empty. The smallest case is $|S| = 1$ and $|R| = 1$. In this case, if $S = \{a\}$, then $S \times S = \{(a, a)\}$ and so $R = \{(a, a)\}$. You should be able to verify that R is reflexive, symmetric, antisymmetric and transitive.

The next simplest case is $|S| = 2$. Things are suddenly more complicated. There are four elements in $S \times S$. Thus $2^4 = 16$ choices for R, fifteen of which are nonempty. To get a feeling for the situation, we look at some *incidence matrices* for R. These are 2×2 matrices whose rows and columns are labeled with the elements of S. The entry (x, y) is 1 if $(x, y) \in R$ and is 0 if $(x, y) \notin R$. Here are six of the fifteen possible incidence matrices for $S = \{a, b\}$:

$$
\begin{array}{cc} & a\ \ b \\ \begin{array}{c} a \\ b \end{array} & \left|\begin{array}{cc} 1 & 1 \\ 0 & 0 \end{array}\right| \end{array}
\quad
\begin{array}{cc} & a\ \ b \\ \begin{array}{c} a \\ b \end{array} & \left|\begin{array}{cc} 1 & 0 \\ 0 & 1 \end{array}\right| \end{array}
\quad
\begin{array}{cc} & a\ \ b \\ \begin{array}{c} a \\ b \end{array} & \left|\begin{array}{cc} 1 & 1 \\ 1 & 0 \end{array}\right| \end{array}
\quad
\begin{array}{cc} & a\ \ b \\ \begin{array}{c} a \\ b \end{array} & \left|\begin{array}{cc} 1 & 0 \\ 1 & 0 \end{array}\right| \end{array}
\quad
\begin{array}{cc} & a\ \ b \\ \begin{array}{c} a \\ b \end{array} & \left|\begin{array}{cc} 1 & 0 \\ 1 & 1 \end{array}\right| \end{array}
\quad
\begin{array}{cc} & a\ \ b \\ \begin{array}{c} a \\ b \end{array} & \left|\begin{array}{cc} 1 & 1 \\ 1 & 1 \end{array}\right| \end{array}
$$
$$\ \ \ \text{A} \qquad\qquad \text{B} \qquad\qquad \text{C} \qquad\qquad \text{D} \qquad\qquad \text{E} \qquad\qquad \text{F}$$

The first matrix $\mathbf{A}$ describes the binary relation $\{(a, a), (a, b)\}$ because $\mathbf{A}(a, a) = \mathbf{A}(a, b) = 1$ and $\mathbf{A}(b, a) = \mathbf{A}(b, b) = 0$. The entries in positions (a, a) and (b, b) are called the entries on the "main diagonal" of $\mathbf{A}$. Positions (a, b) and (b, a) are "symmetric off-diagonal positions" about the main diagonal and their entries are "symmetric off-diagonal entries" with respect to $\mathbf{A}$. The relation defined by $\mathbf{A}$ (or more simply, "the matrix $\mathbf{A}$") is not reflexive, is antisymmetric, is not symmetric, and is transitive. With regard to the last statement, note that, since $R = \{(a, a), (a, b)\}$, there is not really anything to check since combining (a, a) with (a, b) using transitivity just gives us (a, b) again. This is always true: we never need to use a diagonal entry like (a, a) in checking transitivity. On the other hand, matrix $\mathbf{C}$ is *not* transitive because $(b, a) \in R$ and $(a, b) \in R$ would give us $(b, b) \in R$, which is not true. Note that there is already a lesson here. If symmetric off-diagonal entries are both 1, but either of their corresponding diagonal entries is not 1, then the relation is not transitive. This is true for the incidence matrix of any relation. For the various matrices we have

	reflexive	symmetric	antisymmetric	transitive
A	no	no	yes	yes
B	yes	yes	yes	yes
C	no	yes	no	no
D	no	no	yes	yes
E	yes	no	yes	yes
F	yes	yes	no	yes

You should make sure you understand the reasons for all of these statements. □

Example 14 (Counting relations)   As we have seen in the previous example, relations on a set S correspond to matrices of zeroes and ones (the incidence matrices). A relation on a four element set, for example, corresponds to a 4×4 matrix of zeroes and ones (an incidence matrix). The matrix below, with appropriate substitutions of 0's and 1's for the symbolic entries, is such an incidence matrix. The rows and columns should be labeled with the four elements of the set (as in the previous example, where we worked with 2×2 incidence matrices), but we omit that here for simplicity.

Since an $n \times n$ matrix has n^2 entries and each entry can be either a 0 or a 1, there are 2^{n^2} such matrices. This number grows *very* rapidly; for example, when $n = 4$ we have $2^{4^2} = 2^{16} = 65,536$.

Let's look at 4×4 matrices We could think of starting with a matrix, such as the one below, labeled with symbols d_i (for diagonal), u_i (for upper), and l_i (for lower), and then, in some manner (just how is up to us) substituting zeroes and ones for the sixteen symbols.

$$
\begin{array}{cccc}
d_1 & u_1 & u_2 & u_3 \\
l_1 & d_2 & u_4 & u_5 \\
l_2 & l_4 & d_3 & u_6 \\
l_3 & l_5 & l_6 & d_4
\end{array}
$$

Equivalence and Order

- If we want the relation to be reflexive, we must make all the $d_i = 1$.

- If we want the relation to be symmetric, then we must have $u_i = l_i$ for all i.

- If we want the relation to be antisymmetric, then we can never have $u_i = l_i = 1$ for any i (but $u_i = l_i = 0$ is allowed).

You should make sure you understand the reasons for these three statements. You may have noticed that transitivity was not mentioned. It cannot be described in such simple terms.

The 4×4 matrix above can easily be extended to a general $n \times n$ matrix. As we continue to describe certain properties of the 4×4 case, you should think about how these descriptions extend to the $n \times n$ case. As far as the situations just described, we do exactly the same thing in the $n \times n$ case.

From what we have just said, there are 2^{12} reflexive relations on a four element set. Why? We must set all of the $d_i = 1$ and then we can choose freely the u_i and l_i to be 0 or 1. There are 12 total u_i and l_i, giving 2^{12} choices. In general, there are $n^2 - n$ entries in an $n \times n$ matrix which are not d_i's. Thus there are $2^{n^2-n} = 2^{n(n-1)}$ reflexive relations on a set with n elements.

Let's try one more example. How many relations are both reflexive and antisymmetric? All of the $d_i = 1$. For each pair (l_i, u_i), we have three choices: $(l_i, u_i) = (0,0)$, $(l_i, u_i) = (1,0)$, $(l_i, u_i) = (0,1)$. In the 4 element case there are 3^6 such relations. What is the formula for general n? We have seen that there are $n^2 - n$ elements l_i and u_i and so there are $(n^2 - n)/2$ pairs (l_i, u_i). Thus there are $3^{(n^2-n)/2}$ relations on an n-set which are both reflexive and antisymmetric. $\square$

Example 15 (Partitions of a set) The collection $\Pi(S)$ of all partitions of a set S can be made into a poset. Let $S = \underline{15}$. Consider the following partition of S:

$$\alpha = \Big\{ \{1\}, \{2\}, \{9\}, \{3,5\}, \{4,7\}, \{6,8,10,15\}, \{11,12,13,14\} \Big\}.$$

We can *refine* the partition α by taking any block or blocks with at least two elements and splitting each of them into two or more blocks. For example, we could choose the block $\{6,8,10,15\}$ and split it into two blocks: $\{6,15\}, \{8,10\}$. We could also choose the block $\{11,12,13,14\}$ and split it into (for example) three blocks: $\{13\}, \{14\}, \{11,12\}$. The resulting partition β is called a *refinement* of α (we write $\beta \preceq \alpha$):

$$\beta = \Big\{ \{1\}, \{2\}, \{9\}, \{3,5\}, \{4,7\}, \{6,15\}, \{8,10\}, \{13\}, \{14\}, \{11,12\} \Big\}.$$

The set of all partitions of S, $\Pi(S)$, together with the refinement relation is a poset — the lattice of partitions of S. By definition, $\alpha \preceq \alpha$ for any $\alpha \in \Pi(S)$. We leave it to you to check antisymmetry and transitivity. $\square$

New Posets from Old Ones

We now examine restrictions, direct products and lexicographic order, which are three ways of forming new posets from old ones.

Definition 5 (Restriction of a poset) *Let $(S, \preceq)$ be a poset and let X be a subset of S. For $u, v \in X$, define the order relation $\preceq_X$ by $u \preceq_X v$ if and only if $u \preceq v$ in $(S, \preceq)$. We call $(X, \preceq_X)$ a subposet of S or the restriction of S to X. Instead of introducing a new symbol $\preceq_X$ for the order relation, one usually uses $\preceq$, writing $(X, \preceq)$.*

Example 16 (Restrictions of posets) Consider the divides poset on $S = \underline{20}$. Let $X = \{2, 4, \ldots, 20\}$ — the even numbers in S. The set X with the same divides relation is a subposet of S or a restriction of S to X. Alternatively, let Y be the divisors of 20, namely $\{1, 2, 4, 5, 10, 20\}$ with the same divides relation. This is also a subposet.

As another example, consider the subset lattice $\mathcal{P}(S)$, $S = \{a, b, c\}$. Remove from this poset the empty set and the set S. This gives a new poset $\mathcal{P}'(S)$ with six elements (six subsets of S) ordered by set inclusion. The poset $\mathcal{P}'(S)$ is a subposet of $\mathcal{P}(S)$:

$$\mathcal{P}'(S) = \Big\{ \{a\}, \{b\}, \{c\}, \{a, b\}, \{a, c\}, \{b, c\} \Big\}.$$

Since this is a restriction of $\mathcal{P}(S)$, it still has "subset of" as the order relation. $\square$

Example 17 (Direct products of posets) Suppose we have two posets P and Q. Let $\preceq_P$ be the relation on P and $\preceq_Q$ be the relation on Q. The *direct product* of the posets $(P, \preceq_P)$ and $(Q, \preceq_Q)$ is the poset $(P \times Q, \preceq)$ where $(p_1, q_1) \preceq (p_2, q_2)$ if $p_1 \preceq_P p_2$ and $q_1 \preceq_Q q_2$. Sometimes this product order is called "coordinate order." Just as we can define Cartesian product of several sets, we can define the direct product of several posets.

A simple application of this idea is to take $P = Q = \{0, 1\}$. Then,

$$P \times Q = \{(0, 0), (0, 1), (1, 0), (1, 1)\}$$

has just four elements. Suppose $S = \{a, b\}$ is a two-element set. We can think of the elements of $P \times Q$ as the one-line forms for the functions $f : S \to \{0, 1\}$. With each of these functions is associated the subset $f^{-1}(1)$ of S. Using one-line notation for functions with S in the order (a, b), we see that

- $(0, 0)$ corresponds to the empty set,
- $(0, 1)$ corresponds to the set $\{b\}$,
- $(1, 0)$ corresponds to the set $\{a\}$, and
- $(1, 1)$ corresponds to the set $\{a, b\}$.

[129]

Equivalence and Order

In this way the four functions $f : S \to \{0,1\}$ become alternative descriptions of the four subsets of the subset lattice $\mathcal{P}(S)$. Thought of in this way, they are called the *characteristic functions* of the subsets of S. The poset of characteristic functions with coordinate order is just another way to describe the subset lattice with set inclusion.

Instead of a two element subset, we could consider an n-set S. In this case, we form the direct product of n copies of $P = (\{0,1\}, \leq)$. The characteristic function is a bijection from $\mathcal{P}(S)$ to n-long vectors of zeroes and ones. We leave it to you to fill in the details. $\blacksquare$

Is the direct product of posets again a poset? Yes. In fact we now define the direct product of binary relations and prove that properties are "inherited."

Definition 6 (Direct product of binary relations)   *Let $S_1, S_2, \ldots, S_n$ be sets and let R_i be a binary relation on S_i for $i = 1, \ldots, n$. The direct product is the Cartesian product $S = S_1 \times S_2 \times \cdots \times S_n$ with the binary relation R defined by*

$$(a_1, a_2, \ldots, a_n) \, R \, (b_1, b_2, \ldots, b_n) \quad \text{if and only if } a_i \, R_i \, b_i \text{ for } i = 1, 2, \ldots, n.$$

You should verify that this definition gives the definition for the direct product of posets when the (S_i, R_i) are all posets.

The following theorem implies that the direct product of posets is again a poset. There are four statements in one — choose any property in the first { } and then choose the same property in the second { }

Theorem 4 (Properties of direct products)   *If each of the binary relations R_i on the set S_i is* $\left\{ \begin{array}{c} \text{reflexive} \\ \text{symmetric} \\ \text{antisymmetric} \\ \text{transitive} \end{array} \right\}$ *, then the direct product is also* $\left\{ \begin{array}{c} \text{reflexive} \\ \text{symmetric} \\ \text{antisymmetric} \\ \text{transitive} \end{array} \right\}$.

Proof: We prove transitivity and leave the rest to you. Suppose that

$$(a_1, a_2, \ldots, a_n) \, R \, (b_1, b_2, \ldots, b_n) \quad \text{and} \quad (b_1, b_2, \ldots, b_n) \, R \, (c_1, c_2, \ldots, c_n)$$

From the definition of R, $a_i \, R_i \, b_i$ and $b_i \, R_i \, c_i$ for $i = 1, \ldots, n$. Since R_i is transitive, $a_i \, R_i \, c_i$ for $i = 1, \ldots, n$. By the definition of R, $(a_1, a_2, \ldots, a_n) \, R \, (c_1, c_2, \ldots, c_n)$. This proves transitivity of R. $\blacksquare$

Definition 7 (Isomorphism of posets)   *Let $(S, \preceq_S)$ and $(T, \preceq_T)$ be posets. We say the posets are isomorphic if we have a bijection $f : S \to T$ such that $x \preceq_S y$ if and only if $f(x) \preceq_T f(y)$. We then call f an isomorhism between the posets.*

In Example 17, we used the characteristic function to construct an isomorphism between the subset lattice and the direct product $P \times \cdots \times P$ where $P = (\{0,1\}, \leq)$. We now look at another example.

Example 18 (The divisibility relation again) Let $T = \{1, 2, 3, 4, 6, 12\}$, the set of divisors of 12, and let the order relation be "divides." Consider the two chains (linear orders) $C_2 = \{1, 2, 4\}$ and $C_3 = \{1, 3\}$ where the ordering can be thought of as either divisibility or ordinary $\leq$ since it gives the same ordering.

The posets T and $S = C_2 \times C_3$ describe the same situation. To see this, let $(a, b) \in C_2 \times C_3$ correspond to $ab \in S$. In this case $f((a, b)) = ab$ and the posets are isomorphic.

The previous idea can be applied to the set of divisors of n for any $n > 0$. The number of chains will equal the number of different primes dividing n.

We looked at the divisors of 12. What about the set $\underline{12}$ of positive integers less than or equal to 12 ordered by divisibility? This is not isomorphic to a direct product of chains. However, it is isomorphic to a restriction of a direct product of chains. Here is one way to do this. Let

$$V = (\{1, 2, 4, 8\}, |) \times (\{1, 3, 9\}, |) \times (\{1, 5\}, |) \times (\{1, 7\}, |) \times (\{1, 11\}, |).$$

The map $f((a, b, c, d, e)) = abcde$ shows that V is isomorphic to W the poset of divisors of $8 \times 9 \times 5 \times 7 \times 11$ with the divides relation. Since $\underline{12} \subset W$, $\underline{12}$ is a subposet W. ◻

The next concept, when applied to products of linearly ordered sets, is one of the most useful elementary ideas to be found in computer science. It's found in every subdiscipline of computer science and in almost every program of any length or complexity.

Definition 8 (Strings and lexicographic order) Let $(S, \preceq)$ be a poset. We use S^* to denote the set of all "strings" over S; that is S^* contains

- for each $k > 0$, the set of k-long strings $(x_1, \dots, x_k)$ of elements in S, which is denoted $S^k = \times^k S$;

- the empty string ϵ.

We now define a relation $\preceq_L$ on S^*. This relation is called lexicographic order or, more briefly, lex order. Let $(a_1, a_2, \dots, a_m)$ and $(b_1, b_2, \dots, b_n)$ be two elements of S^* with $m, n > 0$. We say that

$$(a_1, a_2, \dots, a_m) \preceq_L (b_1, b_2, \dots, b_n)$$

if either of the following two conditions hold:

(1) $m \leq n$ and $a_i = b_i$ for $i = 1, \dots, m$.

(2) For some $k < \min(m, n)$, $a_i = b_i$, $i = 1, \dots, k$, $a_{k+1} \neq b_{k+1}$, and $a_{k+1} \preceq b_{k+1}$.

In addition we have a third condition:

(3) For the empty string ϵ, we have $\epsilon \preceq_L x$ for every string $x \in S^*$.

Notice that we said "define a relation $\preceq_L$ on S^*" rather than "define a partial order $\preceq_L$ on S^*." Why is that? By its definition, $\preceq_L$ is obviously a relation. The fact that it is a partial order requires proof. We'll give a proof after the next example.

[131]

Equivalence and Order

The term "lexicographic" comes from the listing of words in a dictionary. One writes $S_0 = \{\epsilon\}$ and so $S^* = \cup_{k=0}^{\infty} S^k$. Other notations for the k-tuple $(x_1, x_2, \ldots, x_k)$ are $x_1, x_2, \ldots, x_k$ (leave off the parentheses) or $x_1 x_2 \ldots x_k$ (leave off the parentheses and the commas). Each of these latter notations leaves off information and can be confusing (for example, $112131212131414 = (11, 21, 312, 121, 31414)$, or does it?). Mostly, we will stick to the full notation (vector notation). Such k-tuples are sometimes referred to as *strings of length k over S*. The term "words" over S is also used in this context as meaning the same thing as "strings."

Example 19 (Lexicographic order) At this stage, all we know is that $\preceq_L$ is a relation on strings. Consider the lattice of subsets of $\{1, 2, 3\}$; that is, $(\mathcal{P}(\{1, 2, 3\}), \subseteq)$. By Condition (1),

$$(\{1\}, \{2\}, \{1, 3\}) \preceq_L (\{1\}, \{2\}, \{1, 3\}, \{1, 2\}).$$

By Condition (2),

$$(\{1\}, \{2\}, \{1\}, \{1, 2\}) \preceq_L (\{1\}, \{2\}, \{1, 3\}).$$

On the other hand,

$$(\{1\}, \{2\}, \{1, 2\}) \npreceq_L (\{1\}, \{2\}, \{1, 3\})$$

and

$$(\{1\}, \{2\}, \{1, 3\}) \npreceq_L (\{1\}, \{2\}, \{1, 2\}).$$

The two strings $(\{1\}, \{2\}, \{1, 2\})$ and $(\{1\}, \{2\}, \{1, 3\})$ are incomparable in lex order because $\{1, 2\}$ and $\{1, 3\}$ are incomparable in the subset lattice. If, on the other hand, S is linearly ordered (it is not in this example where S is the lattice of subsets) then S^* with lex order is linearly ordered. $\square$

We now prove that the lexicographic relation in Definition 8 is an order relation on S^*.

Theorem 5 (The relation $\preceq_L$ is an order relation) *Let S be a set with order relation $\preceq$. Let $\preceq_L$ be the lexicographic relation on the strings S^*. Then, $\preceq_L$ is an order relation on S^*. If the poset S is linearly ordered, so is S^*.*

*Proof: In the proof, we refer to the conditions (1), (2) and (3) of that discussion. We also omit commas and parentheses in the strings as there is no possibility of confusion.

First we show that the **reflexive** property is true. Let $w \in S^*$. If $w = \epsilon$ is the null string, then the reflexive property follows from Condition (3). If $w = a_1 a_2 \cdots a_m$ with $m \geq 1$ then the reflexive property follows immediately from Condition (1).

Next we show the **antisymmetric** property. Suppose that

$$a_1 a_2 \cdots a_m \preceq_L b_1 b_2 \cdots b_n \quad \text{and} \quad b_1 b_2 \cdots b_n \preceq_L a_1 a_2 \cdots a_m.$$

This would be impossible if either of these relations were due to Condition (2) since $a_{k+1} \preceq b_{k+1}$ and $a_{k+1} \neq b_{k+1}$ implies $b_{k+1} \npreceq a_{k+1}$. We have used here the fact that $\preceq$ is a partial order relation on S and hence is antisymmetric. Thus both relations are due to

Condition (1). Hence, $m \leq n$ and $n \leq m$, so $m = n$, and $a_i = b_i$ for $1 \leq i \leq m$. This proves the antisymmetric property.

Next we show the **transitive** property. Suppose that

$$a_1 a_2 \cdots a_m \preceq_L b_1 b_2 \cdots b_n \quad \text{and} \quad b_1 b_2 \cdots b_n \preceq_L c_1 c_2 \cdots c_t.$$

If $m = 0$ (the empty string), transitivity is trivial by Condition (3). Thus we may suppose $m, n, t > 0$. We consider cases.

- If both $\preceq_L$ relations are due to Condition (1) then $a_1 a_2 \cdots a_m \preceq_L c_1 c_2 \cdots c_t$ from Condition (1) also.

- If $a_1 a_2 \cdots a_m \preceq_L b_1 b_2 \cdots b_n$ is due to Condition (1) but $b_1 b_2 \cdots b_n \preceq_L c_1 c_2 \cdots c_t$ is due to Condition (2), let k be the smallest integer such that $b_{k+1} \neq c_{k+1}$. If $k < m$ then, using the fact that $\preceq$ is a partial order, $a_1 a_2 \cdots a_m \preceq_L c_1 c_2 \cdots c_t$ follows from Condition (2), otherwise (i.e., $k \geq m$) it follows from Condition (1).

- If $a_1 a_2 \cdots a_m \preceq_L b_1 b_2 \cdots b_n$ is due to Condition (2) but $b_1 b_2 \cdots b_n \preceq_L c_1 c_2 \cdots c_t$ is due to Condition (1), the proof is similar to the preceding case.

- Finally, if $a_1 a_2 \cdots a_m \preceq_L b_1 b_2 \cdots b_n$ and $b_1 b_2 \cdots b_n \preceq_L c_1 c_2 \cdots c_t$ are both due to Condition (2), then, using the fact that $\preceq$ is a partial order relation, $a_1 a_2 \cdots a_m \preceq_L c_1 c_2 \cdots c_t$ follows from Condition (2) also.

We have proved that $\preceq_L$ is an order relation.

It remains to show that if the poset S is **linearly ordered**, so is S^*. Suppose that we are given any two strings $a_1 a_2 \cdots a_m$ and $b_1 b_2 \cdots b_n$. Suppose, without loss of generality, that $m \leq n$. Then either

(a) $a_1 = b_1, \ldots, a_m = b_m$ or

(b) there is a smallest $k < m$ such that $a_{k+1} \neq b_{k+1}$.

We consider cases.

- If (a) holds, then $a_1 a_2 \cdots a_m \preceq_L b_1 b_2 \cdots b_n$ by Condition (1).

- If (b) holds, then either $a_1 a_2 \cdots a_m \preceq_L b_1 b_2 \cdots b_n$ or $b_1 b_2 \cdots b_n \preceq_L a_1 a_2 \cdots a_m$ by Condition (2). This follows since either $a_{k+1} \preceq b_{k+1}$ or $b_{k+1} \preceq a_{k+1}$, because $\preceq$ is a linear order.

We have shown that, given any two strings, either the first is less than or equal to the second in lex order or the reverse. Thus, S^* is linearly ordered by lex order. $\square$

There is a variation on lex order on S^* which first orders the strings by length. Strings of the same length are then ordered lexicographically by restricting the above definition to subsets of S^* of the same length. This order relation is called *short lexicographic order* or *length-first lexicographic order*.

We now present an alternative proof of the theorem. You may find this proof harder to follow because it is somewhat more abstract. So why give it? It illustrates some techniques that are often used by people doing mathematics. Specifically, we will reduce an infinite

Equivalence and Order

problem (S^*) to a finite one (S_m below). We then embed this in another problem (T_m below) that lets us use previous results (Theorem 4).

Proof: Review Definitions 5 and 6 on restrictions and direct products of posets.

Let $S_m = \cup_{k=0}^m S^k$. We claim it suffices to prove the theorem for the restriction of $(S^*, \preceq_L)$ to $(S_m, \preceq_L)$ for each $m = 1, 2, \ldots$. Why? The conditions for being a partial order and for being a linear order refer to at most three elements of the poset. Suppose, for example, we want to verify transitivity. Someone gives us $r \preceq_L s$ and $s \preceq_L t$. Let m be the longest length of r, s and t. Then $r, s, t \in S_m$. By the definition of a restriction $\preceq_L$ is the same in the restriction as it is in the full poset, so we can work in the restriction. In other words, if the theorem is true with S^* replaced by S_m, then it is true for S^*.

We begin by adding a "blank" to S. As we shall see, this lets us use a restriction of a direct product. Let $\sqcup$ be something that is not in S. You can think $\sqcup$ as a blank space. Let $T = S \cup \{\sqcup\}$ and define $\preceq$ on T to be the same as it is on S together with $\sqcup \preceq x$ for all $x \in S$. We'll still call it $\preceq$. Let $(T_m, \preceq_m)$ be the direct product of m copies of $(T, \preceq)$. This is a poset by Theorem 4.

Let U_m be the restriction of T_m to those strings in which $\sqcup$ is never followed by an element of S, but may be followed by more $\sqcup$'s. In other words, blanks appear only at the end of a string.

Define $f(x)$ for $x \in S_m$ to be x "padded out" at the end with enough blanks to give a string of length m. You should have no trouble verifying that f is a bijection from S_m to U_m. You should also check that $x \preceq_L y$ if and only if $f(x) \preceq_m f(y)$. This gives us an isomorphism between $(S_m, \preceq_L)$ and the poset $(T_m, \preceq_m)$. Thus $(S_m, \preceq_L)$ is a poset. □

Example 20 (Lexicographic bucket sort) Let $S = \{1, 2, 3\}$ be ordered in the usual way (as integers). Consider all strings of length three, $\times^3 S = S^3$. Take some subset of S^3, say the set

$$A = \big\{(2,1,3),\ (3,2,3),\ (1,2,1),\ (2,3,2),\ (1,1,3),\ (3,1,1),\ (3,3,1),\ (2,2,2)\big\}.$$

We are interested in an algorithm for sorting the elements of A so that they are in lexicographic order. The topic of sorting is very important for computer science. The literature on sorting methods is vast.

One type of sorting algorithm involves comparisons only. Imagine a bin of bananas that are to be sorted by weight using only a beam balance that tells which of two bananas weighs the most. No numerical values are recorded. Start with one banana. Get another and compare it with the first, laying them on a table in order of weight, left to right. Each new banana is compared with the bananas already sorted until all bananas are sorted by weight. In this manner you can sort the bananas by weight without actually knowing the numerical value of the weight of any banana. Such a sorting algorithm is called a *comparison sort*.

As another approach to sorting bananas by weight, suppose we have a scale that returns the weight of a banana to the nearest one-tenth of an ounce. Suppose we know that the bananas in the bin weigh between 4.0 oz. and 6.0 oz. We put 21 buckets on the table, the buckets labeled with the number 4.0, 4.1, 4.2, ..., 5.7, 5.8, 5.9, 6.0. Take the bananas from the bin one by one and weigh them, accurate to 0.1 oz. Put each banana in the bucket corresponding to its weight. This type of sorting is called a *bucket sort*.

These two basic types of sorting, as well as hybrid forms of these two types, occur in many computer related applications. Comparison sorts are used with tree-type data structures, linked lists, etc. Bucket sorts are implemented by arrays where the buckets correspond to index references into an array.

Returning to the set A, we can sort the elements of A lexicographically using a variation on the bucket sort. We use three buckets, labeled 1, 2, and 3. On the first "pass" through the set A we place elements of

$$A = \{(2,1,3), (3,2,3), (1,2,1), (2,3,2), (1,1,3), (3,1,1), (3,3,1), (2,2,2)\}$$

into buckets according to the rightmost entry in that element:

PASS 1
Bucket 1: (1, 2, 1), (3, 1, 1), (3, 3, 1)
Bucket 2: (2, 3, 2), (2, 2, 2)
Bucket 3: (2,1,3), (3, 2, 3), (1, 1, 3)

Although A is a set and, technically, has no order, we now have imposed a linear order on A. This linear order is gotten, from the placement in the buckets, by reading the elements in the buckets from left to right, first from Bucket 1, then from Bucket 2 and finally from Bucket 3. We obtain the Pass 1 *concatenated order*:

$$(1,2,1), (3,1,1), (3,3,1), (2,3,2), (2,2,2), (2,1,3), (3,2,3), (1,1,3).$$

Notice that if you just read the third elements of each vector (string) you obtain, in order left to right, $1,1,1,2,2,3,3,3$. These strings of length one are in order.

We now do PASS 2. We go through the Pass 1 concatenated order, left to right, putting strings into buckets based on the value of their second-from-the-right coordinate (middle coordinate). In carrying this out, it is essential that the the order of the strings in each bucket is the correct order relative to PASS 1 concatenated order. The term "bucket" is not suggestive of order. Perhaps "sublist" would be better here, but we follow conventional terminology. Here is the composition of the buckets after PASS 2:

PASS 2
Bucket 1: (3, 1, 1), (2, 1, 3), (1, 1, 3)
Bucket 2: (1, 2, 1), (2, 2, 2), (3, 2, 3)
Bucket 3: (3, 3, 1), (2, 3, 2)

PASS 2 concatenated order is

$$(3,1,1), (2,1,3), (1,1,3), (1,2,1), (2,2,2), (3,2,3), (3,3,1), (2,3,2).$$

Note now that the list of all last two elements is in lexicographic order:

$$(1,1), (1,3), (1,3), (2,1), (2,2), (2,3), (3,1), (3,2).$$

Equivalence and Order

Finally, we do PASS 3, putting strings into buckets according to first elements, retaining within each bucket the PASS 2 concatenated order.

PASS 3
Bucket 1: (1, 1, 3), (1, 2, 1)
Bucket 2: (2, 1, 3), (2, 2, 2), (2, 3, 2)
Bucket 3: (3, 1, 1), (3, 2, 3), (3, 3, 1)

PASS 3 concatenated order is the lexicographic order on A.

$$(1,1,3), (1,2,1), (2,1,3), (2,2,2), (2,3,2), (3,1,1), (3,2,3), (3,3,1).$$

This process extends to a general algorithm called the *lexicographic bucket sort*. The correctness of the algorithm can be proved by induction on the number of passes (the length of the strings).

The lexicographic bucket sort was used in the early days of computers to sort "punched cards" on which data was stored. ☐

Example 21 (Lexicographic order and domino coverings) Lexicographic order is used in many computer applications where geometric objects are being manipulated as input data. We give an example from a class of problems called "tiling problems." Below we show a 3×4 "chess board" or grid. The grid is to be tiled or "covered" with horizontal and vertical dominoes (little 1×2 size rectangles):

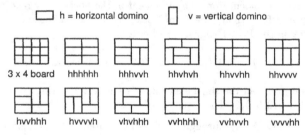

If the squares of the board are numbered systematically, left to right, top to bottom, from 1 to 12, we can describe any placement of dominoes by a sequence of 6 h's and v's: Each of the domino placements in the above picture has such a description just below it. Take as an example, hhvhvh (the third domino covering in the picture). We begin with no dominoes on the board. None of the squares, numbered 1 to 12 are covered. The list of "unoccupied squares" is as follows:

1   2   3   4
5   6   7   8
9   10   11   12

Thus, the smallest unoccupied square is 1. The first symbol in hhvhvh is the h. That means that we take a horizontal domino and cover the square 1 with it. That forces us to cover square 2 also. The list of unoccupied squares is as follows:

    3   4
5   6   7   8
9   10   11   12

Now the smallest unoccupied square is 3. The second symbol in hhvhhv is also an h. Cover square 3 with a horizontal domino, forcing us to cover square 4 also. The list of unoccupied squares is as follows:

$$
\begin{array}{cccc}
5 & 6 & 7 & 8 \\
9 & 10 & 11 & 12
\end{array}
$$

At this point, the first row of the board is covered with two horizontal dominoes (check the picture). Now the smallest unoccupied square is 5 (the first square in the second row). The third symbol in hhvhvh is v. Thus we cover square 5 with a vertical domino, forcing us to cover square 9 also. The list of unoccupied squares is as follows:

$$
\begin{array}{ccc}
6 & 7 & 8 \\
10 & 11 & 12
\end{array}
$$

We leave it to you to continue this process to the end and obtain the domino covering shown in the picture.

Here is the general description of the process. Place dominoes sequentially as follows. If the first unused element in the sequence is h, place a horizontal domino on the first (smallest numbered) unoccupied square and the square to its right. If the first unused element in the sequence is v, place a vertical domino on the first unoccupied square and the square just below it. Not all sequences correspond to legal placements of dominoes (try hhhhhv). For a 2×2 board, the only legal sequences are hh and vv. For a 2×3 board, the legal sequences are hvh, vhh and vvv. For a 3×4 board, there are eleven legal sequences as shown in the above picture.

Having developed this correspondence between tiling of a rectangular board and strings of letters from the set $S = \{h, v\}$, we can now list the strings that represent coverings of the board in lexicographic order. This order is useful for generating, storing, retrieving, and comparing domino coverings. □

More Poset Concepts

In the next example we introduce a useful pictorial or geometric way of visualizing a relation. The only problem with this method of thinking about relations is that the picture can become much too complicated. These pictures are combined with the idea of "transitive closure" to deal with this growth of complexity.

Equivalence and Order

Example 22 (Transitive closure and directed graph diagrams)   Consider the relation $R = \{(1,2),\ (2,3),\ (2,4),\ (4,3),\ (4,2)\}$ on the set $S = \{1,2,3,4\}$. Figure (a) below is another way of representing R using a *directed graph diagram*. The elements of S are written down in some manner, in this case one after the other in a straight line (any way will do). An arrow is drawn between i and j if and only if $(i,j) \in R$. Note that R is not transitive. For example, $(1,2) \in R$ and $(2,3) \in R$, but $(1,3) \notin R$. Suppose we add the missing pair $(1,3)$ to R as indicated in Figure (b) below. This gives a new relation $\{(1,2),\ (2,3),\ (2,4),\ (4,3),\ (4,2),\ (1,3)\}$. This new relation is still not transitive. For example, $(2,4) \in R$ and $(4,2) \in R$, but $(2,2) \notin R$. So, add $(2,2)$ to R. Keep repeating this process until no violations of transitivity can be found. The directed graph diagram of this final transitive relation is shown in Figure (b). This relation is the smallest transitive relation that contains R. It is called the *transitive closure* of R.

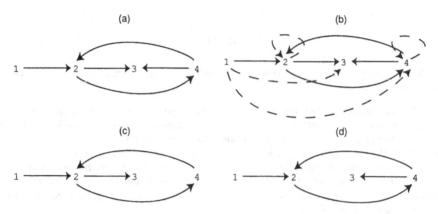

Note that there are even smaller (fewer elements) relations than R whose transitive closure is the same as R. The directed graph diagrams of two such relations are shown in Figures (c) and (d). By the way, we use the terminology "directed graph diagram," so you might naturally wonder what is a directed graph (the thing that is being "diagrammed" here)? The answer is that directed graphs and binary relations are, mathematically, the same thing. The terminology "directed graph diagram" is standard in this context, rather than the more natural "relation diagram." ☐

*Example 23 (Transitive closure and matrices)   There is another way to compute transitive closure. We start the same relation, $R = \{(1,2),\ (2,3),\ (2,4),\ (4,3),\ (4,2)\}$ on the set $S = \{1,2,3,4\}$, used in the previous example. Here is the "incidence matrix" of this relation:

$$A = \begin{pmatrix} 0 & 1 & 0 & 0 \\ 0 & 0 & 1 & 1 \\ 0 & 0 & 0 & 0 \\ 0 & 1 & 1 & 0 \end{pmatrix}$$

The interpretation is that if $A(i,j) = 1$ then $(i,j) \in R$, else $(i,j) \notin R$. If we compute the square of A, then, by definition of matrix multiplication, $A^2(i,j) = \sum_{k=0}^{4} A(i,k)A(k,j)$. Note that $A^2(i,j) \neq 0$ if and only if there is a pair $(i,t) \in R$ and $(t,j) \in R$. In other words, for this pair, $A(i,t)A(t,j) = 1$ so $A^2(i,j) > 0$. For our purposes, we don't care how big

$A^2(i,j)$ is only whether or not it is zero. So, we replace all nonzero entries in $A^2(i,j)$ by 1. This is called the "Boolean product" of A with A. We just use the same notation for this Boolean product as for the square. Here is this Boolean product:

$$A^2 = \begin{pmatrix} 0 & 0 & 1 & 1 \\ 0 & 1 & 1 & 0 \\ 0 & 0 & 0 & 0 \\ 0 & 0 & 1 & 1 \end{pmatrix}$$

You can think of the Boolean product as follows: Multiply matrices in the usual way be replace "plus" with $\vee$ and "times" with $\wedge$. Thus

$$A^2(i,j) = \big(A(i,1) \wedge A(1,j)\big) \vee \big(A(i,2) \wedge A(1,2)\big) \vee \cdots \vee \big(A(i,n) \wedge A(1,n)\big).$$

If we now form the "Boolean sum" $A + A^2$ (Again, replace "plus" with $\vee$.) we get the following matrix:

$$A + A^2 = \begin{pmatrix} 0 & 1 & 1 & 1 \\ 0 & 1 & 1 & 1 \\ 0 & 0 & 0 & 0 \\ 0 & 1 & 1 & 1 \end{pmatrix}$$

This matrix has a 1 in position (i,j) if and only if either $(i,j) \in R$ or $(i,t) \in R$ and $(t,j) \in R$ for at least one t (perhaps both of these conditions hold). In terms of the directed graph diagram, this matrix has a 1 in position (i,j) if and only if there is a directed arrow joining i to j or a sequence of two directed arrows that you can follow to go from i to j. Such a sequence is called a directed path of length two from i to j.

We could continue this process to compute the Boolean matrix $A + A^2 + A^3$, but, if we do, we see that this latter matrix is the same as $A + A^2$. A little thought should tell you that this means $A + A^2$ is the incidence matrix of the transitive closure of R.

This idea can be applied to any binary relation. For large relations a computer helps. You start with the incidence matrix A and keep forming Boolean partial sums $S_k = \sum_{i=1}^{k} A^i$ until, for some $k = t$, $S_t = S_{t+1}$. Then we'll have $S_{t+i} = S_t$ for all $i > 0$. At this point S_t is the incidence matrix of the transitive closure. Note that $S_1 = A$, $S_2 = A S_1 + A$, $S_3 = A S_2 + A$, and, in general $S_{k+1} = A S_k + A$. This is a convenient way to carry out these computations. It also makes it easy to prove the earlier claim that $S_{t+i} = S_t$: Use induction in i and note that $S_{t+i} = A S_{t+i-1} + A$, which equals $A S_t + A$ by the induction hypothesis.

The story doesn't end here. We can find a similar algorithm that is much faster for large problems. We claim that $S_t S_t + A = S_{2t}$. Why is this? If you consider ordinary multiplication, you should see that $S_t S_t$ consists of all the powers $A^2, A^3, \ldots, A^{2t}$ added together, some of them many times. Notice that when you do Boolean addition for any matrix B, you have $B + \cdots + B = B$. Thus $S_t S_t = A^2 + \cdots + A^{2t}$ when we do Boolean addition and multiplication. Here's our new algorithm:

$$P_0 = A \quad \text{and} \quad P_{k+1} = P_k P_k + A \quad \text{for} \quad k \geq 0,$$

and we stop when $P_k = P_{k+1}$. You should be able to prove by induction that $P_k = S_{2^k}$.

Equivalence and Order

Is this algorithm really faster? Yes. The simplest example of this is a chain:

$$S = \{0, 1, \ldots, n\} \quad \text{and} \quad R = \{(0,1), (1,2), \ldots, (n-1,n)\}.$$

We claim that $S_{n+1} = S_n$ and the n matrices $S_1, S_2, \ldots, S_n$ are all different.[4] Thus we must compute $S_2, \ldots, S_{n+1}$ to obtain the transitive closure if we use the first algorithm. For the second algorithm, we compute $P_1, \ldots, P_{m+1}$ where $m = \lceil \log_2 n \rceil$. Why this value of m? We will not have $P_k = P_{k+1}$ until $2^k \geq n$ and m is the smallest such k. For large values of n, $\log_2 n$ is *much* smaller than n.

Which algorithm should you use in the problems? It doesn't make much difference because the sets S we look at are small. □

Example 24 (Covering relations and Hasse diagrams) Let S be a finite poset with relation $\preceq$. We define a new relation on S called the *covering relation*, denoted by $\prec_c$. For $x, y \in S$, we say $x \prec_c y$ if

(a) $x \neq y$ and $x \preceq y$, and

(b) $x \preceq z \preceq y$ implies that either $x = z$ or $y = z$.

In words, $x \prec_c y$ if x and y are different and there is no third element of S "between" x and y. In this case, we say that "y covers x" or "x is covered by y."

The condition $x \neq y$ means that the covering relation of a nonempty set S is never reflexive. In fact, $x \not\prec_c x$ for all $x \in S$. Thus, the covering relation fails badly the test of being reflexive. The covering relation is always antisymmetric. If there are three distinct elements $a \preceq b \preceq c$ in S then the covering relation is not transitive; otherwise, it is trivially transitive.

If you recall the discussion of Example 22 you can easily see that any order relation is *almost* the transitive closure of its covering relation. Missing, when we take the transitive closure of the covering relation, are all of the relations of the form $x \preceq x$. If you add those at the end, after taking the transitive closure of the covering relation, then you recover $\preceq$. Or, start with the covering relation, add in all pairs (x, x), $x \in S$, and then take the transitive closure.[5]

Take for example a set $A = \{1, 2, 3\}$ and the subset lattice $\mathcal{P}(A)$. Let $x = \{1\}$ and $y = \{1, 2, 3\}$. In this example, $x \subseteq y$ but y does not cover x, written $x \not\subset_c y$. To see why, note that there is a third element $z = \{1, 2\}$ between x and y: $\{1\} \subseteq \{1, 2\} \subseteq \{1, 2, 3\}$. In this example, x is covered by z and z is covered by y. Here is the directed graph diagram

---

[4] You are encouraged to experiment with small values of n to convince yourself that this is true.

[5] If S is not finite, this process may not work. For example, the covering relation for the real numbers, $(\mathbb{R}, \leq)$, is empty!

for the covering relation:

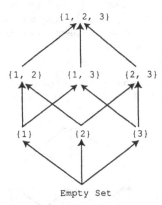

The directed graph diagram of the covering relation of a poset is called the *Hasse diagram* of the poset. The Hasse diagram is a very useful geometric way to picture posets. The transitive closure of the relation represented by the Hasse diagram, plus all pairs (x, x), x in the poset, is the order relation. □

Example 25 (Poset terminology) Here are the Hasse diagrams of two posets, the second a subposet of the first:

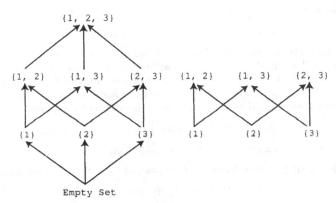

The subset $\{\{1\}, \{1,3\}, \{1,2,3\}\}$, of the first poset is called a *chain* in that poset. It is a chain because, as a subposet, it is linearly ordered: $\{1\} \subseteq \{1,3\} \subseteq \{1,2,3\}$. The *length* of this chain is two (one less than the number of elements in the chain). The longest chain in this poset has length three. There are six such "maximal" chains. You should try to find them all. In this first poset, the empty set $x = \emptyset$ is special in that for all y in the poset, $x \subseteq y$. Such an element is called the *least element* in the poset. Correspondingly, the element $t = \{1,2,3\}$, is the *greatest element* in the poset because $y \subseteq t$ for all y in the poset. There can be at most one greatest element and at most one least element in a poset.

Consider now the second poset. There is no least element and no greatest element in this poset. The element $x = \{2\}$ has the property that there is no y in the poset with $y \neq x$

and $y \subseteq x$. Such an element x is called a *minimal element* of a poset. A least element is a minimal element, but not necessarily the other way around. Similarly, $\{1, 2\}$ is a *maximal element* of this poset (but not a greatest element). Confused? Read it over again and look at the pictures. It is not that bad! ◻

Example 26 (Linear extensions — topological sorts) For this example, we shall return to the lattice of subsets of the set $\{1, 2, 3\}$. Recall its Hasse diagram:

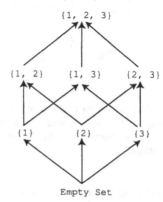

Here is a special listing of the elements of the lattice of subsets in which every element occurs exactly once:

$$\emptyset, \{1\}, \{2\}, \{1,2\}, \{3\}, \{1,3\}, \{2,3\}, \{1,2,3\}.$$

What is special about this listing? If you scan the list from left to right, you will find that for each set in the list, all of its supersets are to the right of it. Or, said in another way, if you scan from right to left, no set is a subset of some other set located to its left. Such a listing is called a *linear extension* of the poset (by mathematicians) or a *topological sort* (by computer scientists). Here is another linear extension of the poset:

$$\emptyset, \{3\}, \{2\}, \{1\}, \{1,3\}, \{1,2\}, \{2,3\}, \{1,2,3\}.$$

This poset has 48 linear extensions. Can you list them all?

Here is a listing of the elements of the lattice of subsets that is not a linear extension:

$$\emptyset, \{1\}, \{2\}, \{1,2\}, \{1,3\}, \{3\}, \{2,3\}, \{1,2,3\}.$$

Scanning from left to right, the set $\{3\}$ does not have all of its supersets to the right of it (the superset $\{1,3\}$ is to the left). Scanning from right to left, we again see that $\{3\}$ is a subset of $\{1,3\}$ which is to the left.

In general, let S be a poset with n elements and with relation $\preceq$. A *linear extension* of S is a listing of the elements of S, $s_1, s_2, \ldots, s_n$, such that for any $1 \leq i, j \leq n$, if $s_i \preceq s_j$ then $i \leq j$.

It is usually difficult to count the number of linear extensions of an arbitrary poset except by listing them. There are no easy formulas for many common posets. An easy case is an n-set with the empty relation (the "discrete" poset): There are $n!$ linear extensions. ◻

Exercises for Section 2

2.1. In each case a binary relation R on a set S is specified directly as a subset of $S \times S$. Determine, for each property, whether the relation R is reflexive, symmetric, or transitive. Explain your answers.

    (a) $R = \{(0,0), (0,1), (0,3), (1,0), (1,1), (2,3), (3,3)\}$ where $S = \{0,1,2,3\}$.

    (b) $R = \{(1,3), (3,1), (0,3), (3,0), (3,3)\}$ where $S = \{0,1,2,3\}$.

    (c) $R = \{(a,a), (a,b), (b,c), (a,c)\}$ where $S = \{a,b,c\}$.

    (d) $R = \{(a,a), (b,b)\}$ where $S = \{a,b,c\}$.

    (e) $R = \emptyset$ where $S = \{a\}$.

2.2. Define a binary relation on $\mathbb{R}$ (the reals) by $x\,R\,y$ if $\exists\, n \in \mathbb{Z}$ (the integers) such that $x^2 + y^2 = n^2$. Determine, for each property, whether the relation R is reflexive, symmetric, or transitive. Explain your answers.

2.3. Define a binary relation on $\mathbb{Z}$ by $x\,R\,y$ if $x = y$ or if $x - y = 2k + 1$ for some integer k. Determine, for each property, whether the relation R is reflexive, symmetric, or transitive. Explain your answers.

2.4. Let $S = \mathbb{R}$, the real numbers. Define a binary relation on S by $x\,R\,y$ if $x^2 = y^2$. Determine, for each property, whether the relation R is reflexive, symmetric, or transitive. Explain your answers.

2.5. Define a binary relation on $\mathbb{N}^+$ (the positive integers) by $x\,R\,y$ if $\gcd(x,y) > 1$. Determine, for each property, whether the relation R is reflexive, symmetric, or transitive. Explain your answers.

2.6. Let $S = \mathcal{P}(\underline{4}) - \{\emptyset\}$, the power set of $\underline{4} = \{1,2,3,4\}$ with the empty set discarded. Define a binary relation on S by $X\,R\,Y$ if $X \cap Y \neq \emptyset$. Determine, for each property, whether the relation R is reflexive, symmetric, or transitive. Explain your answers.

2.7. Let $S = \mathcal{P}(T)$, be the power set of $T = \{1,2,3,4\}$. Define a binary relation on S by $X\,R\,Y$ if either $X \subseteq Y$ or $Y \subseteq X$. Determine, for each property, whether the relation R is reflexive, symmetric, or transitive. Explain your answers.

2.8. Let S be a set with n elements. How many binary relations on S are reflexive? How many are not reflexive?

2.9. Let S be a set with n elements. How many binary relations on S are symmetric? How many are not symmetric?

Equivalence and Order

2.10. Let S be a set with n elements. How many binary relations on S are reflexive and symmetric?

2.11. Let S be a set with n elements. How many binary relations on S are antisymmetric? How many are reflexive and antisymmetric?

2.12. Let $R = \{(0,0),(0,3),(1,0),(1,2),(2,0),(3,2)\}$ be a binary relation on $\{0,1,2,3\}$. Find the transitive closure of R.

2.13. Let $R = \{(a,c),(b,c),(c,d)\}$ be a binary relation on $\{a,b,c,d\}$. Find the transitive closure of R by experimentation and by the matrix method.

2.14. Let S be the set of composite integers n, $4 \le n \le 20$. Order S with the divides relation. What is the covering relation? Draw the Hasse diagram. List the minimal and maximal elements. Specify a chain of longest length.

2.15. Let $S = \{1,2,3,4,5\}$. Let $\mathcal{P}^{(2)}(S)$ denote the subset of $\mathcal{P}(S)$ consisting of all subsets A such that if $i, j \in A$, then $i \ne j$ implies that $|i - j| \ge 2$. Order the elements of $\mathcal{P}^{(2)}(S)$ by set inclusion. What is the cardinality of the covering relation of $\mathcal{P}^{(2)}(S)$? How many chains are there of length three? What are the maximal elements? the minimal elements? Is there a greatest element? a least element?

2.16. Give an example of a poset with no maximal element.

2.17. Let $S_2 = \{0,1\} \times \{0,1\} = \times^2\{0,1\}$. Use coordinate order: $(x_1,x_2) \le (y_1,y_2)$ if $x_1 \le y_1$ and $x_2 \le y_2$. What is the covering relation? Compare this covering relation with $\mathcal{P}(X)$, $|X| = 2$, and set inclusion as the order relation. How do these ideas extend to $\times^3\{0,1\}$? To $\times^n\{0,1\}$?

2.18. Let S be the set of composite integers n, $4 \le n \le 20$. Order S with the divides relation. Let S^* denote the set of all finite strings (words) over S ordered lexicographically based on the poset S. Answer the following by stating whether or not the pair of strings (read left to right) is in order, in reverse order, or incomparable lexicographically.

(a) 4 6 18 and 4 6 9

(b) 4 6 8 and 4 6 8 9

(c) 4 16 8 and 4 6 10 9

2.19. Let S be an n-element set where $n \ge 3$ and let $x, y \in S$ where $x \ne y$. Suppose that S is made into a poset in such a way that all pairs of elements are comparable except x and y.

(a) What is the covering relation for S and does the Hasse diagram of S look like? (Describe *all* possible answers.)

(b) Let the poset T be $S \times S$ with the lex order. How many pairs $\{(a_1, a_2), (b_1, b_2)\}$ of incomparable elements does T have?

2.20. List in lexicographic order all ways of placing six dominoes on a 2×6 board.

2.21. Sort the following list into lexicographic order using a three-pass bucket sort: 321, 441, 143, 312, 422, 221, 214, 311, 234, 111. (Each element in the list is a sequence of three digits — not a 3-digit number. Thus 321 is the list (3,2,1).) Show the composition of the buckets after each pass.

2.22. Let S be the set of composite integers n, $4 \leq n \leq 20$. Order S with the divides relation. Let $x_1, x_2, \ldots, x_{11}$ be a topological sort of this poset. A pair (i, j), where $i < j$ and the integer x_i is smaller than the integer x_j will be called an "in-order pair." Find a topological sort where the number of in order pairs is less than or equal to 26.
Hint: First draw the Hasse diagram.

2.23. Let $S = \{a, b, c\}$, a set with three elements. Let $\mathcal{P}(S)$ be the set of all subsets of S ordered by set inclusion. Find 48 different topological sorts of $\mathcal{P}(S)$. You need not list them all if you can describe them in a convincing way.

Multiple Choice Questions for Review

In each case there is one correct answer (given at the end of the problem set). Try to work the problem first without looking at the answer. Understand both why the correct answer is correct and why the other answers are wrong.

1. Let $S = \{0, 1, 2, 3, 4, 5, 6, 7, 8, 9\}$. What is the smallest integer K such that any subset of S of size K contains two disjoint subsets of size two, $\{x_1, x_2\}$ and $\{y_1, y_2\}$, such that $x_1 + x_2 = y_1 + y_2 = 9$?

(a) 8   (b) 9   (c) 7   (d) 6   (e) 5

2. There are K people in a room, each person picks a day of the year to get a free dinner at a fancy restaurant. K is such that there must be at least one group of six people who select the same day. What is the smallest such K if the year is a leap year (366 days)?

(a) 1829   (b) 1831   (c) 1830   (d) 1832   (e) 1833

3. A mineral collection contains twelve samples of Calomel, seven samples of Magnesite, and N samples of Siderite. Suppose that the smallest K such that choosing K samples from the collection guarantees that you have six samples of the same type of mineral is $K = 15$. What is N?

(a) 6   (b) 2   (c) 3   (d) 5   (e) 4

4. What is the smallest $N > 0$ such that any set of N nonnegative integers must have two distinct integers whose sum or difference is divisible by 1000?

(a) 502   (b) 520   (c) 5002   (d) 5020   (e) 52002

5. Let $S = \{1, 3, 5, 7, 9, 11, 13, 15, 17, 19, 21\}$. What is the smallest integer $N > 0$ such that for any set of N integers, chosen from S, there must be two distinct integers that divide each other?

(a) 10   (b) 7   (c) 9   (d) 8   (e) 11

6. The binary relation $R = \{(0, 0), (1, 1)\}$ on $A = \{0, 1, 2, 3, \}$ is

(a) Reflexive, Not Symmetric, Transitive

(b) Not Reflexive, Symmetric, Transitive

(c) Reflexive, Symmetric, Not Transitive

(d) Reflexive, Not Symmetric, Not Transitive

(e) Not Reflexive, Not Symmetric, Not Transitive

7. Define a binary relation $R = \{(0, 1), (1, 2), (2, 3), (3, 2), (2, 0)\}$ on $A = \{0, 1, 2, 3\}$. The directed graph (including loops) of the transitive closure of this relation has

(a) 16 arrows

(b) 12 arrows

(c) 8 arrows

(d) 6 arrows

(e) 4 arrows

8. Let $\mathbb{N}^+$ denote the nonzero natural numbers. Define a binary relation R on $\mathbb{N}^+ \times \mathbb{N}^+$ by $(m,n)R(s,t)$ if $\gcd(m,n) = \gcd(s,t)$. The binary relation R is

(a) Reflexive, Not Symmetric, Transitive

(b) Reflexive, Symmetric, Transitive

(c) Reflexive, Symmetric, Not Transitive

(d) Reflexive, Not Symmetric, Not Transitive

(e) Not Reflexive, Not Symmetric, Not Transitive

9. Let $\mathbb{N}_2^+$ denote the natural numbers greater than or equal to 2. Let mRn if $\gcd(m,n) > 1$. The binary relation R on $\mathbb{N}_2$ is

(a) Reflexive, Symmetric, Not Transitive

(b) Reflexive, Not Symmetric, Transitive

(c) Reflexive, Symmetric, Transitive

(d) Reflexive, Not Symmetric, Not Transitive

(e) Not Reflexive, Symmetric, Not Transitive

10. Define a binary relation R on a set A to be *antireflexive* if xRx doesn't hold for any $x \in A$. The number of symmetric, antireflexive binary relations on a set of ten elements is

    (a) 2^{10} (b) 2^{50} (c) 2^{45} (d) 2^{90} (e) 2^{55}

11. Let R and S be binary relations on a set A. Suppose that R is reflexive, symmetric, and transitive and that S is symmetric, and transitive but is **not** reflexive. Which statement is always true for any such R and S?

(a) $R \cup S$ is symmetric but not reflexive and not transitive.

(b) $R \cup S$ is symmetric but not reflexive.

(c) $R \cup S$ is transitive and symmetric but not reflexive

(d) $R \cup S$ is reflexive and symmetric.

(e) $R \cup S$ is symmetric but not transitive.

12. Define an equivalence relation R on the positive integers $A = \{2,3,4,\ldots,20\}$ by $m\ R\ n$ if the largest prime divisor of m is the same as the largest prime divisor of n. The number of equivalence classes of R is

    (a) 8   (b) 10   (c) 9   (d) 11   (e) 7

Equivalence and Order

13. Let $R = \{(a, a), (a, b), (b, b), (a, c), (c, c)\}$ be a partial order relation on $\Sigma = \{a, b, c\}$. Let $\preceq$ be the corresponding lexicographic order on Σ^*. Which of the following is true?

    (a) $bc \preceq ba$

    (b) $abbaaacc \preceq abbaab$

    (c) $abbac \preceq abb$

    (d) $abbac \preceq abbab$

    (e) $abbac \preceq abbaac$

14. Consider the divides relation, $m \mid n$, on the set $A = \{2, 3, 4, 5, 6, 7, 8, 9, 10\}$. The cardinality of the covering relation for this partial order relation (i.e., the number of edges in the Hasse diagram) is

    (a) 4   (b) 6   (c) 5   (d) 8   (e) 7

15. Consider the divides relation, $m \mid n$, on the set $A = \{2, 3, 4, 5, 6, 7, 8, 9, 10\}$. Which of the following permutations of A is **not** a topological sort of this partial order relation?

    (a) 7,2,3,6,9,5,4,10,8

    (b) 2,3,7,6,9,5,4,10,8

    (c) 2,6,3,9,5,7,4,10,8

    (d) 3,7,2,9,5,4,10,8,6

    (e) 3,2,6,9,5,7,4,10,8

16. Let $A = \{2, 3, 4, 5, 6, 7, 8, 9, 10, 11, 12, 13, 14, 15, 16\}$ and consider the divides relation on A. Let C denote the length of the maximal chain, M the number of maximal elements, and m the number of minimal elements. Which is true?

    (a) $C = 3$, $M = 8$, $m = 6$

    (b) $C = 4$, $M = 8$, $m = 6$

    (c) $C = 3$, $M = 6$, $m = 6$

    (d) $C = 4$, $M = 6$, $m = 4$

    (e) $C = 3$, $M = 6$, $m = 4$

Answers: 1 (c), **2** (b), **3** (e), **4** (a), **5** (d), **6** (b), **7** (a), **8** (b), **9** (a), **10** (c), **11** (d), **12** (a), **13** (b), **14** (e), **15** (c), **16** (a).

Induction, Sequences and Series

Section 1: Induction

Suppose $A(n)$ is an assertion that depends on n. We use *induction* to prove that $A(n)$ is true when we show that

- it's true for the smallest value of n and

- if it's true for everything less than n, then it's true for n.

In this section, we will review the idea of proof by induction and give some examples. Here is a formal statement of proof by induction:

Theorem 1 (Induction)   *Let $A(m)$ be an assertion, the nature of which is dependent on the integer m. Suppose that we have proved $A(n_0)$ and the statement*

"*If $n > n_0$ and $A(k)$ is true for all k such that $n_0 \leq k < n$, then $A(n)$ is true.*"

Then $A(m)$ is true for all $m \geq n_0$.[1]

Proof: We now prove the theorem. Suppose that $A(n)$ is false for some $n \geq n_0$. Let m be the least such n. We cannot have $m = n_0$ because one of our hypotheses is that $A(n_0)$ is true. On the other hand, since m is as small as possible, $A(k)$ is true for $n_0 \leq k < m$. By the inductive step, $A(m)$ is also true, a contradiction. Hence our assumption that $A(n)$ is false for some n is itself false; in other words, $A(n)$ is never false. This completes the proof. $\square$

Definition 1 (Induction terminology)   "*$A(k)$ is true for all k such that $n_0 \leq k < n$*" *is called the induction assumption or induction hypothesis and proving that this implies $A(n)$ is called the inductive step. $A(n_0)$ is called the base case or simplest case.*

---

[1] This form of induction is sometimes called *strong* induction. The term "strong" comes from the assumption "$A(k)$ is true for all k such that $n_0 \leq k < n$." This is replaced by a more restrictive assumption "$A(k)$ is true for $k = n - 1$" in *simple* induction. Actually, there are many intermediate variations on the nature of this assumption, some of which we shall explore in the exercises (e.g., "$A(k)$ is true for $k = n - 1$ and $k = n - 2$," "$A(k)$ is true for $k = n - 1$, $k = n - 2$, and $k = n - 3$," etc.).

Induction, Sequences and Series

Example 1 (Every integer is a product of primes) A positive integer $n > 1$ is called a *prime* if its only divisors are 1 and n. The first few primes are 2, 3, 5, 7, 11, 13, 17, 19, 23. In another unit, we proved that every integer $n > 1$ is a product of primes. We now redo the proof, being careful with the induction.

We adopt the terminology that a single prime p is a product of one prime, itself. We shall prove $A(n)$:

"Every integer $n \geq 2$ is a product of primes."

Our proof that $A(n)$ is true for all $n \geq 2$ will be by induction. We start with $n_0 = 2$, which is a prime and hence a product of primes. The induction hypothesis is the following:

"Suppose that for some $n > 2$, $A(k)$ is true for all k such that $2 \leq k < n$."

Assume the induction hypothesis and consider $A(n)$. If n is a prime, then it is a product of primes (itself). Otherwise, $n = st$ where $1 < s < n$ and $1 < t < n$. By the induction hypothesis, s and t are each a product of primes, hence $n = st$ is a product of primes. This completes the proof of $A(n)$; that is, we've done the inductive step. Hence $A(n)$ is true for all $n \geq 2$. $\square$

In the example just given, we needed the induction hypothesis "for all k such that $2 \leq k < n$." In the next example we have the more common situation where we only need "for $k = n - 1$." We can still make the stronger assumption "for all k such that $1 \leq k < n$" and the proof is valid.

Example 2 (Sum of first n integers) We would like a formula for the sum of the first n integers. Let us write $S(n) = 1 + 2 + \ldots + n$ for the value of the sum. By a little calculation,

$$S(1) = 1, \quad S(2) = 3, \quad S(3) = 6, \quad S(4) = 10, \quad S(5) = 15, \quad S(6) = 21.$$

What is the general pattern? It turns out that $S(n) = \frac{n(n+1)}{2}$ is correct for $1 \leq n \leq 6$. Is it true in general? This is a perfect candidate for an induction proof with

$$n_0 = 1 \quad \text{and} \quad A(n): \quad \text{``} S(n) = \frac{n(n+1)}{2}.\text{''}$$

Let's prove it. We have shown that A(1) is true. In this case we need only the restricted induction hypothesis; that is, we will prove the formula for $S(n)$ by assuming the formula for for $k = n - 1$. Thus, we assume only $S(n-1)$ is true. Here it is (the inductive step):

$$\begin{aligned}
S(n) &= 1 + 2 + \cdots + n \\
&= \left(1 + 2 + \cdots + (n-1)\right) + n \\
&= S(n-1) + n \\
&= \frac{(n-1)\big((n-1)+1\big)}{2} + n \qquad &\text{by } A(n-1), \\
&= \frac{n(n+1)}{2} \qquad &\text{by algebra.}
\end{aligned}$$

This completes the proof. $\square$

Example 3 (Intuition behind the sum of first n integers) Whenever you prove something by induction you should try to gain an intuitive understanding of why the result is true. Sometimes a proof by induction will obscure such an understanding. In the following array, you will find one 1, two 2's, three 3's, etc. The total number of entries is $1 + 2 + \cdots + 8$. On the other hand, the array is a rectangle with $4 \times 9 = 36$ entries. This verifies that $1 + 2 + \ldots + n = \frac{n(n+1)}{2}$ is correct for $n = 8$. The same way of laying out the integers works for any n (if n is odd, it is laid out along the bottom row, if n is even, it is laid out in the last two columns).

$$
\begin{array}{ccccccccc}
1 & 2 & 2 & 4 & 4 & 6 & 6 & 8 & 8 \\
3 & 3 & 3 & 4 & 4 & 6 & 6 & 8 & 8 \\
5 & 5 & 5 & 5 & 5 & 6 & 6 & 8 & 8 \\
7 & 7 & 7 & 7 & 7 & 7 & 7 & 8 & 8
\end{array}
$$

This argument, devised by a fourth-grade girl, has all of the features of a powerful intuitive image.

Here is another proof based on adding columns

$$
\begin{array}{ccccccc}
S(n) = & 1 & + & 2 & + \cdots + & n \\
S(n) = & n & + (n-1) & + \cdots + & 1 \\
\hline
2S(n) = & (n+1) + (n+1) & + \cdots + & (n+1) \\
= & n(n+1) &
\end{array}
$$

Here is geometric view of this approach for $n = 8$.

$$
\begin{array}{ccccccccc}
O & X & X & X & X & X & X & X & X \\
O & O & X & X & X & X & X & X & X \\
O & O & O & X & X & X & X & X & X \\
O & O & O & O & X & X & X & X & X \\
O & O & O & O & O & X & X & X & X \\
O & O & O & O & O & O & X & X & X \\
O & O & O & O & O & O & O & X & X \\
O & O & O & O & O & O & O & O & X
\end{array}
$$

$\square$

Example 4 (Bounding the terms of a recursion) Consider the recursion

$$f_k = f_{k-1} + 2f_{k-2} + f_{k-3}, \ k \geq 3, \text{ with } f_0 = 1, \ f_1 = 2, \ f_2 = 4.$$

We would like to obtain a bound on the f_k, namely $f_k \leq r^k$ for all $k \geq 0$. Thus there are two problems: (a) what is the best (smallest) value we can find for r and (b) how can we prove the result?

Since the recursion tells us how to compute f_k from previous values, we expect to give a proof by induction. The inequality $f_k \leq r^k$ tells us that $f_1 \leq r^1 = r$. Since $f_1 = 2$, maybe $r = 2$ will work. Let's try giving a proof with $r = 2$. Thus $\mathcal{A}(n)$ is the statement "$f_n \leq 2^n$" and $n_0 = 0$. In order to use the recursion for f_n, we need $n \geq 3$. Thus we must treat $n = 0, 1, 2$ separately

- Since $f_0 = 1$ and $2^0 = 1$, we've done $n = 0$.

- We've already done $n = 1$.

- Since $f_2 = 4 = 2^2$, we've done $n = 2$.

- Suppose $n \geq 3$. By our induction hypothesis, $f_{n-1} \leq 2^{n-1}$, $f_{n-2} \leq 2^{n-2}$, and $f_{n-3} \leq 2^{n-3}$. Thus

$$f_n = f_{n-1} + 2f_{n-2} + f_{n-3} \leq 2^{n-1} + 2 \times 2^{n-2} + 2^{n-3} = 2^n + 2^{n-3}.$$

This won't work because we wanted to conclude that $f_n \leq 2^n$.

What is wrong? Either our guess that $f_n \leq 2^n$ wrong or our guess is right and we need to look for another way to prove it. Since it's easier to compute values of f_n than it is to find proofs, let's compute. We have $f_3 = f_2 + 2f_1 + f_0 = 4 + 2 \times 2 + 1 = 9$. Thus $f_3 \leq 2^3$ is false! This illustrates an important idea: Often computing a few values can save a lot of time.

Since 2 won't work, what will? Let's pretend we know the answer and call it r. We already know that we need to have $r > 2$.

- Since $r > 2$, $f_n \leq r^n$ for $n = 0, 1, 2$.

- Suppose $n \geq 3$. Working just as we did for the case $r = 2$, we have

$$f_n \leq r^{n-1} + 2r^{n-2} + r^{n-3}.$$

We want this to be less than r^n; that is, we want $r^{n-1} + 2r^{n-2} + r^{n-3} \leq r^n$. Dividing both sides by r^{n-3}, we see that we want $r^2 + 2r + 1 \leq r^3$. The smallest $r \geq 2$ that satisfies this inequality is an irrational number which is approximately 2.148.

For practice, you should go back and write a formal induction proof when $r = 2.2$. $\square$

---

*More Advanced Examples of Induction

The next two examples are related, first because they both deal with polynomials, and second because the theorem in one is used in the other. They also illustrate a point about proof by induction that is sometimes missed: Because exercises on proof by induction are chosen to give experience with the inductive step, students frequently assume that the inductive step will be the hard part of the proof. The next example fits this stereotype — the inductive step is the hard part of the proof. In contrast, the base case is difficult and the inductive step is nearly trivial in the second example. A word of caution: these examples are more complicated than the preceding ones.

Example 5 (Sum of kth powers of integers) Let $S_k(n)$ be the sum of the first n k^{th} powers of integers. In other words,

$$S_k(n) = 1^k + 2^k + \cdots + n^k \quad \text{for } n \text{ a positive integer.}$$

In particular $S_k(0) = 0$ (since there is nothing to add up) and $S_k(1) = 1$ (since $1^k = 1$) for all k. We have
$$S_0(n) = 1^0 + 2^0 + \cdots + n^0 = 1 + 1 + \cdots + 1 = n.$$

In Example 2 we showed that $S_1(n) = n(n+1)/2$. Can we observe any patterns here? Well, it looks like $S_k(n)$ might be $\frac{n(n+1)\cdots(n+k)}{k+1}$ A little checking shows that this is wrong since $S_2(2) = 5$. Well, maybe we shouldn't be so specific. If you're familiar with integration, you might notice that $S_k(n)$ is a Riemann sum for $\int_0^n x^k \, dx = n^{k+1}/(k+1)$. Maybe $S_k(n)$ behaves something like $n^{k+1}/(k+1)$. That's rather vague. We'll prove

Theorem 2 (Sum of kth powers) If $k \geq 0$ is an integer, then $S_k(n)$ is a polynomial in n of degree $k + 1$. The constant term is zero and the coefficient of n^{k+1} is $1/(k+1)$.

Two questions may come to mind. First, how can we prove this since there is no formula to prove? Second, what good is the theorem since it doesn't give us a formula for $S_k(n)$?

Let's start with second question. We can use the theorem to find $S_k(n)$ for any particular k. To illustrate, suppose we don't know what $S_1(n)$ is. According to the theorem $S_1(n) = n^2/2 + An$ for some A since it says that $S_1(n)$ is a polynomial of degree two with no constant term and leading term $n^2/2$. With $n = 1$ we have $S_1(1) = 1^2/2 + A \times 1 = 1/2 + A$. Since $S_1(1) = 1^1 = 1$, it follows that $A = 1/2$. We have our formula: $S_1(n) = n^2/2 + n/2$.

Let's find $S_2(n)$. By the theorem $S_2(n) = n^3/3 + An^2 + Bn$. With $n = 1$ and $n = 2$ we get

n	direct calculation	polynomial
1	$S_2(1) = 1^2 = 1$	$S_2(1) = 1^3/3 + A \times 1^2 + B \times 1$
2	$S_2(2) = 1^2 + 2^2 = 5$	$S_2(2) = 2^3/3 + A \times 2^2 + B \times 2$

After a little algebra, we obtain the two equations

$$n = 1: \quad A + B = 2/3$$
$$n = 2: \quad 4A + 2B = 7/3$$

Solving these equations, we find that $A = 1/2$ and $B = 1/6$. Thus $S_2(n) = \frac{n^3}{3} + \frac{n^2}{2} + \frac{n}{6}$.

Okay, enough examples — on with the proof! We are going to use induction on k and a couple of tricks. The assertion we want to prove is

$$\mathcal{A}(k) \quad = \quad \begin{array}{l} S_k(n) \text{ is a polynomial in } n \text{ of degree } k+1 \\ \text{with constant term zero and leading term } \frac{1}{k+1}. \end{array}$$

The base case, $k = 0$ is easy: $1^0 + 2^0 + \cdots + n^0 = 1 + 1 + \cdots + 1 = n$, which has no constant term and has leading coefficient $\frac{1}{0+1} = 1$.

[153]

Induction, Sequences and Series

Now for the inductive step. We want to prove $\mathcal{A}(k)$. To do so, we will need $\mathcal{A}(t)$ for $0 \le t < k$.

The first trick uses the binomial theorem $(x+y)^m = \sum_{t=0}^m \binom{m}{t} x^t y^{m-t}$ with $m = k+1$, $x = j$ and $y = -1$: We have

$$j^{k+1} - (j-1)^{k+1} = j^{k+1} - \sum_{t=0}^{k+1} \binom{k+1}{t} j^t (-1)^{k+1-t} = -\sum_{t=0}^{k} \binom{k+1}{t} j^t (-1)^{k+1-t}.$$

Sum both sides over $1 \le j \le n$. When we sum the right side over j we get

$$-\sum_{t=0}^{k} \binom{k+1}{t} S_t(n)(-1)^{k+1-t}.$$

The second trick is what happens when we sum $j^{k+1} - (j-1)^{k+1}$ over j: Almost all the terms cancel:

$$(1^{k+1} - 0^{k+1}) + (2^{k+1} - 1^{k+1}) + \cdots + ((n-1)^{k+1} - (n-2)^{k+1}) + (n^{k+1} - (n-1)^{k+1})$$
$$= -0^{k+1} + n^{n+1} = n^{k+1}.$$

Thus we have

$$n^{k+1} = -\sum_{t=0}^{k} \binom{k+1}{t} S_t(n)(-1)^{k+1-t}$$

$$= -\binom{k+1}{k} S_k(n)(-1)^{k+1-k} - \sum_{t=0}^{k-1} \binom{k+1}{t} S_t(n)(-1)^{k+1-t}$$

$$= (k+1)S_k(n) - \sum_{t=0}^{k-1} \binom{k+1}{t} S_t(n)(-1)^{k+1-t}.$$

We can solve this equation for $S_k(n)$:

$$S_k(n) = \frac{n^{k+1}}{k+1} + \sum_{t=0}^{k-1} \frac{1}{k+1} \binom{k+1}{t} (-1)^{k+1-t} S_t(n).$$

By the induction hypothesis, $S_t(n)$ is a polynomial in n with no constant term and degree $t+1$. Since $0 \le t \le k-1$, it follows that each term in the messy sum is a polynomial in n with no constant term and degree at most k Thus the same is true of the entire sum. We have proved that

$$S_k(n) = \frac{n^{k+1}}{k+1} + P_k(n),$$

where $P_k(n)$ is a polynomial in n with no constant term and degree at most k. This completes the proof of the theorem. $\square$

Definition 2 (Forward difference) *Suppose $S : \mathbb{N} \to \mathbb{R}$. The forward difference of S is another function denoted by ΔS and defined by $\Delta S(n) = S(n+1) - S(n)$. In this context, Δ is called a difference operator.*

IS-6   [154]

We can iterate Δ. For example, $\Delta^2 S = \Delta(\Delta S)$. If we let $T = \Delta S$, then $T(n) = S(n+1) - S(n)$ and

$$\begin{aligned}
\Delta^2 S(n) = \Delta T(n) &= T(n+1) - T(n) \\
&= (S(n+2) - S(n+1)) - (S(n+1) - S(n)) \\
&= S(n+2) - 2S(n+1) + S(n).
\end{aligned}$$

The operator Δ has properties similar to the derivative operator d/dx. For example $\Delta(S + T) = \Delta S + \Delta T$. In some subjects, "differences" of functions play the role that derivatives play in other subjects. Derivatives arise in the study of rates of change in continuous situations. Differences arise in the study of rates of change in discrete situations. Although there is only one type of ordinary derivative, there are three common types of differences: backward, central and forward.

The next example gives another property of the difference operator that is like the derivative. You may know that the general solution of the differential equation $f^{(k)}(x) = $ constant is a polynomial of degree $k + 1$. In the next example we prove that the same is true for the difference equation $\Delta^k f(x) = $ constant.

Example 6 (Differences of polynomials) Suppose $S(n) = an + b$ for some constants a and b. You should be able to check that $\Delta S(n) = a$, a constant. With a little more work, you can check that $\Delta^2(an^2 + bn + c) = 2a$. We now state and prove a general converse of these results.

Theorem 3 (Polynomial differences) If $\Delta^k S$ is a polynomial of degree j, then $S(n)$ is a polynomial of degree $j + k$ in n.

We'll prove this by induction on k. $\mathcal{A}(k)$ is simply the statement of the theorem.

We now do the base case. Suppose $k = 1$. Let $T = \Delta S$. We want to show that, if T is a polynomial of degree j, then S is a polynomial of degree $j + 1$. We have

$$\begin{aligned}
S(n+1) =& \Big(S(n+1) - S(n)\Big) + \Big(S(n) - S(n-1)\Big) + \Big(S(n-1) - S(n-2)\Big) \\
& + \cdots + \Big(S(2) - S(1)\Big) + S(1) \\
=& T(n) + T(n-1) + T(n-2) + \cdots + T(1) + S(1) \\
=& \sum_{t=1}^{n} T(t) + S(1).
\end{aligned}$$

What have we gained by this manipulation? We've expressed an unknown function $S(n+1)$ as the sum of a constant $S(1)$ and the sum of a function T which is known to be a polynomial of degree j. Now we need to make use of our knowledge of T to say something about $\sum T(t)$.

[155]

Induction, Sequences and Series

By assumption, T is a polynomial of degree j. Let $T(n) = a_j n^j + \cdots + a_1 n + a_0$, where $a_0, \ldots, a_j$ are constants. Then

$$\sum_{t=1}^{n} T(t) = \sum_{t=1}^{n} (a_j t^j + \cdots + a_1 t + a_0)$$

$$= \sum_{t=1}^{n} a_j t^j + \cdots + \sum_{t=1}^{n} a_1 t + \sum_{t=1}^{n} a_0$$

$$= a_j \sum_{t=1}^{n} t^j + \cdots + a_1 \sum_{t=1}^{n} t + n a_0.$$

By Theorem 2,

$$\sum_{t=1}^{n} t^j \qquad \text{is a polynomial of degree } j+1,$$

$$\sum_{t=1}^{n} t^{j-1} \qquad \text{is a polynomial of degree less than } j+1,$$

$$\cdots$$

$$\sum_{t=1}^{n} t \qquad \text{is a polynomial of degree less than } j+1.$$

Thus $\sum_{t=1}^{n} T(t)$ is a polynomial of degree $j+1$. Since $S(n+1) = \sum_{t=1}^{n} T(t) + S(1)$, it is a polynomial of degree $j+1$ in n.

Let's see where we are with the base case. We've proved that $S(n+1)$ is a polynomial of degree $j+1$ in n. But we want to prove that $S(n)$ is a polynomial of degree $j+1$ in n, so we have a bit more work.

We can write $S(n+1) = b_{j+1} n^{j+1} + b_j n^j + \cdots + b_1 n + b_0$. Replace n by $n-1$:

$$S(n) = b_{j+1}(n-1)^{j+1} + b_j(n-1)^j + \cdots + b_1(n-1) + b_0.$$

Using the binomial theorem in the form $(n-1)^k = \sum_{i=0}^{k} \binom{k}{i} n^i (-1)^{k-i}$, you should be able to see that $(n-1)^k$ is a polynomial of degree k in n. Using this in the displayed equation, you can see that $S(n)$ is a polynomial of degree $j+1$ in n. The base case is done. Whew!

The induction step is easy: We are given that $\Delta^k S$ is a polynomial of degree j. We want to show that S is a polynomial of degree $j+k$. By definition, $\Delta^k S = \Delta(\Delta^{k-1} S)$. Let $T = \Delta^{k-1} S$. We now take three simple steps.

- By the definition of T, $\Delta T = \Delta^k S$, which is a polynomial of degree j by the hypothesis of $\mathcal{A}(k)$.

- By $\mathcal{A}(1)$, T is a polynomial of degree $j+1$; that is, $\Delta^{k-1} S$ is a polynomial of degree $j+1$.

- By $\mathcal{A}(k-1)$ with j replaced by $j+1$, it now follows that S is a polynomial of degree $(j+1) + (k-1) = j+k$.

The proof is done. $\square$

The best way, perhaps the only way, to understand induction and inductive proof technique is to work lots of problems. That we now do!

IS-8

[156]

Exercises for Section 1

1.1. In each case, express the given infinite series or product in summation or product notation.

(a) $1^2 - 2^2 + 3^2 - 4^2 \cdots$

(b) $(1^3 - 1) + (2^3 + 1) + (3^3 - 1) \cdots$

(c) $(2^2 - 1)(3^2 + 1)(4^2 - 1) \cdots$

(d) $(1 - r)(1 - r^3)(1 - r^5) \cdots$

(e) $\frac{1}{2!} + \frac{2}{3!} + \frac{3}{4!} + \cdots$

(f) $n + \frac{n-1}{2!} + \frac{n-2}{3!} + \cdots$

1.2. In each case give a formula for the nth term of the indicated sequence. Be sure to specify the starting value for n.

(a) $1 - \frac{1}{2}, \frac{1}{2} - \frac{1}{3}, \frac{1}{3} - \frac{3}{4}, \ldots$

(b) $\frac{1}{4}, \frac{2}{9}, \frac{3}{16}, \ldots$

(c) $\frac{1}{2}, -\frac{2}{3}, \frac{3}{4}, \ldots$

(d) $2, 6, 12, 20, 30, 42, \ldots$

(e) $0, 0, 1, 1, 2, 2, 3, 3, \ldots$

1.3. In each case make the change of variable $j = i - 1$.

(a) $\displaystyle\prod_{i=2}^{n+1} \frac{(i-1)^2}{i}$

(b) $\displaystyle\sum_{i=1}^{n-1} \frac{i}{(n-i)^2}$

(c) $\displaystyle\prod_{i=n}^{2n} \frac{n-i+1}{i}$

(d) $\displaystyle\prod_{i=1}^{n} \frac{i}{i+1} \prod_{i=1}^{n} \frac{i+1}{i+2}$

1.4. Prove by induction that $\displaystyle\sum_{k=1}^{n} k^2 = \frac{n(n+1)(2n+1)}{6}$ for $n \geq 1$.

1.5. Prove twice, once using Theorem 2 and once by induction, that $\displaystyle\sum_{k=1}^{n} k^3 = \left(\frac{n(n+1)}{2}\right)^2$ for $n \geq 1$.

Induction, Sequences and Series

1.6. Prove by induction that $\displaystyle\sum_{i=1}^{n} \frac{1}{i(i+1)} = \frac{n}{n+1}$ for $n \geq 1$.

1.7. Prove by induction that $\displaystyle\sum_{i=1}^{n+1} i2^i = n2^{n+2} + 2$ for $n \geq 0$.

1.8. Prove by induction that $\displaystyle\prod_{i=2}^{n} \left(1 - \frac{1}{i^2}\right) = \frac{n+1}{2n}$ for $n \geq 2$.

1.9. Prove by induction that $\displaystyle\sum_{i=1}^{n} i\,i! = (n+1)! - 1$ for $n \geq 1$.

1.10. Prove by induction that $\displaystyle\prod_{i=0}^{n} \frac{1}{2i+1}\frac{1}{2i+2} = \frac{1}{(2n+2)!}$ for $n \geq 0$.

1.11. Prove without using induction that $\displaystyle\sum_{k=1}^{n} 5k = 2.5n(n+1)$.

1.12. Prove that, for $a \neq 1$ and $n \geq t$,

$$\sum_{k=t}^{n} a^k = a^t \left(\frac{a^{n-t+1} - 1}{a-1}\right).$$

1.13. Prove twice, once with induction and once without induction, that $3 \mid (n^3 - 10n + 9)$ for all integers $n \geq 0$; that is, $n^3 - 10n + 9$ is a multiple of 3.

1.14. Prove by induction that $(x - y) \mid (x^n - y^n)$ where $x \neq y$ are integers, $n > 0$.

1.15. Prove twice, once with induction and once without induction, that $6 \mid n(n^2 + 5)$ for all $n \geq 1$.

1.16. Prove by induction that $n^2 \leq 2^n$ for all $n \geq 0$, $n \neq 3$.

1.17. Prove by induction that

$$\sqrt{n} < \sum_{i=1}^{n} \frac{1}{\sqrt{i}} \quad \text{for } n \geq 2.$$

1.18. Consider the Fibonacci recursion $f_k = f_{k-2} + f_{k-1}$, $k \geq 2$, with $f_0 = 3$ and $f_1 = 6$. Prove by induction that $3 \mid f_k$ for all $k \geq 0$.

1.19. Consider the recursion $F_k = F_{k-1} + F_{k-2}$, $k \geq 2$, with $F_0 = 0$ and $F_1 = 1$. Prove that F_k is even if and only if $3 \mid k$. In other words, prove that, modulo 2, $F_{3t} = 0$, $F_{3t+1} = 1$, and $F_{3t+2} = 1$ for $t \geq 0$.

1.20. Consider the recursion $f_k = 2f_{\lfloor \frac{k}{2} \rfloor}$, $k \geq 2$, with $f_1 = 1$. Prove by induction that $f_k \leq k$ for all $k \geq 1$.

1.21. We wish to prove by induction that for any real number $r > 0$, and every integer $n \geq 0$, $r^n = 1$. For $n = 0$, we have $r^n = 1$ for all $r > 0$. This is the base case. Assume that for $k > 0$, we have that, for $0 \leq j \leq k$, $r^j = 1$ for all $r > 0$. We must show that for $0 \leq j \leq k + 1$, $r^j = 1$ for all $r > 0$. Write $r^{k+1} = r^s r^t$ where $0 \leq s \leq k$ and $0 \leq t \leq k$. By the induction hypothesis, $r^s = 1$ and $r^t = 1$ for all $r > 0$. Thus, $r^{k+1} = r^s r^t = 1$ for all $r > 0$. Combining this with the induction hypothesis gives that for $0 \leq j \leq k + 1$, $r^j = 1$ for all $r > 0$. Thus the theorem is proved by induction. What is wrong?

1.22. We wish to prove by induction the proposition $\mathcal{A}(n)$ that all positive integers j, $1 \leq j \leq n$, are equal. The case $\mathcal{A}(1)$ is true. Assume that, for some $k \geq 1$, $\mathcal{A}(k)$ is true. Show that this implies that $\mathcal{A}(k+1)$ is true. Suppose that p and q are positive integers less than or equal to $k + 1$. By the induction hypothesis, $p - 1 = q - 1$. Thus, $p = q$. Thus $\mathcal{A}(n)$ is proved by induction. What is wrong?

*1.23. Let $a \in \mathbb{R}$, $f : \mathbb{N} \to \mathbb{R}$ and $g : \mathbb{N} \to \mathbb{R}$. Prove the following.

(a) $\Delta(af) = a\Delta f$; that is, for all $n \in \mathbb{N}$, the function $\Delta(af)$ evaluated at n equals a times the function Δf evaluated at n.

(b) $\Delta(f + g) = \Delta f + \Delta g$.

(c) $\Delta(fg) = f\Delta g + g\Delta f + (\Delta f)(\Delta g)$; that is, for all $n \in \mathbb{N}$,
$(\Delta(fg))(n) = f(n)(\Delta g)(n) + g(n)(\Delta f)(n) + (\Delta f)(n) \, (Deltag)(n)$.

*1.24. Prove by induction on k that, for $k \geq 1$,

$$(\Delta^k f)(n) = \sum_{j=0}^{k} \binom{k}{j} (-1)^{k-j} f(n + j).$$

Hint: You may find it useful to recall that $\binom{k-1}{j-1} + \binom{k-1}{j} = \binom{k}{j}$ for $k \geq j > 0$.

Section 2: Infinite Sequences

Our purpose in this section and the next is to present the intuition behind infinite sequences and series. It is our experience, however, that the development of this intuition is greatly aided by an exposure to a small amount of the precise formalism that lies behind the mathematical study of sequences and series. This exposure takes away much of the mystery of the subject and focuses the intuition on what really matters.

Recall that a function f with domain D and range (codomain) R is a rule which, to every $x \in D$ assigns a unique element $f(x) \in R$. Sequences are a special class of functions.

Definition 3 (Infinite sequence) *Let $n_0 \in \mathbb{N} = \{0, 1, 2, \ldots\}$ A function f whose domain is $D = \mathbb{N} + n_0 = \{n \mid n \in \mathbb{N} \text{ and } n \geq n_0\}$ and whose range is the set $\mathbb{R}$ of real numbers is called an infinite sequence.*

An infinite sequence is often written in subscript notation; for example, $a_2, a_3, a_4, \ldots$ corresponds to a function f with domain $\mathbb{N} + 2$, $f(2) = a_2$, $f(3) = a_3$ and so on.

Each value of the function is a term of the sequence. Thus $f(4)$ is a term in functional notation and a_7 is a term in subscript notation.

If f is an infinite sequence with domain $\mathbb{N} + n_0$ and $k \geq n_0$, the f restricted to $\mathbb{N} + k$ is called a tail of f. For example, $a_7, a_8, \ldots$ is a tail of $a_2, a_3, \ldots$.

Example 7 (Specifying sequences) People specify infinite sequences in various ways. The function is usually given by subscript notation rather than parenthetic notation; that is, a_n instead of $f(n)$. Let's look at some examples of sequence specification.

- "Consider the sequence $1/n$ for $n \geq 1$." This is a perfectly good specification of the function. Since the sequence starts at $n = 1$, we have $n_0 = 1$ and $a_n = 1/n$.

- "Consider the sequence $1/n$." Since the domain of n has not been specified this is not a function; however, specifying a sequence in this manner is common. What should the domain be? The convention is that $n_0 \geq 0$ be chosen as small as possible. Since $1/0$ is not defined, $n_0 = 1$.

- "Consider the sequence $1/1$, $1/2$, $1/3$, $1/4, \ldots$" It's clear what the terms of this sequence are, however no domain has been specified. There are an infinite number of possibilities. Here are three.

$$n_0 = 0 \text{ and } a_n = \tfrac{1}{n+1} \qquad n_0 = 1 \text{ and } a_n = \tfrac{1}{n} \qquad n_0 = 37 \text{ and } a_n = \tfrac{1}{n-36}$$

The first choice makes n_0 as small as possible. The second choice makes a_n as simple as possible, which may be convenient. The third choice is because we like the number 37. Which is correct? They all are — but use one of the first two approaches since the third only confuses people. Since we haven't specified a function by saying $1/1$, $1/2$, $1/3$, $1/4, \ldots$, why do we consider this to be a sequence? Often it's the terms in the sequence that are important, so any way you make it into a function is okay.

People sometimes define an infinite sequence to be an infinite list.

Sometimes, we will specify an infinite sequence that way, too.

• "Given the sequence a_n, consider the sequence $a_0, a_2, a_4, \ldots$ of the even terms." As just discussed, $a_0, a_2, a_4, \ldots$ specifies a sequence from the list point of view. We should have said "the terms with even subscripts" rather than "the even terms;" however, people seldom do that. ◻

The next definition may sound strange at first, but you will get used to it.

Definition 4 (Limit of a sequence) *Let a_n, $n \geq n_0$, be an infinite sequence. We say that a real number A is the* limit *of a_n as n goes to infinity and write*

$$\lim_{n \to \infty} a_n = A$$

if, for every real number $\epsilon > 0$, there exists N_ϵ such that for all $n \geq N_\epsilon$, $|a_n - A| \leq \epsilon$.

We often omit "as n goes to infinity and simply say "A is the limit of the sequence a_n."

If a sequence a_n has a real number A as a limit, we say that the sequence *converges* to A. If a sequence does not converge, we say that it *diverges*.

Since Definition 4 refers only to a_n with $n \geq N_\epsilon$ and since N_ϵ can be as large as we wish, we only need to look at tails of sequences. We state this as a theorem and omit the proof.

Theorem 4 (Convergence and tails) *Let a_n, $n \geq n_0$, be an infinite sequence. The following are equivalent*

• *The sequence a_n converges.*

• *Every tail of the sequence a_n converges.*

• *Some tail of the sequence a_n converges.*

The theorem tells us that we can ignore any "inconvenient" terms at the beginning of a sequence when we are checking for convergence.

Example 8 (What does the Definition 4 mean?) It helps to have some intuitive feel for the definition of the limit of a sequence. We'll explore it here and in the next example.

The definition says a_n will be as close as you want to A if n is large enough. Note that the definition does not say that A is unique — perhaps a sequence could have two limits A and A^*. Since a_n will be as close as you want to A and also to A^* at the same time if n is large, we must have $A = A^*$. (If you don't see this, draw a picture where a_n is within $|A - A^*|/3$ of both A and A^*.) Since $A = A^*$ whenever A and A^* are limits of the same sequence, the limit is unique. We state this as a theorem:

Theorem 5 (The limit is unique) *An infinite sequence has at most one limit. In other words, if the limit of an infinite sequence exists, it is unique.*

Induction, Sequences and Series

Here's another way to picture the limit of an infinite sequence. Imagine that you are in a room sitting at a desk. You have with you a sequence a_n, $n = 0, 1, 2, \ldots$, that you have announced converges to a number A. Every now and then, there is a knock on the door and someone enters the room and gives you positive real number ϵ (like $\epsilon = 0.001$). You must give that person an integer $N_\epsilon > 0$ such that for all $n \geq N_\epsilon$, $|a_n - A| \leq \epsilon$. If you can do that, the person will go away contented. If you are able to convincingly prove that for any such $\epsilon > 0$ there is such an N_ϵ, then they will leave you alone because you are right in asserting that A is the limit of the sequence a_n, as n goes to infinity.

We can phrase the condition for A to be the limit of the sequence in logic notation:

$$\forall \, \epsilon > 0, \; \exists \, N_\epsilon, \; \forall \, n \geq N_\epsilon, \; |a_n - A| \leq \epsilon. \qquad \square$$

Suppose we know that a sequence $a_0, a_1, \ldots$ has a limit A and we want to estimate A. We can do this by computing a_n for large values of n. Of course, estimating the limit A only makes sense if we know the sequence has a limit. How can we know that the sequence has a limit? By Definition 4 of course! Unfortunately, Definition 4 requires that we know the value of A.

What can we do about this? We'd like to know that a limit exists without knowing the value of that limit. How can that be? Let's look at it intuitively. The definition says all the values of a_n are near A when n is large. But if they are all near A, then a_n and a_m must be near each other when n and m are large. (You should be able to see why this is so.) What about the converse; that is, if all the values of a_n and a_m are near each other when n and m are large are they near some A which is the limit of the sequence? We state the following theorem without proof.

Theorem 6 (Second "definition" of a convergent sequence)   *Let a_n, $n \geq n_0$, be an infinite sequence.*

The sequence a_n, $n \geq n_0$, converges to some limit A
if and only if
for every real number $\epsilon > 0$ there is an N_ϵ such that for all $n, m \geq N_\epsilon$, $|a_n - a_m| \leq \epsilon$.

In other words, if the terms far out in the sequence are as close together as we wish, then the sequence converges.

Some students misunderstand the definition and think we only need to show that $|a_n - a_{n+1}| \leq \epsilon$ for $n \geq N_\epsilon$. *Don't fall into this trap.* The sequence $a_n = \log n$ shows that we can't do that because $\log n$ grows without limit but $|\log n - \log(n+1)| = \log(1 + 1/n)$ which can be made as close to zero as you want by making n large enough.

Most beginning students have little patience with the formal precision of Definition 4 and Theorem 6. If you look at a particular example such as the sequence $\frac{2n+1}{n+1}$, $n = 0, 1, 2, \ldots$, it is obvious that, as n goes to infinity, this sequence approaches $A = 2$ as a limit. So why confuse the obvious with such formality? The reason is that we need the precise definition of a limit is to enable us to discuss convergent sequences in general, independent of particular examples such as $\frac{2n+1}{n+1}$, $n \geq 0$. Without such formal definitions, we couldn't state general theorems precisely and proofs would be impossible.

Example 9 (Convergence from three viewpoints) Let's take a look at the convergence of $a_n = \frac{2n+1}{n+1}$, $n = 0, 1, 2, \ldots$ from three different points of view.

- First, we can manipulate the terms to see that they converge: Since

$$\frac{2n+1}{n+1} = \frac{2+1/n}{1+1/n}, \quad \lim_{n\to\infty}(2+1/n) = 2 \quad \text{and} \quad \lim_{n\to\infty}(1+1/n) = 1,$$

we have

$$\lim_{n\to\infty}\frac{2n+1}{n+1} = \lim_{n\to\infty}\frac{2+1/n}{1+1/n} = \frac{\lim_{n\to\infty}(2+1/n)}{\lim_{n\to\infty}(1+1/n)} = 2/1 = 2.$$

- Second, using Definition 4, given $\epsilon > 0$, choose $N_\epsilon = 1/\epsilon$. Then, if $n \geq N_\epsilon$,

$$|a_n - 2| = \left|\frac{2n+1}{n+1} - 2\right| = \left|\frac{-1}{n+1}\right| = \frac{1}{n+1} < \frac{1}{n} \leq \frac{1}{N_\epsilon} = \epsilon.$$

- Third, using Theorem 6, given $\epsilon > 0$, choose $N_\epsilon = \frac{2}{\epsilon}$. We have

$$|a_n - a_m| = \left|\frac{2n+1}{n+1} - \frac{2m+1}{m+1}\right| = \left|\left(2 - \frac{1}{n+1}\right) - \left(2 - \frac{1}{m+1}\right)\right| = \left|\frac{1}{m+1} - \frac{1}{n+1}\right|.$$

But, since $|x - y| \leq |x| + |y|$,

$$\left|\frac{1}{m+1} - \frac{1}{n+1}\right| \leq \frac{1}{m+1} + \frac{1}{n+1} < \frac{1}{N_\epsilon} + \frac{1}{N_\epsilon} = \frac{2}{N_\epsilon} = \epsilon.$$

The easiest method for showing convergence of a particular sequence is usually the first method. You may wonder about our values of N_ϵ in the other two methods:

- *How did we find them?* We found them by working from both ends. To illustrate, consider the third method. Suppose $n \geq N_\epsilon$ and $m \geq N_\epsilon$ but we don't know what to choose for N_ϵ. We found that $|a_n - a_m| < 2/N_\epsilon$. We want to know how to choose N_ϵ so that $|a_n - a_m| \leq \epsilon$. You should be able to see that it will be okay if $2/N_\epsilon \leq \epsilon$. Thus we need $N_\epsilon \geq 2/\epsilon$.

- *Would other values work?* Yes. If someone comes up with a value that works, then any larger value of N_ϵ would also work because it tells us to ignore more of the earlier values in the sequence. □

In Definition 4, we said that, if a sequence a_n, $n \geq n_0$, does not converge then it is said to diverge. So far we haven't looked at any examples. Here are two.

- The infinite sequence is $a_n = (-1)^n$ alternates between $+1$ and -1. It clearly fails our definition and theorem on convergence. For example, the theorem fails with any $0 < \epsilon < 2$. There is no N_ϵ such that for all $m, n \geq N_\epsilon$, $|a_n - a_m| \leq \epsilon$, since $|a_n - a_{n+1}| = 2$ for all $n \geq 0$.

- Another example of a divergent sequence is $b_n = \log n$, $n \geq 1$. Although

$$\lim_{n\to\infty}|b_n - b_{n+1}| \to 0,$$

Induction, Sequences and Series

$|b_n - b_{2n}| = \log 2$ and so the theorem fails for any $\epsilon < \log 2$.

The sequences a_n and b_n of the previous paragraph differ in a fundamental way, as described by the following definition.

Definition 5 (Bounded sequence)   *A sequence a_n, $n = 0, 1, 2, \ldots$ is bounded if there exists a positive number B such that $|a_n| \leq B$ for $n = 0, 1, 2, \ldots$.*

The sequence $a_n = (-1)^n$ is an example of a bounded divergent sequence. The sequence $b_n = \log n$ is an example of an unbounded divergent sequence. All the convergent sequences we have looked at are bounded. The next theorem shows that there are no unbounded convergent sequences.

Theorem 7 (Boundedness)   *Convergent sequences are bounded.*

Proof: Let a_n, $n \geq n_0$, be convergent with limit A. Take $\epsilon = 1$. Then there is an N_1 such that for all $n \geq N_1$, $|a_n - A| \leq 1$. Since a_n is within 1 of A, it follows that $|a_n| \leq |A| + 1$ for all $n \geq N_1$. Let B be the maximum of $|a_{n_0}|, |a_{n_0+1}|, |a_{n_0+2}|, \ldots, |a_{N_1-1}|$, and $|A| + 1$. Then, $|a_n| \leq B$ for $n \geq n_0$. $\square$

The converse of the previous theorem is, "Bounded sequences are convergent." This statement is false ($a_n = (-1)^n$ for example).

The next theorem gives some elementary rules for working with sequences.

Theorem 8 (Algebraic rules for sequences)   *Suppose that a_n, $n \geq n_0$ and b_n, $n \geq n_0$ are convergent sequences and that*

$$\lim_{n \to \infty} a_n = A \quad \text{and} \quad \lim_{n \to \infty} b_n = B.$$

Define sequences t_n, r_n, s_n, p_n and q_n, $n \geq n_0$, by

$$t_n = \alpha a_n + \beta, \quad \alpha, \beta \in \mathbb{R}; \qquad s_n = a_n + b_n;$$
$$p_n = a_n b_n; \qquad\qquad\qquad \text{and, if } b_n \neq 0 \text{ for all } n \geq n_0, \quad q_n = a_n/b_n.$$

Then

$$\lim_{n \to \infty} t_n = \alpha A + \beta, \quad \lim_{n \to \infty} s_n = A + B, \quad \lim_{n \to \infty} p_n = AB$$

and, if $B \neq 0$, $\lim_{n \to \infty} q_n = A/B$.

Proof: All we are given is that the sequences a_n and b_n converge. This means that $|a_n - A|$ and $|b_n - B|$ are small when n is large. The proof technique is to use that fact to show that other values are small. We illustrate the technique by proving the assertion about p_n. We omit the proofs for t_n, s_n and q_n.

We must show that we can make $|a_n b_n - AB|$ small. Thus, we need to relate $a_n - A$ and $b_n - B$ to $a_n b_n - AB$. An obvious idea is to try multiplying $a_n - A$ and $b_n - B$.

Unfortunately, the product is not of the right form, so we need to be more clever. After some experimentation, you might notice that

$$a_n b_n - AB = a_n(b_n - B) + B(a_n - A)$$

and that the parenthesized expressions are small. This is the key! We have

$$|a_n b_n - AB| = |a_n(b_n - B) + B(a_n - A)| \le |a_n||b_n - B| + |B||a_n - A|.$$

By Theorem 7, there is a constant A^* such that $|a_n| \le A^*$ for all n. Thus

$$|a_n b_n - AB| \le A^*|b_n - B| + |B||a_n - A|.$$

This says that, for all large n, $|a_n b_n - AB|$ is at most a constant (A^*) times a small number $(|b_n - B|)$ plus a constant times another small number. If we were being informal in our proof, we could stop here. However, a formal proof requires that we tell how to compute N_ϵ for the sequence $a_n b_n$.

We find the rule for N_ϵ by, in effect, working backwards. For $\delta > 0$, let N_δ^* be such that $|a_n - A| \le \delta$ and $|b_n - B| \le \delta$ for all $n \ge N_\delta^*$. We can do this because a_n and b_n converge. Now we have

$$|a_n b_n - AB| \le A^*|b_n - B| + |B||a_n - A| \le A^*\delta + |B|\delta = (A^* + |B|)\delta.$$

Since we want this to be at most epsilon, we define δ by $(A^* + |B|)\delta = \epsilon$. Thus $\delta = \epsilon/(A^* + |B|)$ and so $N_\epsilon = N_{\epsilon/(A^* + |B|)}^*$. $\quad\blacksquare$

An important class of sequences are those which are "eventually monotone," a concept we now define.

Definition 6 (Monotone sequence)   *A sequence a_n, $n \ge n_0$, is*

- *increasing if $a_{n_0} < a_{n_0+1} < a_{n_0+2} < \cdots$,*
- *decreasing if $a_{n_0} > a_{n_0+1} > a_{n_0+2} > \cdots$,*
- *nondecreasing if $a_{n_0} \le a_{n_0+1} \le a_{n_0+2} \le \cdots$,*
- *nonincreasing if $a_{n_0} \ge a_{n_0+1} \ge a_{n_0+2} \ge \cdots$,*
- *monotone if it is either nonincreasing or nondecreasing.*

If a tail of the sequence is monotone, we say the sequence is eventually monotone. We define "eventually increasing" and so on similarly.

Nonincreasing is also called "weakly decreasing" and nondecreasing is also called "weakly increasing." If you understand the definition, you should see the reason for this terminology.

Eventually monotone sequences are fairly common and have nice properties. The following theorem gives one property.

Theorem 9 (Convergence of bounded monotone sequences)   *If an infinite sequence is bounded and eventually monotone, then it converges.*

Induction, Sequences and Series

We won't prove this theorem. It is, in a very basic sense, a fundamental property of real numbers. We leave the understanding of this theorem to your intuition. The power of the theorem is in its generality so that it can be applied in discussing sequences in general as well as to discussing specific examples.

We now study three common classes of eventually monotone functions and their relative rates of growth.

Example 10 (Polynomials, exponentials and logarithms) Consider the sequence a_n, $n = 0, 1, 2, \ldots$, where $a_n = n/1.1^n$. It is a fact that you probably learned in high school, and certainly learned if you have had a course in calculus, that any exponential function $f(x) = b^x$, $b > 1$, "grows faster" than any polynomial function $g(x) = c_k x^k + \ldots + c_1 x + c_0$. By this we mean that

$$\lim_{x \to \infty} g(x)/f(x) = 0 \quad \text{when} \quad g(x) = c_k x^k + \ldots + c_1 x + c_0, \quad f(x) = b^x \quad \text{and} \quad b > 1.$$

If for example, we take the sequence a_n, $n = 0, 1, 2, \ldots$, where $a_n = n^3/2^n$, we get $a_0 = 0$, $a_1 = 1/5$, $a_2 = 2.25$, $a_3 = 3.375$, $a_4 = 4$, and $a_5 = 3.90625$. Some calculations may convince you that $a_4 > a_5 > a_6 > \cdots$, and so the sequence is eventually decreasing.

Recall from high school that the inverse function of the function b^x is the function $\log_b(x)$. That these functions are inverses of each other means that $b^{\log_b(x)} = \log_b(b^x) = x$ for all $x > 0$. It is particularly important that all computer science students understand the case $b = 2$ as well as the usual $b = e$ (the "natural log") and $b = 10$. You should graph 2^x, for $-1 \le x \le 5$ and $\log_2(x)$ for $0.5 \le x \le 32$. You can compute $\log_2(x)$ on your calculator using the LN key: $\log_2(x) = LN(x)/LN(2)$ (or you can use the LOG key instead of LN). Note that $\log_2(x)$ is also written $\lg(x)$. Here are typical graphs for $b > 1$.

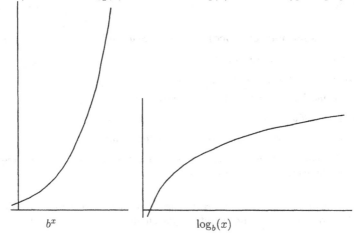

b^x $\log_b(x)$

Notice that, although both b^x and $\log_b(x)$ get arbitrarily large as x gets arbitrarily large, b^x grows *much* more rapidly than $\log_b(x)$. In fact, $\log_b(x)$ grows so slowly that, for any $\alpha > 0$

$$\lim_{x \to \infty} \log_b(x)/x^\alpha = 0.$$

For example,

$$\lim_{x \to \infty} \log_b(x)/x^{0.01} = 0.$$

For those of you who have had some calculus, you can prove the above limit is correct by using l'Hospital's Rule. If you haven't had calculus, you can do some computations with your computer or calculator to get a feeling for this limit. For example, if $b = 2$ then

$$\log_2(2^{10})/2^{0.10} = 9.33033 \quad \log_2(2^{100})/2^{1.0} = 50 \quad \log_2(2^{1000})/2^{10.0} = 0.976563.$$

If for example, we take the sequence a_n, $n = 0, 1, 2, \ldots$, where $a_n = \log_2(n)/n^{0.01}$, we will find that the sequence increases at first. But, starting at some (rather large) m, we have $a_m > a_{m+1} > a_{m+2} > \cdots$. These terms will continue to get smaller and smaller and approach zero as a limit. The sequence is eventually decreasing.

These are examples of general results such as:

If $b > 1$, $c > 0$ and $d > 0$, then n^c/b^{n^d} and $(\log_b(n^d))/n^c$ are eventually monotonic sequences that converge to zero.

We omit the proof. One can replace n^c and n^d by more general functions of n.

People may write log without specifying a base as in $\log_b$. What do they mean? Some people mean $b = 10$ and others mean $b = e$. Still others mean that it doesn't matter what value you choose for b as long as it's the same throughout the discussion. That's what we mean — if there's no base on the logarithm, choose your favorite $b > 1$. $\square$

We conclude our discussion of sequences with a discussion of "converges to infinity."

In Definition 4, we defined what it means for a sequence to have a real number A as its limit. We also find in many mathematical discussions, the statement that "a_n, $n = 0, 1, 2, \ldots$ has limit $+\infty$" or "a_n, $n = 0, 1, 2, \ldots$ has limit $-\infty$." Alternatively, one sees "a_n, $n = 0, 1, 2, \ldots$ tends to $+\infty$, converges to $+\infty$, or diverges to $+\infty$." In symbols,

$$\lim_{n \to \infty} a_n = +\infty \quad \text{or} \quad \lim_{n \to \infty} a_n = -\infty.$$

This use of "limit" is really an abuse of the term. Such sequences are actually divergent sequences, but they diverge with a certain consistency. Thus, $a_n = n$, $n = 0, 1, 2, \ldots$ or $a_n = -n$, $n = 0, 1, 2, \ldots$, though divergent, are said to "have limit $+\infty$" or "have limit $-\infty$," respectively. Compare this with the divergent sequence $a_n = (-1)^n n$, $n = 0, 1, 2, \ldots$, which hops around between ever increasing positive and negative values. Here is a formal definition.

Definition 7 (Diverges to infinity) *Let a_n, $n \geq n_0$ be an infinite sequence. We say that the sequence converges to $+\infty$ or that it diverges to $+\infty$ and write*

$$\lim_{n \to \infty} a_n = +\infty$$

if, for every real number $r > 0$, there exists N_r such that for all $n \geq N_r$, $a_n \geq r$.

Similarly, we say that the sequence converges to $-\infty$ or that it diverges to $-\infty$ and write

$$\lim_{n \to \infty} a_n = -\infty$$

if, for every real number $r < 0$, there exists N_r such that for all $n \geq N_r$, $a_n \leq r$.

Exercises for Section 2

2.1. For each of the following sequences, answer the following questions.
- Is the sequence bounded?
- Is the sequence monotonic?
- Is the sequence eventually monotonic?

(a) $a_n = n$ for all $n \geq 0$.

(b) $a_n = 1$ for all $n \geq 0$.

(c) $a_n = 2n + (-1)^n$ for all $n \geq 0$.

(d) $a_n = n + (-1)^n 2$ for all $n \geq 0$.

(e) $a_n = 2^n - 10n$ for all $n \geq 0$.

(f) $a_n = 10 - 2^{-n}$ for all $n \geq 0$.

2.2. Discuss the convergence or divergence of the following sequences:

(a) $\frac{2n^3 + 3n + 1}{3n^3 + 2}$, $n = 0, 1, 2, \ldots$

(b) $\frac{-n^3 + 1}{2n^2 + 3}$, $n = 0, 1, 2, \ldots$

(c) $\frac{(-n)^n + 1}{n^n + 1}$, $n = 0, 1, 2, \ldots$

(d) $\frac{n^n}{(n/2)^{2n}}$, $n = 1, 2, \ldots$

2.3. Discuss the convergence or divergence of the following sequences:

(a) $\frac{\log_2(n)}{\log_3(n)}$, $n = 1, 2, \ldots$

(b) $\frac{\log_2(\log_2(n))}{\log_2(n)}$, $n = 2, 3, \ldots$

*Section 3: Infinite Series

We now look at infinite series. Every infinite series is associated with two infinite sequences. Thus the study of infinite series can be thought of as the study of sequences. However, the viewpoint is different.

Definition 8 (Infinite series)   Let a_n, $n \geq n_0$, be an infinite sequence. Define a new sequence s_n, $n \geq n_0$, by

$$s_n = a_{n_0} + a_{n_0+1} + \cdots + a_n = \sum_{k=n_0}^{n} a_k.$$

The infinite sequence s_n is called the *sequence of partial sums* of the sequence a_n. We call a_n a *term* of the series.

If $\lim_{n \to \infty} s_n$ exists, we write

$$\sum_{k=n_0}^{\infty} a_k = \lim_{n \to \infty} s_n.$$

We call $\sum_{k=n_0}^{\infty} a_k$ the infinite series whose terms are the a_k and whose sum is $\lim_{n \to \infty} s_n$. We say the infinite series converges to $\lim_{n \to \infty} s_n$.

If $\lim_{n \to \infty} s_n$ does not exist, we still speak of the infinite series $\sum_{k=n_0}^{\infty} a_k$, but now we say that the series diverges and that it has no sum. If s_n diverges to $+\infty$ or to $-\infty$, we say that the infinite series diverges to $+\infty$ or to $-\infty$.

The infinite series associated with a tail of a sequence, is a tail of the infinite series associated with the sequence. In this case, mathematical notation is clearer than words: If $t \geq n_0$, then

$$\sum_{k=t}^{\infty} a_k \quad \text{is a tail of} \quad \sum_{k=n_0}^{\infty} a_k.$$

So where are we? Given an infinite sequence a_n, $n \geq n_0$, we can ask whether the infinite series $\sum_{k=n_0}^{\infty} a_k$ converges. This is the same as asking whether the sequence of partial sums converges. So what's new? There are often situations where we know something about the terms a_n and are interested in the sum of the series. For example, what can be said about the value of $\sum_{k=1}^{\infty} 1/k$? the value of $\sum_{k=0}^{\infty} (-1)^k/k!$? We get to see the terms, but we're interested in the sum. Thus, we want to use information about the infinite sequence a_n to say something about the infinite sequence s_n of partial sums. This presence of *two* sequences is what makes the study of infinite series different from the study of a single sequence. Here's a simple example of that interplay:

Theorem 10 (Terms are small)   *If the infinite series $\sum_{n=n_0}^{\infty} a_n$ converges, then* $\lim_{n \to \infty} a_n = 0$.

Proof: We are given that the infinite series converges, which means that the sequence $s_n = \sum_{k=n_0}^{n} a_n$ converges. We use Theorem 6 with $m = n - 1$ and a_n in the theorem replaced by s_n. By Theorem 6, whenever n is large enough

$$\epsilon \geq |s_n - s_m| = |s_n - s_{n-1}| = |a_n| = |a_n - 0|.$$

Since ϵ can be made as close to zero as we wish, this proves that $\lim_{n \to \infty} |a_n - 0| = 0$. Therefore a_n converges to zero. $\square$

Induction, Sequences and Series

Example 11 (Geometric series) For $r \in \mathbb{R}$, let $a_n = r^n$, $n \geq 0$. The partial sum s_n associated with a_n is called a *geometric series*. Note that, from high school mathematics,

$$s_n = \sum_{k=0}^{n} r^k = \begin{cases} \frac{r^{n+1}-1}{r-1} & \text{if } r \neq 1, \\ n+1 & \text{if } r = 1. \end{cases}$$

If $|r| \geq 1$, the infinite series $\sum_{k=0}^{\infty} r^k$ diverges by Theorem 10. If $|r| < 1$ then

$$\lim_{n \to \infty} s_n = \sum_{k=0}^{\infty} r^k = \frac{1}{1-r}.$$

For example, when $r = 2/3$, we have $\sum_{k=0}^{\infty} (2/3)^k = 3$. $\blacksquare$

Example 12 (Harmonic series) A basic infinite series, denoted by H_n, is the one that is associated with the sequence $a_n = 1/n$, $n = 1, 2, \ldots$. Let $H_n = a_1 + \cdots + a_n$ denote the partial sums of this series. The sequence H_n, $n = 1, 2, \ldots$, is called the *harmonic series* (for reasons that any of you who have studied music will know). In infinite series notation, this series can be represented by

$$\sum_{n=1}^{\infty} \frac{1}{n}.$$

We can visualize this series by grouping its terms as follows:

$$\underbrace{\frac{1}{1}}_{b_0} + \underbrace{\frac{1}{2} + \frac{1}{3}}_{b_1} + \underbrace{\frac{1}{4} + \frac{1}{5} + \frac{1}{6} + \frac{1}{7}}_{b_2} + \underbrace{\frac{1}{8} + \frac{1}{9} + \frac{1}{10} + \frac{1}{11} + \frac{1}{12} + \frac{1}{13} + \frac{1}{14} + \frac{1}{15}}_{b_3} + \cdots.$$

Note that b_k contains the terms

$$\frac{1}{2^k} \quad \frac{1}{2^k + 1} \quad \cdots \quad \frac{1}{2^{k+1} - 1} = \frac{1}{2^k + (2^k - 1)}$$

and so contains 2^k terms. Which b_k is $\frac{1}{11}$ in? Easy. Just take $\lfloor \log_2(11) \rfloor = 3$ and you get the answer, b_3. In general, $\frac{1}{n}$ is in b_k where $k = \lfloor \log_2(n) \rfloor$.

What is a lower bound for the sum of all the numbers in b_3? Easy. They are all bigger than $\frac{1}{16}$, the first number in b_4. There are 8 numbers in b_3, all bigger than $\frac{1}{16}$, so a lower bound is $b_3 > 8 \times \frac{1}{16} = \frac{1}{2}$. You can do this calculation in general for group b_k, getting $b_k > \frac{1}{2^{k+1}} \times 2^k = \frac{1}{2}$. Now that you are getting a feeling for this grouping, you can see that an upper bound for the sum of the terms in b_k is $\frac{1}{2^k} \times 2^k = 1$. Thus

$$\frac{1}{2} \leq b_k \leq 1.$$

Now suppose you pick an integer n and want to get an estimate on the size of H_n. To get a lower bound just find the k such that b_k contains the term $1/n$. (By our earlier work, $k = \lfloor \log_2(n) \rfloor$.) Then

$$H_n > b_0 + b_1 + \cdots + b_{k-1} > k/2 \quad \text{and} \quad H_n \leq b_0 + b_1 + \cdots + b_k \leq k+1.$$

Using our value for k and the fact that $x - 1 < \lfloor x \rfloor \leq x$, we have

$$\frac{\log_2(n) - 1}{2} < H_n \leq \log_2(n) + 1.$$

We learned in Example 10 that $\log_2(n)$ is a very slowly growing function of n. But it does get arbitrarily large (has limit $+\infty$). Thus, H_n grows very slowly and diverges.

There is more to the story of the harmonic series. Although the derivations are beyond the scope of our study, the results are worth knowing. Here is a very interesting way of representing H_n:

$$H_n = \ln(n) + \gamma + \frac{1}{2n} - \frac{1}{12n^2} + \frac{\epsilon_n}{120n^4} \quad \text{where} \quad 0 < \epsilon_n < 1.$$

The "ln" refers to the natural logarithm. It is a special function key on all scientific calculators. To ten decimal places, $\gamma = 0.5772156649$.

Your first reaction might be, "What good is this formula, we don't know ϵ_n exactly?" Since $\epsilon_n > 0$, we'll get a number that is less than H_n if we throw away $\epsilon_n/120n^4$. Since $\epsilon_n < 1$, we'll get a number that is greater than H_n if we replace $\epsilon_n/120n^4$ with $1/120n^4$. These upper and lower bounds for H_n are quite close together — they differ by $1/120n^4$. With $n = 10$ we have upper and lower bounds that differ by only $1/1200000 = 0.0000008333\ldots$. For example, by adding up the terms we get $H_{10} = 2.928968254$ to nine decimal places. The lower bound gotten with $\epsilon_{10} = 0$ is 2.928967425 and the upper bound gotten with $\epsilon_{10} = 1$ is 2.928968258. Get the idea? No matter what value ϵ_n takes in the interval from 0 to 1, the denominator $120n^4$ grows rapidly with n, so the error is small. $\square$

Example 13 (Alternating harmonic series)   Let h_n be the sequence of partial sums associated with the sequence $(-1)^{n-1}/n$ for $n \geq 1$. The series h_n is called the *alternating harmonic* series. What about the infinite series

$$\sum_{n=1}^{\infty} \frac{(-1)^{n-1}}{n}?$$

It converges. To see why, imagine that you are standing in a room with your back against the wall. Imagine that you step forward 1 meter, then backwards 1/2 meter, then forwards 1/3 meter, etc. After n such steps, your distance from the wall is h_n meters. By the time you are stepping backwards one millimeter, forwards 0.99 millimeter, etc., an observer in the room (who by now has decided that you are crazy) would conclude that you are standing still. In other words, you have converged. It turns out your position doesn't converge to infinity because your forward and backward motions practically cancel each other out. How can we see this? Each pair of forward–backward steps moves you a little further from the wall; e.g., $1 - \frac{1}{2} = \frac{1}{2}$, $\frac{1}{3} - \frac{1}{4} = \frac{1}{12}$. Thus you never have to step through the wall. (All partial sums are positive.) On the other hand, after first stepping forward 1 meter, each following pair of backward–forward steps moves you a little closer to the the wall; e.g., $-\frac{1}{2} + \frac{1}{3} = \frac{-1}{6}$, $-\frac{1}{4} + \frac{1}{5} = \frac{-1}{20}$. Thus you are never further than 1 meter from the wall.

This argument works just as well for any size steps as long as they are decreasing in size towards zero and are alternating forward and backwards. In the case of the alternating harmonic series, your distance from the wall will converge to $\ln(2)$, meters, where "ln" is the natural logarithm. We won't prove this fact, as it is best proved using calculus. You can check this out on your calculator or computer by adding up a lot of terms in the series. □

A series is called *alternating* if the terms alternate in sign; that is, the sign pattern of terms is $+ - + - \cdots$ or $- + - + \cdots$.

Example 14 (Some particular alternating series and variations) By taking particular sequences a_n that converge monotonically to zero, you get particular alternating series. Here are some examples of alternating convergent series:

$$\sum_{n=1}^{\infty}(-1)^n\frac{1}{\sqrt{n}} \qquad \sum_{n=2}^{\infty}(-1)^n\frac{1}{\ln(n)} \qquad \sum_{n=3}^{\infty}(-1)^n\frac{1}{\ln(\ln(n))}.$$

It is an interesting fact about such series that the sequence $(-1)^n$ in the above examples can be replaced by any sequence b_n which has bounded partial sums. Of course, $(-1)^n$, $n = k, k+1, \ldots$, has bounded partial sums for any starting value k (bounded by $B = 1$). For example, it can be shown that $b_n = \sin(n)$ and $b_n = \cos(n)$ are sequences with bounded partial sums.[2] Thus,

$$\sum_{n=1}^{\infty}\sin(n)\frac{1}{n} \quad \text{and} \quad \sum_{n=0}^{\infty}\cos(n)\frac{1}{\ln(n)}$$

are convergent generalized "alternating" series. The fact that these generalized "alternating" series converge is proved in more advanced courses and called *Dirichlet's Theorem*. □

Example 15 (Series and the integral test) Suppose we have a function $f(x)$ that is defined for all $x \geq m$ where $m \geq 0$ is an integer. Then we can associate with $f(x)$ a sequence $a_n = f(n)$, $n \geq m$. In summation notation, $\sum_{n=m}^{\infty} a_n$ is an infinite series, and we are interested in the divergence or convergence of this series. Suppose that $f(x)$ is weakly decreasing for all $x \geq t$ where $t \geq m$. Study the pictures shown below. If the area under the curve is infinite, as intended in the first picture, then the summation $\sum_{k=t}^{\infty} a_k$, which represents the sum of the areas of the rectangles, must also be infinite.

If the area under the curve is finite, as in the second picture, then the summation $\sum_{k=t}^{\infty} a_k$, which represents the sum of the areas of the rectangles, must also be finite.

---

[2] Here is how it's done for those of you who are familiar with complex numbers and Euler's relation. From Euler's relation, $\cos(n) = \Re(e^{in})$ and so

$$\sum_{n=0}^{N}\cos(n) = \Re\left(\sum_{n=0}^{N}(e^i)^n\right) = \Re\left(\frac{e^{i(N+1)} - 1}{e^i - 1}\right).$$

Since the numerator is bounded and the denominator is constant, this is bounded.

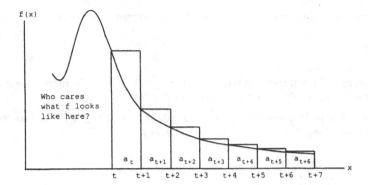

If $a_k = f(k)$, $k \geq t$, then $\displaystyle\int_t^\infty f(x)\,dx = +\infty$ implies $\displaystyle\sum_{k=t}^\infty a_k$ diverges.

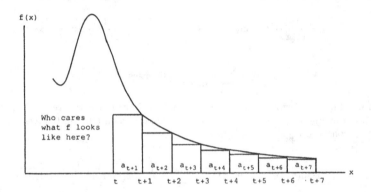

If $a_k = f(k)$, $k \geq t$, then $\displaystyle\int_t^\infty f(x)\,dx < +\infty$ implies $\displaystyle\sum_{k=t}^\infty a_k$ converges.

In one or the other of the two cases, we conclude that a tail of the given series diverges or converges and, thus, that the given series diverges or converges.

This way of checking for convergence and/or divergence is called the *integral test*. $\square$

Example 16 (General harmonic series)   We can extend the harmonic series H_n with terms $\frac{1}{n}$ to a series $H_n^{(r)}$ based on the sequence $\frac{1}{n^r}$, where r is a real number. We call the series the *general harmonic series with parameter r*. In summation notation, this series is

$$\sum_{n=1}^\infty \frac{1}{n^r}.$$

[173]

Induction, Sequences and Series

If $r \leq 0$ then it is obvious that $H_n^{(r)}$, $n = 1, 2, \ldots$, diverges. For example, $r = -1$ gives the series

$$\sum_{n=1}^{\infty} n,$$

which diverges. If $r > 0$ then the function $f_r(x) = \frac{1}{x^r}$ is strictly decreasing for $x \geq 1$. This means that we can apply the integral test with $t = 1$.

From calculus, it is known that $\int_1^{\infty} (1/x^r)\, dx = +\infty$ if $r \leq 1$. It is also known that $\int_1^{\infty} (1/x^r))\, dx = \frac{1}{r-1}$ if $r > 1$. Thus, by the integral test,

$$\sum_{n=1}^{\infty} \frac{1}{n^r} \quad \text{diverges if } 0 < r \leq 1 \text{ and converges if } r > 1. \qquad \square$$

The integral test can produce some surprises. The harmonic series H_n, based on $\frac{1}{n}$, $n = 1, 2, \ldots$, diverges. But what about the series s_n, $n = 2, 3, \ldots$, based on $\frac{1}{n\ln(n)}$? The terms of that series get smaller faster, so maybe it converges? Applying the integral test gives

$$\int \frac{1}{x\ln(x)}\, dx = \ln(\ln(x)) + C \quad \text{so} \quad \int_2^{\infty} \frac{1}{x\ln(x)}\, dx = +\infty.$$

Thus

$$\sum_{n=2}^{\infty} \frac{1}{n\ln(n)} \quad \text{diverges}.$$

It looks like $\ln(n)$ just doesn't grow fast enough to help make the terms $1/n\ln(n)$ small enough for convergence. So using $\ln(n)$ twice probably won't help. It gives the series s_n, $n = 2, 3, \ldots$, based on $\frac{1}{n(\ln(n))^2}$. We have

$$\int \frac{1}{x(\ln(x))^2}\, dx = \frac{-1}{\ln(x)} + C \quad \text{so} \quad \int_2^{\infty} \frac{1}{x(\ln(x))^2}\, dx < \frac{1}{\ln(2)}.$$

Thus

$$\sum_{n=2}^{\infty} \frac{1}{n(\ln(n))^2} \quad \text{converges!}$$

In fact, if $\delta > 0$, then

$$\sum_{n=2}^{\infty} \frac{1}{n(\ln(n))^{1+\delta}} \quad \text{converges}.$$

You should prove this by using the integral test.

Definition 9 (Absolute convergence) *Let s_n, $n = 0, 1, 2, \ldots$, be a series based on the sequence a_n, $n = 0, 1, 2, \ldots$. Let t_n, $n = 0, 1, 2, \ldots$, be a series based on the sequence $|a_n|$, $n = 0, 1, 2, \ldots$. If the series t_n converges then the series s_n is said to converge absolutely or to be absolutely convergent. In other words,*

$$\sum_{n=0}^{\infty} a_n \quad \text{converges absolutely if} \quad \sum_{n=0}^{\infty} |a_n| \quad \text{converges}.$$

IS-26 [174]

If a series is convergent, but not absolutely convergent, then it is called *conditionally convergent*.

Any geometric series with $|r| < 1$ is absolutely convergent. The alternating harmonic series is convergent but not absolutely convergent (since the harmonic series diverges).

Theorem 11 (Absolute convergence and bounded sequences)   *Suppose that s_n, $n \geq n_0$ is an absolutely convergent series based on the sequence a_n, $n \geq n_0$. Let b_n, $n \geq n_0$ be a bounded sequence. Then the series p_n, $n \geq n_0$, based on the sequence $a_n b_n$, $n \geq n_0$, is absolutely convergent. In other words,*

$$\sum_{n=n_0}^{\infty} a_n \text{ converges absolutely and } b_n \text{ bounded}\quad \text{implies}\quad \sum_{n=n_0}^{\infty} a_n b_n \text{ converges absolutely.}$$

Proof: Let $M > 0$ be a bound for b_n. Thus, $M \geq |b_n|$, $n \geq n_0$. Since a_n is absolutely convergent, given $\epsilon > 0$, there exists $N_{\epsilon/M}$ such that for all $i \geq j \geq N_{\epsilon/M}$, $|a_{j+1}| + |a_{j+2}| + \cdots |a_i| \leq \epsilon/M$. Given $\epsilon > 0$, let $N_\epsilon = N_{\epsilon/M}$. Then, for all $i \geq j \geq N_\epsilon$,

$$|a_{j+1}||b_{j+1}| + |a_{j+2}||b_{j+2}| + \cdots |a_i||b_i| \leq (|a_{j+1}| + |a_{j+2}| + \cdots |a_i|)M \leq (\epsilon/M)M = \epsilon.$$

This shows that p_n is absolutely convergent. $\square$

Example 17 (Series convergence and using your intuition) Based on the ideas we have studied thus far, you can develop some very powerful intuitive ideas that will correctly tell you whether or not a series converges. We discuss these without proof. The basic idea is to look at a constant C times a convergent or divergent series: $C \sum_{n=0}^{\infty} a_n = \sum_{n=0}^{\infty} C a_n$. Then think about what conditions on a sequence b_n, will allow you to replace the constant C on the right hand side by b_n to get $\sum_{n=0}^{\infty} b_n a_n$ without changing the convergence or divergence of the series. Here are some specific examples:

(1) Suppose the series $\sum_{n=0}^{\infty} a_n$ converges absolutely. An example is $a_n = r^n$, $0 \leq r < 1$ (i.e., the geometric series). If you have a bounded sequence b_n, $n = 0, 1, 2, \ldots$ then you can replace C to get $\sum_{n=0}^{\infty} b_n a_n$ and still retain absolute convergence. This was proved in Theorem 11. An example is

$$\sum_{n=0}^{\infty} (1 + \sin(n)) r^n, 0 \leq r < 1.$$

Note that, b_n can be any convergent sequence (which is necessarily bounded). One way this situation arises in practice is that you are given a series such as

$$\sum_{n=1}^{\infty} \frac{2n+1}{n^3+1}.$$

You notice that the terms $\frac{2n+1}{n^3+1}$ can be written $\frac{2+1/n}{n^2+1/n}$ and thus, for large n, the original series should be very similar to the terms of the series

$$\sum_{n=1}^{\infty} \frac{2}{n^2},$$

[175]

which converges absolutely (general harmonic series with parameter 2). Thus, the original series with terms $\frac{2n+1}{n^3+1}$ converges absolutely. Here is an explanation based on absolute convergence and bounded sequences. Start with the absolutely convergent series $\sum_{n=1}^{\infty} \frac{1}{n^2}$. Here, $a_n = n^{-2}$. Let $c_n = (2n+1)/(n^3+1)$. The $\lim_{n\to\infty} c_n/a_n = 2$. By our previous discussion, with $b_n = c_n/a_n$,

$$\sum_{n=0}^{\infty} b_n a_n = \sum_{n=1}^{\infty} \frac{2n+1}{n^3+1}$$

converges absolutely.

(2) Suppose the series $\sum_{n=n_0}^{\infty} a_n$ converges (but perhaps only conditionally). In that case, you can replace the constant C by any eventually monotonic convergent sequence b_n. In this case, $\sum_{n=n_0}^{\infty} a_n b_n$ converges. This result is proved in more advanced courses and called *Abel's Theorem*. For example, take the alternating series

$$\sum_{n=1}^{\infty} C \frac{(-1)^n}{\sqrt{n}},$$

which converges by Example 14. Replace C with $b_n = (1 + 1/\sqrt{n})$ which is weakly decreasing, converging to 1:

$$\sum_{n=1}^{\infty} \left(1 + \frac{1}{\sqrt{n}}\right) \frac{(-1)^n}{\sqrt{n}}.$$

The monotonicity of b_n is *important*. If we replace C with $b_n = (1 + (-1)^n/\sqrt{n})$ which converges to 1 but is not monotonic. We obtain

$$S = \sum_{n=1}^{\infty} \left(1 + \frac{(-1)^n}{\sqrt{n}}\right) \frac{(-1)^n}{\sqrt{n}} = \sum_{n=1}^{\infty} \left(\frac{(-1)^n}{\sqrt{n}} + \frac{1}{n}\right).$$

Since

$$\sum_{n=1}^{\infty} \frac{(-1)^n}{\sqrt{n}} \quad \text{converges and} \quad \sum_{n=1}^{\infty} \frac{1}{n} \quad \text{diverges,}$$

S diverges. $\square$

We conclude this section by looking at the question "How common are primes?" What does this mean? Suppose the primes are called p_n so that $p_1 = 2$, $p_2 = 3$, $p_3 = 5$, $p_4 = 7$, $p_5 = 11$ and so on. We might ask for an estimate of p_n. It turns out that p_n is approximately $n \ln n$. In fact, the *Prime Number Theorem*, states that the ratio of p_n and $n \ln n$ approaches 1 as n goes to infinity. The proof of the theorem requires much more background in number theory and much more time than is available in this course.

It might be easier to look at $p_1 + \cdots + p_n$. Indeed it is, but it is still too hard for this course.

It turns out that things are easier if we can work with an infinite sum. Of course $p_1 + \cdots + p_n + \cdots$ diverges to infinity because there are an infinite number of primes, so that sum is no help. What about summing the reciprocals:

$$\sum_{n=1}^{\infty} \frac{1}{p_n}?$$

Now we're onto something useful that is within our abilities!

- If the primes are not very common, we might expect $p_n \geq Cn^{1+\delta}$ for some $\delta > 0$ and some C. In that case, $\sum 1/p_n$ converges because $\sum 1/p_n \leq C \sum 1/n^{1+\delta}$ and this general harmonic series converges by Example 16.

- On the other hand, if the primes are fairly common, then $\sum 1/p_n$ might diverge because $\sum 1/n$ diverges.[3]

How can we study $\sum 1/p_n$? The key is unique factorization.

Imagine that it made sense to talk about the infinite sum $1 + 2 + 3 + 4 + \cdots$. We claim that then

$$1 + 2 + 3 + 4 + \cdots = (1 + 2 + 2^2 + 2^3 + \cdots)(1 + 3 + 3^2 + \cdots)(1 + 5 + 5^2 + \cdots)\cdots,$$

where the factors on the right are sums of powers of primes. Why is this? Imagine that you multiply this out using the distributive law. Let's look at some number, say $300 = 2^2 \times 3 \times 5^2$. We get it by taking 2 from $1 + 2 + 2^2 + \cdots$, 3^2 from $1 + 3 + 3^2 + \cdots$, 5^2 from $1 + 5 + 5^2 + \cdots$ and 1 from each of the remaining factors $1 + p + p^2 + \cdots$ for $p = 7, 11, \ldots$. This is the only way to get 300 as a product. In fact, by unique factorization, each positive integer is obtained exactly once this way.

Instead, suppose we do this with the reciprocals, remembering that $p^0 = 1$. We have

$$\frac{1}{1} + \frac{1}{2} + \frac{1}{3} + \frac{1}{4} + \cdots = \left(\frac{1}{p_1^0} + \frac{1}{p_1^1} + \frac{1}{p_1^2} + \cdots\right)\left(\frac{1}{p_2^0} + \frac{1}{p_2^1} + \frac{1}{p_2^2} + \cdots\right)\cdots.$$

Each of the series in parentheses is a geometric series and so it can be summed. In fact

$$\frac{1}{p_n^0} + \frac{1}{p_n^1} + \frac{1}{p_n^2} + \cdots = \frac{1}{1 - 1/p_n} = \frac{p_n}{p_n - 1} = 1 + \frac{1}{p_n - 1}.$$

We can't give a proof in this way because the series we started with is the harmonic series, which diverges, and we don't have tools for dealing with divergent series. As a result, we work backwards and give a proof by contradiction.

Suppose that $\sum 1/p_n$ converges. Since the terms are positive, it converges absolutely. We now introduce a mysterious sequence b_n. (Actually the values of b_n were found by continuing with the incorrect approach in the previous paragraph.) Let

$$b_n = p_n \log\left(1 + \frac{1}{p_n - 1}\right).$$

---

[3] In fact, $\sum 1/p_n$ diverges because p_n behaves like $n \ln n$ (which we can't prove) and $\sum 1/n \ln n$ diverges by Example 16.

We claim b_n is bounded. This can be proved easily by l'Hôpital's Rule, but we omit the proof since we have not discussed l'Hôpital's Rule. Let $a_n = 1/p_n$. Remember that we are assuming $\sum 1/p_n$ converges. By Theorem 11, $\sum a_n b_n$ converges. By the previous paragraph, $a_n b_n$ is the logarithm of

$$\frac{1}{p_n^0} + \frac{1}{p_n^1} + \frac{1}{p_n^2} + \cdots.$$

Hence, again by the previous paragraph, $\sum a_n b_n$ is the logarithm of the harmonic series. Since $\sum a_n b_n$ converges, so does the harmonic series. This is a contradiction.

Since we reached a contradiction by assuming that $\sum 1/p_n$ converges, it follows that $\sum 1/p_n$ diverges and so the primes are fairly common. How close are we to the Prime Number Theorem (p_n behaves like $n \ln n$)? If p_n grew much faster than this, say $p_n > Cn(\ln n)^{1+\delta}$ for some C and some $\delta > 0$, then $\sum 1/p_n$ would converge because $\sum 1/n(\ln n)^{1+\delta}$ converges by Example 16. But we've just shown that $\sum 1/p_n$ diverges.

Exercises for Section 3

3.1. Discuss the convergence or divergence of the following series:

$$\text{(a)} \sum_{n=1}^{\infty} \frac{2^{n/2}}{n^2 + n + 1} \qquad \text{(b)} \sum_{n=1}^{\infty} \frac{n+1}{2n+1}$$

3.2. Discuss the convergence or divergence of the following series:

$$\text{(a)} \sum_{n=1}^{\infty} \frac{n^5}{5^n} \qquad \text{(b)} \sum_{n=1}^{\infty} \frac{1}{n^2 - 150}$$

3.3. Discuss the convergence or divergence of the following series:

$$\text{(a)} \sum_{n=1}^{\infty} \frac{1}{(n^3 - n^2 - 1)^{1/2}} \qquad \text{(b)} \sum_{n=1}^{\infty} \frac{(n+1)^{1/2} - (n-1)^{1/2}}{n}$$

3.4. Discuss the convergence or divergence of the following series:

$$\text{(a)} \sum_{n=1}^{\infty} \frac{(-1)^n}{n} \left(1 + \frac{1}{2^2} + \frac{1}{3^2} + \cdots + \frac{1}{n^2} \right)$$

$$\text{(b)} \sum_{n=1}^{\infty} \frac{(-1)^n}{n} \left(1 + \frac{1}{2} + \frac{1}{3} + \cdots + \frac{1}{n} \right)$$

3.5. Discuss the convergence or divergence of the following series:

$$\text{(a)} \sum_{n=0}^{\infty} \frac{\sin(n)}{|n - 99.5|} \qquad \text{(b)} \sum_{n=0}^{\infty} (-1)^n \frac{-9n^2 - 5}{n^3 + 1}$$

Multiple Choice Questions for Review

In each case there is one correct answer (given at the end of the problem set). Try to work the problem first without looking at the answer. Understand both why the correct answer is correct and why the other answers are wrong.

1. Which of the following sequences is described, as far as it goes, by an explicit formula $(n \geq 0)$ of the form $g_n = \lfloor \frac{n}{k} \rfloor$?

   (a) 0000111122222

   (b) 001112223333

   (c) 000111222333

   (d) 0000011112222

   (e) 0001122233444

2. Given that $k > 1$, which of the following sum or product representations is **WRONG**?

   (a) $(2^2 + 1)(3^2 + 1) \cdots (k^2 + 1) = \prod_{j=2}^{k}[(j+1)^2 - 2j]$

   (b) $(1^3 - 1) + (2^3 - 2) + \cdots + (k^3 - k) = \sum_{j=1}^{k-1}[(k-j)^3 - (k-j)]$

   (c) $(1 - r)(1 - r^2)(1 - r^3) \cdots (1 - r^k) = \prod_{j=0}^{k-1}(1 - r^{k-j})$

   (d) $\frac{1}{2!} + \frac{2}{3!} + \frac{3}{4!} + \cdots + \frac{k-1}{k!} = \sum_{j=2}^{k} \frac{j-1}{j!}$

   (e) $n + (n-1) + (n-2) + \cdots + (n-k) = \sum_{j=1}^{k+1}(n - j + 1)$

3. Which of the following sums is gotten from $\sum_{i=1}^{n-1} \frac{i}{(n-i)^2}$ by the change of variable $j = i + 1$?

   (a) $\sum_{j=2}^{n} \frac{j-1}{(n-j+1)^2}$

   (b) $\sum_{j=2}^{n} \frac{j-1}{(n-j-1)^2}$

   (c) $\sum_{j=2}^{n} \frac{j}{(n-j+1)^2}$

   (d) $\sum_{j=2}^{n} \frac{j}{(n-j-1)^2}$

   (e) $\sum_{j=2}^{n} \frac{j+1}{(n-j+1)^2}$

4. We are going to prove by induction that $\sum_{i=1}^{n} Q(i) = n^2(n+1)$. For which choice of $Q(i)$ will induction work?

   (a) $3i^2 - 2$ (b) $2i^2$ (c) $3i^3 - i$ (d) $i(3i - 1)$ (e) $3i^3 - 7i$

5. The sum $\sum_{k=1}^{n}(1 + 2 + 3 + \cdots + k)$ is a polynomial in n of degree

   (a) 3   (b) 1   (c) 2   (d) 4   (e) 5

6. We are going to prove by induction that for all integers $k \geq 1$,

$$\sqrt{k} \leq \frac{1}{\sqrt{1}} + \frac{1}{\sqrt{2}} + \cdots + \frac{1}{\sqrt{k}}.$$

[179]

IS-31

Induction, Sequences and Series

Clearly this is true for $k = 1$. Assume the Induction Hypothesis (IH) that $\sqrt{n} \leq \frac{1}{\sqrt{1}} + \frac{1}{\sqrt{2}} + \cdots \frac{1}{\sqrt{n}}$. Which is a correct way of concluding this proof by induction?

(a) By IH, $\frac{1}{\sqrt{1}} + \frac{1}{\sqrt{2}} + \cdots \frac{1}{\sqrt{n+1}} \geq \sqrt{n} + \frac{1}{\sqrt{n+1}} = \sqrt{n+1} + 1 \geq \sqrt{n+1}$.

(b) By IH, $\frac{1}{\sqrt{1}} + \frac{1}{\sqrt{2}} + \cdots \frac{1}{\sqrt{n+1}} \geq \sqrt{n+1} + \frac{1}{\sqrt{n+1}} \geq \sqrt{n+1}$.

(c) By IH, $\frac{1}{\sqrt{1}} + \frac{1}{\sqrt{2}} + \cdots \frac{1}{\sqrt{n+1}} \geq \sqrt{n} + 1 \geq \sqrt{n+1}$.

(d) By IH, $\frac{1}{\sqrt{1}} + \frac{1}{\sqrt{2}} + \cdots \frac{1}{\sqrt{n+1}} \geq \sqrt{n} + \frac{1}{\sqrt{n}} \geq \frac{\sqrt{n}\sqrt{n+1}}{\sqrt{n}} \geq \frac{n+1}{\sqrt{n+1}} = \sqrt{n+1}$.

(e) By IH, $\frac{1}{\sqrt{1}} + \frac{1}{\sqrt{2}} + \cdots \frac{1}{\sqrt{n+1}} \geq \sqrt{n} + \frac{1}{\sqrt{n+1}} = \frac{\sqrt{n}\sqrt{n+1}+1}{\sqrt{n+1}} \geq \frac{\sqrt{n}\sqrt{n+1}}{\sqrt{n+1}} = \frac{n+1}{\sqrt{n+1}} = \sqrt{n+1}$.

7. Suppose $b_1, b_2, b_3, \cdots$ is a sequence defined by $b_1 = 3$, $b_2 = 6$, $b_k = b_{k-2} + b_{k-1}$ for $k \geq 3$. Prove that b_n is divisible by 3 for all integers $n \geq 1$. Regarding the induction hypothesis, which is true?

(a) Assuming this statement is true for $k \leq n$ is enough to show that it is true for $n+1$ and no weaker assumption will do since this proof is an example of "strong induction."

(b) Assuming this statement is true for n and $n-1$ is enough to show that it is true for $n+1$.

(c) Assuming this statement is true for n, $n-1$, and $n-3$ is enough to show that it is true for $n+1$ and no weaker assumption will do since you need three consecutive integers to insure divisibility by 3.

(d) Assuming this statement is true for n is enough to show that it is true for $n+1$.

(e) Assuming this statement is true for n and $n-3$ is enough to show that it is true for $n+1$ since 3 divides n if and only if 3 divides $n-3$.

8. Evaluate $\lim\limits_{n\to\infty} \dfrac{(-1)^{n^3} n^3 + 1}{2n^3 + 3}$.

    (a) $-\infty$ (b) $+\infty$ (c) Does not exist.   (d) $+1$ (e) -1

9. Evaluate $\lim\limits_{n\to\infty} \dfrac{\log_5(n)}{\log_9(n)}$.

    (a) $\ln(9)/\ln(5)$ (b) $\ln(5)/\ln(9)$ (c) $5/9$ (d) $9/5$ (e) 0

10. Evaluate $\lim\limits_{n\to\infty} \dfrac{\cos(n)}{\log_2(n)}$.

    (a) Does not exist.   (b) 0 (c) $+1$ (d) -1 (e) $+\infty$

*11. The series $\displaystyle\sum_{n=1}^{\infty} \dfrac{(-1)^n n^{500}}{(1.0001)^n}$

IS-32   [180]

(a) converges absolutely.

(b) converges conditionally, but not absolutely.

(c) converges to $+\infty$

(d) converges to $-\infty$

(e) is bounded but divergent.

*12. The series $\displaystyle\sum_{n=1}^{\infty} \frac{(-1)^n}{\sqrt{n}}\left(1+\frac{1}{n^2}\right)$.

(a) is bounded but divergent.

(b) converges absolutely.

(c) converges to $+\infty$

(d) converges to $-\infty$

(e) converges conditionally, but not absolutely.

Answers: **1** (c), **2** (b), **3** (a), **4** (d), **5** (a), **6** (e), **7** (b), **8** (c), **9** (a), **10** (b), **11** (a), **12** (e).

IS-33

Solutions for Boolean Functions and Computer Arithmetic

BF-1.1 The idea of this problem is to show how English phrases are translated into logical expressions.

    (a) $f \wedge s$. In English, "but" is "and" with an underlying message: The use of "but" in this way often (but not always) indicates surprise. You might say, "The animal is a fish *and* it lives in the water." $(F \wedge W)$ Or you might say "The animal is a fish *but* it lives on dry land." $(F \wedge L)$ Logic doesn't make a fuss over surprises.

    (b) Either of the equivalent functions $\sim(f \vee s)$ and $(\sim f) \wedge (\sim s)$. Some people think of "Neither A nor B nor ..." as "None of A and B and ...," which is the first form. Other people think of "Neither A nor B nor ..." as "Not A and not B and ...," which is the second form.

BF-1.2 $r \wedge \sim v$. Same idea as the previous exercise. It may or may not be a surprise since most registered voters don't vote, but we don't need to know that to write it in logic notation.

BF-1.3 It is helpful to include intermediate columns in the table to help with the computation of $f(p,q) = \sim\big((p \wedge q) \vee \sim(p \vee q)\big)$. In this case, we have included columns for $p \wedge q$ $p \vee q$ and $\sim(p \vee q)$. With more practice, less columns are needed.

p	q	$p \wedge q$	$p \vee q$	$\sim(p \vee q)$	f
0	0	0	0	1	0
0	1	0	1	0	1
1	0	0	1	0	1
1	1	1	1	0	0

BF-1.4

p	q	r	$q \vee \sim r$	$\sim p \wedge (q \vee \sim r)$
0	0	0	1	1
0	0	1	0	0
0	1	0	1	1
0	1	1	1	1
1	0	0	1	0
1	0	1	0	0
1	1	0	1	0
1	1	1	1	0

BF-1.5 Before making a truth table, it may help to simplify the expression. Using the associative law $p \vee (\sim p \vee q) = (p \vee \sim p) \vee q = 1 \vee q = 1$. Thus

$$\big(p \vee (\sim p \vee q)\big) \wedge \sim(q \wedge \sim r) = 1 \wedge \sim(q \wedge \sim r) = \sim(q \wedge \sim r) = \sim q \vee r,$$

where the last is by DeMorgan's Law. Now we are ready to make the table. We don't even need to include p since it does not enter into the final function!

q	r	$\sim q \vee r$
0	0	1
0	1	1
1	0	0
1	1	1

Solutions for Boolean Functions and Computer Arithmetic

BF-1.6 Let $m =$ "Mary is a musician" and let $c =$ "Mary plays chess." The statement is $m \wedge c$ and its negation is $\sim(m \wedge c) = \sim m \vee \sim c$ by DeMorgan's law. Now put the final statement into words: "Either Mary is not a musician or she does not play chess."

BF-1.7 Let $g =$ "The car has gas" and let $f =$ "The fuel line is plugged." The statement is $\sim g \vee f$. Its negation is $\sim(\sim g \vee f) = g \wedge \sim f$. In words, "The car has gas and the fuel line isn't plugged."

You could have taken $g =$ "The car is out of gas". The statement and its negation would be $g \vee f$ and $\sim(g \vee f) = \sim g \wedge \sim f$. In words, "The car is not out of gas and the fuel line isn't plugged." There is a double negative in "is not out of gas," which you could simplify to "has gas."

BF-1.8 Here is the standard beginners way of doing this: $p \vee (p \wedge q) = (p \vee p) \wedge (p \vee q)$ by the distributive rule. This is $p \wedge (p \vee q)$ by the idempotent rule. This becomes p by the absorption rule. This is perfectly correct.

We assume from now on that you can look up or memorize the names of the rules, and we do not require you to write down the names each time you use a rule. Generally, you need only write the steps, showing the changes in the forms of functions that result from the basic rules. Here is all you need to write for this problem:

$$p \vee (p \wedge q) = (p \vee p) \wedge (p \vee q) = p \wedge (p \vee q) = p.$$

There are often different ways to apply the basic rules to reduce one function to another. If done correctly, they are all "full credit." The goal is clarity. If you feel that certain steps are made clearer by including the names of the rules (distributive, associative, etc.) then include them. If you combine two short steps and want to indicate that (e.g, associative law and distributive law) do so. It is up to you to be clear and correct.

You don't need to make up a truth table for equal functions when you can reduce one to the other using algebraic manipulation. However, unless you are specifically asked for an algebraic proof, you can give a truth table proof.

BF-1.9 No. For $h(p,q,r) = (p \wedge q) \vee r$ and $g(p,q,r) = p \wedge (q \vee r)$ we have $h(0,0,1) = 1$ and $g(0,0,1) = 0$, so the functions are *not* equal. **But wait** — they seem to be equal by the associative law. What's wrong with that?

BF-1.10 No. For $h(p,q,r) = (p \vee q) \vee (p \wedge r)$ and $g(p,q,r) = (p \vee q) \wedge r$ we have $h(1,0,0) = 1$ and $g(1,0,0) = 0$, so the functions are *not* equal. Note also that $h(p,q,r) = p \vee q$. Show this and explain why this makes it easy to see that $h(p,q,r) \neq g(p,q,r)$.

BF-1.11 If no choice of variables comes to mind, one can simplify functions and then either look at them and see the situation or compute truth tables. We leave the truth tables to you and take the algebraic approach. We want to simplify $f(p,q,r) = ((\sim p \vee q) \wedge (p \vee \sim r)) \wedge (\sim p \vee \sim q)$ and then see where we are. Since $f(p,q,r)$ is of the form $A \wedge B \wedge C$ — parentheses not neeeded because of the associative law — where A, B and C involve "ors," a good strategy is to use the distributive laws to rearrange the "ands" and "ors" and use DeMorgan's law as needed. Also note that the order of A, B and C in $A \wedge B \wedge C$ does not matter because of the commutative law. Which two of the three possibilities (namely $\sim p \vee q$, $p \vee \sim r$ and $\sim p \vee \sim q$) should we combine first? In the end, it doesn't matter since it will all lead to the same answer. The easiest is $(\sim p \vee q) \wedge (\sim p \vee \sim q)$, which you should be able to simplify to $\sim p \vee (q \wedge \sim q) = \sim p$ with the distributive law.

Solutions for Boolean Functions and Computer Arithmetic

Thus we have $f(p,q,r) = \sim p \wedge (p \vee \sim r)$, which, with the distributive law, becomes $(\sim p \wedge p) \vee (\sim p \wedge \sim r) = \sim p \wedge \sim r$. This is equal to the other function by DeMorgan's law. The key to solving it this way was to keep using the distributive law and simplifying expressions such as $\sim p \wedge p$ that arose along the way.

BF-1.12 If we want to use the algebraic method, this may be another case for the algebraic method, just like the previous problem. We have $(r \vee p) \wedge (r \vee q) = r \vee (p \wedge q)$. Thus the first function in the problem is

$$\big(\sim r \vee (p \wedge q)\big) \wedge \big(r \vee (p \wedge q)\big) = (\sim r \wedge r) \vee (p \wedge q) = p \wedge q.$$

Thus, they are the same.

BF-1.13 Yes. Write the first function as $(\sim p \wedge q) \vee (\sim p \wedge \sim q)$ with the help of DeMorgan's law. Now you should be able to use the distributive law.

BF-1.14 No. Since there are only two variables, a truth table will have only four rows, so that is probably the quickest way for you to do it. You could try algebraic simplification. You should try those two methods just for the practice. Here's another trick. What happens when $q = 0$? when $q = 1$? With $q = 0$, the first function becomes

$$\sim\big((\sim p \wedge 0) \vee (\sim p \wedge 1)\big) \vee (p \wedge 0) = \sim(0 \vee \sim p) \vee 0 = p,$$

which is not $\sim p$. This leads to a possible choice for p and q: either $(p,q) = (0,0)$ or $(p,q) = (1,0)$. Since the first function simplifies to p and the second function is $\sim p$, we'll get different values. You can try $q = 1$ and see what happens.

BF-1.15 No. The solution to the previous exercise gives three approaches. You should try all three. For the algebraic approach, it's easiest to first use DeMorgan's law on $\sim(\sim p \vee q)$.

BF-2.1 These problems are routine: First write the information in the truth table as a Boolean function as done in the proof of Theorem 1, then perhaps simplify the function, and finally construct a circuit for the function. Many circuits are possible, depending on the final form of the function.

(a) Directly from the truth table we have the function

$$(\sim P \wedge Q \wedge \sim R) \vee (P \wedge \sim Q \wedge \sim R) \vee (P \wedge \sim Q \wedge R).$$

The last two terms can be combined using the distributive law:

$$(\sim P \wedge Q \wedge \sim R) \vee (P \wedge \sim Q).$$

Allowing a 3-input **and** gate, we can represent it with the following circuit.

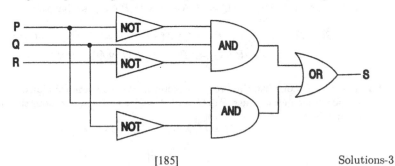

Solutions for Boolean Functions and Computer Arithmetic

(b) From the truth table we have

$$(P \wedge Q \wedge {\sim}R) \vee ({\sim}P \wedge Q \wedge R) \vee ({\sim}P \wedge {\sim}Q \wedge R).$$

Combining the last two parenthesized expressions reduces this to $(P \wedge Q \wedge {\sim}R) \vee ({\sim}P \wedge R)$.

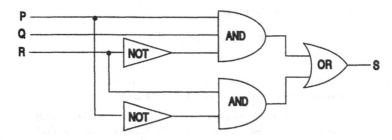

BF-2.2 It's simpler to construct ${\sim}S$ and then negate it. This gives us

$$S = {\sim}\big((P \wedge Q \wedge R) \vee ({\sim}P \wedge {\sim}Q \wedge {\sim}R)\big) = \big({\sim}(P \wedge Q \wedge R)\big) \wedge (P \vee Q \vee R),$$

where we used DeMorgan's rule.

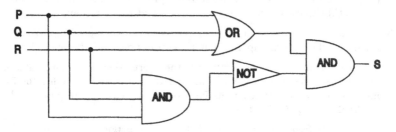

BF-2.3 In general, with k switches such that moving any one changes the state of the lights, the function is $s_1 \oplus s_2 \oplus \cdots \oplus s_k$. The associative and commutative rules hold for $\oplus$ just as they do for $\wedge$ and $\vee$, so we can rearrange and parenthesize this expression any way we wish.

BF-2.4 We could compute tables of the two functions or we could manipulate them algebraically. We'll let you construct tables. The first circuit computes $S(P,Q) = (P \wedge Q) \vee (P \oplus Q)$. Using $P \oplus Q = (P \wedge {\sim}Q) \vee ({\sim}P \wedge Q)$, we have

$$S(P,Q) = (P \wedge Q) \vee (P \wedge {\sim}Q) \vee ({\sim}P \wedge Q) = \big(P \wedge (Q \vee {\sim}Q)\big) \vee ({\sim}P \wedge Q)$$
$$= P \vee ({\sim}P \wedge Q) = (P \vee {\sim}P) \wedge (P \vee Q) = P \vee Q.$$

BF-2.5 This can be done like the previous exercise. We leave the tabular method to you. For the algebraic method, use $P \oplus Q = (P \wedge {\sim}Q) \vee ({\sim}P \wedge Q)$ to note that the first circuit computes

$$(P \vee Q) \wedge \big((P \wedge {\sim}Q) \vee ({\sim}P \wedge Q)\big),$$

which you should simplify to $(P \wedge \sim Q) \vee (\sim P \wedge Q)$.

Here is another approach. Look at the first circuit. When the result of the **or** is 1, the result of the **and** will be the result of the **xor**. Thus the circuit computes $P \oplus Q$ except possibly when $P = Q = 0$. This case is easily checked.

BF-2.6 You are asked to show that $(\sim P \wedge \sim Q) \vee (P \oplus Q) = \sim(P \wedge Q)$. Having done the previous exercises, you should be able to do this. Another way to do it is to notice that, once you write $\sim(P \wedge Q) = \sim P \vee \sim Q$ and $P \oplus Q = \sim P \oplus \sim Q$, this is Exercise 2.4 with P and Q replaced by $\sim P$ and $\sim Q$.

BF-2.7 From the truth table, the function is

$$S = (P \wedge \sim Q \wedge \sim R) \vee (P \wedge \sim Q \wedge R) \vee (P \wedge Q \wedge R)$$
$$= \big((P \wedge \sim Q \wedge \sim R) \vee (P \wedge \sim Q \wedge R)\big) \vee \big((P \wedge \sim Q \wedge R) \vee (P \wedge Q \wedge R)\big)$$
$$= (P \wedge \sim Q) \vee (P \wedge R).$$

This can be built with an **or** gate, an **and** gate and a gate the computes $f(x,y) = x \wedge \sim y$. You might object that this requires a "nonstandard" gate and so you've been tricked. We can get a solution with standard gates:

$$S = (P \wedge \sim Q \wedge \sim R) \vee (P \wedge \sim Q \wedge R) \vee (P \wedge Q \wedge R)$$
$$= P \wedge \big((\sim Q \wedge \sim R) \vee (\sim Q \wedge R) \vee (Q \wedge R)\big)$$
$$= P \wedge (\sim R \vee Q),$$

where the last step omitted some manipulation. We can get this directly from the truth table. First note that $S = P \wedge f(Q, R)$ for some function f. Since the truth table for f contains three ones and only one zero, f can be written as an **or**. Since $f(Q, R) = 0$ only when $Q = 1$ and $R = 0$, it is the **or** of $\sim Q$ and R. The function $P \wedge (\sim R \vee Q)$ requires only three standard gates. If we allow nonstandard gates, we can get by with two — one to compute f and an **and** gate.

BF-2.8 1011101. You can use the standard "human" subtraction procedure, or you can use two's-complement arithmetic, making sure the register is big enough so that 1110100 appears to be positive. We can do that with eight bits. Then 1110100 is 01110100 and the two's-complement of 00010111 is 11101001. Adding these gives 01011101, with a carry of 1 discarded.

BF-2.9 $B7C5_{16} = 133705_8$.

BF-2.10 (a) $61502_8 = 6 \times 8^4 + 1 \times 8^3 + 5 \times 8^2 + 2 = 25410$.

(b) $EB7C5_{16} = 1110\,1011\,0111\,1100\,0101_2 = 11\,101\,011\,011\,111\,000\,101_2 = 3533705_8$.

BF-2.11 (a) *jhecmnwdyh*   (b) *study − hard*

BF-2.12 First method: $67_{10} = 01000111_2$. The two's complement is 10111101.

Second method: By definition, the 8-bit two's complement is $2^8 - 67 = 189$. Covert 189 to binary to obtain 10111101.

BF-2.13 $108_{10} = 01101100_2$ (using 8 bits total). Using our algorithm for computing, we fix the 100 pattern of bits on the right and complement all others to get 10010100.

Solutions for Boolean Functions and Computer Arithmetic

BF-2.14 First method: Start with 10001001. Using the two's complement algorithm this converts to 01110111 which is $64 + 32 + 16 + 4 + 2 + 1 = 119$ and so $k = 119$.
Second method: $10000100l_2 = 128 + 8 + 1 = 137$. It's two's complement in an 8-bit register is $2^8 - 137 = 119$.

BF-2.15 The two's complement of the given number is 01000110, which equals $2^6 + 2^2 + 2^1 = 70$. Thus the original number is -70. Equivalently, 10111010, without considering two's complement, is $2^7 + 2^5 + 2^4 + 2^3 + 2^1 = 186$. Because it is 8-bit two's complement, it represents $2^8 - 186 = 70$.

BF-2.16 $79_{10} = 1001111_2$ and $43_{10} = 101011_2$. The calculations:

Two's complement

1001111 regular		01001111 8-bit
-101011 arithmetic	00101011	11010101 register
100100	11010101	**1**00100100

The boldface **1** is a carry off the end of the register, which we discard because we are combining a positive and negative number.

BF-2.17 This proceeds much like the previous exercise, with a couple of changes. We do just the two's-complement arithmetic. We want $(-15) + (-46)$. Since $15_{10} = 1111_2$, its two's complement is 11110001 Since $46_{10} = 101110_2$, its two's complement is 11010010. Adding gives us 1100011, where we have discarded the leftmost carry bit. (This is the two's complement of 00111101, which is 61.)
This could have been done by writing it as $-(15 + 46)$. We would do the addition and take the two's complement of the result.

Now for $46 + 46 + 46$. From the previous work, the register for 46 is 00101110. Adding this to itself, we get 01011100. Since we added two positive numbers and the result is positive, there has been no overflow. Adding 00101110 to this we obtain 10001010. Since we added two positive numbers and the result is negative, there was overflow.

BF-2.18 The n-bit two's complement of x is $2^n - x$, which is obtained by counting backwards from 2^n. (Remember the clock — time before the hour is $60 - minutes$.) Similarly, the n-bit ten's complement of x is $10^n - x$. Is there a short-cut way to compute it. Yes. Scan from right to left, stopping at the first nonzero digit. Subtract that digit from 10 and all digits to the left from 9. Thus the 8-digit ten's complement of 67834000 is 32166000. To do subtraction such as $71121333 - 67834000$, you can do it in the usual way, or you add the ten's complement: $71121333 + 32166000 = 103287333$. A carry into the ninth digit should be discarded since we're doing 8-digit ten's-complement arithmetic. Thus the answer is 3287333. You should give a more complete explanation of why this works and you should compare it carefully to the two's complement to explain the analogies in more detail.

Solutions for Logic

Lo-1.1 We noted that exclusive or is seldom used in logic. In set theory, it corresponds to the symmetric difference.

Lo-1.2 "But" means "and." It usually indicates that what follows "but" is surprising given what came before "but."

(a) $h \wedge w \wedge \sim s$.

(b) $\sim w \wedge (h \wedge s)$, which can be rearranged by the commutative law if you wish.

(c) $\sim h \wedge \sim w \wedge \sim s$ or $\sim(h \vee w \vee s)$, which is equivalent by DeMorgan's law.

Lo-1.3 $(n \vee k) \wedge \sim(n \wedge k)$. Two other possible forms are $(n \vee k) \wedge (\sim n \vee \sim k)$ and $(k \wedge \sim n) \vee (\sim k \wedge n)$. Can you show symbolically that they are equivalent? What about exclusive or? Can't we also write $n \oplus k$? Yes and no. Thinking in terms of Boolean functions, this is fine; however, $\oplus$ is seldom used in logic.

Lo-1.4 (a) $p \wedge q \wedge r$ (i.e. all three occur).

(b) $p \wedge \sim q$ (ZIP can be anything).

(c) $p \wedge (\sim q \vee \sim r)$ — same as $p \wedge \sim(q \wedge r)$ and $(p \wedge \sim q) \vee (p \wedge \sim r)$. Note that "however" is used in the same way as "but."

(d) $\sim p \wedge q \wedge \sim r$ or $\sim(p \vee \sim q \vee r)$ (Do you see why?)

(e) $\sim p \vee (p \wedge q)$

Lo-1.5 One can construct a truth table or manipulate the statement form algebraically. We choose the latter approach. Note that

$$\sim p \vee (p \wedge \sim q) \Leftrightarrow (\sim p \vee p) \wedge (\sim p \vee \sim q) \Leftrightarrow \sim p \vee \sim q \Leftrightarrow \sim(p \wedge q).$$

With $S = p \wedge q$, the statement form becomes $S \vee \sim S \vee r$, which is a tautology.

Here is another way to look at it. With $r = 1$, the value of the statement form is 1, so it cannot be a contradiction. With $r = 0$, the value of the statement form is $(p \wedge q) \vee (\sim p \vee (p \wedge \sim q))$. One can construct a truth table for this or, use the previous manipulations to see that it is equivalent to $S \vee \sim S$.

Lo-1.6 One can use any of the ideas in the previous problem. Using the algebraic approach:

$$(p \wedge \sim q) \wedge (\sim p \vee q) \Leftrightarrow p \wedge (\sim q \wedge (\sim p \vee q)) \Leftrightarrow p \wedge ((\sim q \wedge \sim p) \vee (\sim q \vee q))$$
$$\Leftrightarrow p \wedge (\sim q \wedge \sim p) \Leftrightarrow \sim q \wedge (p \wedge \sim p) \Leftrightarrow 0.$$

Thus the original statement form is equivalent to $0 \wedge r$, a contradiction.

Lo-1.7 Again we use the algebraic method. You may ask, "Why don't you ever use a truth table?" Truth tables are a mechanical approach and you should be able to use them without any help. With the algebraic method, there are many choices. While all algebraic simplifications will eventually lead to the answer, some choices do so much more quickly than others. By showing which choices lead quickly to solutions, we hope you'll gain some ability to make such choices.

$$((\sim p \wedge q) \wedge (q \vee r)) \wedge \sim q \wedge r \Leftrightarrow \sim p \wedge q \wedge (q \vee r) \wedge \sim q \wedge r.$$

Solutions for Logic

Combining the q and $\sim q$, we obtain 0. Since we have a bunch of things joined by "and," the entire statement form becomes 0.

Lo-1.8 Remember that we can simplify a statement form before constructing a truth table. Note that
$$p \vee (\sim p \wedge q) \Leftrightarrow (p \vee \sim p) \wedge (p \vee q) \Leftrightarrow p \vee q.$$
Thus the statement form of this problem can be written $(p \vee q) \Rightarrow q$, which is 0 if and only if $p = 1$ and $q = 0$. (Incidentally, it can be simplified further to $p \Rightarrow q$.)

Lo-1.9 By the previous exercise, the statement form is equivalent to $(p \vee q) \Rightarrow \sim q$, which is equivalent to $\sim q$.

Lo-1.10 This statement is equivalent to $q \Rightarrow p$.

Lo-1.11 The negation $\sim(p \Rightarrow q)$ of "if p then q" can be written as $p \wedge \sim q$. We use this form.

(a) P is a pentagon, but P is not a polygon.

(b) Let T, J, S and M be statement variables for "Tom is Ann's father," "Jim is Ann's uncle," and so on. The negation is $T \wedge \sim(J \wedge S \wedge M)$. Use DeMorgan's law to move the negation inside. Thus we have "Either Jim is not Ann's uncle or Sue is not her aunt or Mary is not her cousin, but Tom is Ann's father.

Lo-1.12 (a) Converse: If P is a polygon then P is a pentagon.
Inverse: If P is not a pentagon then P is not a polygon.

(b) Converse: If Jim is Ann's uncle and Sue is her aunt and Mary is her cousin, then Tom is Ann's father.
Inverse: If Tom is not Ann's father, then either Jim is not her uncle or Sue is not her aunt or Mary is not her cousin.

Lo-1.13 Of course, one can simply say they are equivalent because they are both contradictions and all contradictions are equivalent. However, we hoped yo would notice that the contrapositive of the converse is the inverse.

Lo-1.14 (a) If P is not a polygon, then P is not a pentagon.

(b) If Jim is not Ann's uncle or Sue is not her aunt or Mary is not her cousin, then Tom is not Ann's father.

Lo-1.15 If Dennis enters the America's Cup, then he is sure of victory.

Lo-1.16 No. The statement "p only if q" means $p \Rightarrow q$; that is, $H \Rightarrow ((M \vee C) \wedge B \wedge A)$. We know that A, B and C are true, but that says nothing about H. Why does it feel like you were lied to? Probably because the requirements were spelled out in such detail. As a result you thought he meant either "if" or "if and only if" when he said "only if." Of course, maybe your high school principal wasn't that familiar with logic and he thought he was lying just to get you to work more.

Lo-1.17 Let L stand for "learning to program in L." The given statement is C++ $\Rightarrow$ C. Thus, "If you learn to program in C++, then you learn to program in C." The other equivalent form is the contrapositive: "If you don't learn to program in C, then you don't learn to program in C++."

Lo-1.18 Using DeMorgan's laws a couple of times, we have
$$\sim(\sim p \vee q) \vee (r \vee \sim q) \Leftrightarrow (p \wedge \sim q) \vee r \vee \sim q$$
$$\Leftrightarrow \sim\left(\sim(p \wedge \sim q \wedge \sim r \wedge q)\right)$$

There are other ways to do this.

Lo-1.19 We have
$$p \Rightarrow (q \Rightarrow r) \Leftrightarrow {\sim}p \vee ({\sim}q \vee r)$$
$$\Leftrightarrow {\sim}p \vee {\sim}q \vee r$$

and
$$(p \wedge q) \Rightarrow r \Leftrightarrow {\sim}(p \wedge q) \vee r$$
$$\Leftrightarrow {\sim}p \vee {\sim}q \vee r.$$

Hence the expression in large parentheses in the problem is always true. Thus the problem reduces to rewriting ${\sim}p \wedge {\sim}q \wedge {\sim}r$. By DeMorgan's law, this equals ${\sim}(p \vee q \vee r)$.

Lo-1.20 "A is a sufficient condition for B" means that A forces B; that is, "If A then B." Applied here: "If I get up when the alarm rings, then I will get to work on time."

Lo-1.21 If the sides of a triangle have lengths 3, 4, and 5 then the triangle is a right triangle.

Lo-1.22 This can be done using either Example 5 or the method in Example 6. We use the latter. The statement is false if Jane doesn't do the programming but passes anyway. By the method in Example 6, we can write either "If Jane passes her Java course, then she did all the programming assignments" or "If Jane does not do all the programming assignments, then she will not pass her Java course."

Lo-1.23 Since all the statements are implications, it is sufficient to check for the single false situation. The given statement is false if the program is running and there is less than 250K of RAM. Therefore, we ask the following question in each of (a)–(f): Is the statement false when the program is running and there is less than 250K of RAM. If "yes," it is false when the given implication is false and so they are equivalent; if "no," they are not equivalent. Here are the answers:

(a) No   (b) Yes   (c) Yes   (d) No   (e) Yes   (f) No

You should fill in the explanations as the why each answer is "yes" or "no."

Lo-2.1 (a) $\forall x \in \mathbb{R}, \ \big((x < 0) \vee (x = 0) \vee (x > 0)\big)$

(b) Let $\mathcal{C}$ be the set of computer scientists, let $\mathcal{U}$ be the set of unemployed people and let $\mathcal{E}$ be the set of employed people. We could say

$$\forall\, x \in \mathcal{C}, \ x \notin \mathcal{U} \quad \text{or} \quad \forall\, x \in \mathcal{C}, \ x \in \mathcal{E}.$$

We could use words instead of the sets $\mathcal{C}$, $\mathcal{U}$ and $\mathcal{E}$:

$\forall$ computer scientists x, x is not unemployed.
$\forall$ computer scientists x, x is employed.

The former is a straight translation of the text. The latter involves knowing that, with the set of people the universal set, $\mathcal{U}^c = \mathcal{E}$. As such, it involves some knowledge of the world and so is not a direct translation of (b) into logic.

Lo-2.2 The original statement is true. Statements (b), (d) and (e) say the same thing and so are also true. However, they are open to misinterpretation. The standard interpretation of (b) and (e) is that they are true regardless of what integer one comes up with. But suppose you tell someone "I noticed that 276^2 is even." He might answer with (b), now thinking of (b) as applying only to 276. Of course, he should have said "If that integer...," not "If a given integer...," but people are often careless in speech. What about (a), (c) and (f)? Since $1^2 = 1$ is odd, (a) is false. While (c) is true, it does not say the same thing as the original statement. Statement (f) may appear to be the same as the original, but it is not. It is the converse. It says "$\forall n \in \mathbb{N}$, if n is even then n^2 is even."

Lo-2.3 (a) $\forall$ correct algorithms A, (A is correctly coded)$\Rightarrow$(A runs correctly).

(b) $\forall s, t \in \mathbb{Z}$, $\big((s \text{ odd}) \wedge (t \text{ odd})\big) \Rightarrow (st \text{ odd})$.

(c) This is the converse of (b): $\forall s, t \in \mathbb{Z}$, $(st \text{ odd}) \Rightarrow \big((s \text{ odd}) \wedge (t \text{ odd})\big)$.

Lo-2.4 (a) "$\forall$ S, (S is a computer science student) $\Rightarrow$ (S needs to take Java programming)."

(b) "$\forall$ computer science students S, S needs to take Java programming."

Note the sets over which quantification takes place (i.e., the sets to which S belongs) is different in the two answers. In (b), "$\forall$ computer science students S" tells us that S runs through the set of computer science students. For (a), common sense tells us that the set can be any set that includes all computer science students; however, the set was not specified. Perhaps it's the set of all fish, in which case the statement is trivial if no fish are computer science students. Thus, the correct form for (a) would be as follows.

(a) "$\forall$ S $\in \mathcal{S}$, (S is a computer science student) $\Rightarrow$ (S needs to take Java programming)," where $\mathcal{S}$ is ... (e.g., the set of all people).

Lo-2.5 (a) "$\exists$ a question Q such that Q is easy."   or
"$\exists$ a question Q, Q is easy."

(b) "$\exists$ S $\in \mathcal{S}$, (S is a question) $\wedge$ (S is easy)," where $\mathcal{S}$ is the set of all sentences.

Lo-2.6 The proposed negation is incorrect. A correct version is "There exists an irrational number x and a rational number y such that the product xy is rational."
The negation is true since we could take x to be any irrational number and $y = 0$. If we start with "The product of any irrational number and any nonzero rational number is irrational," then that statement is true and its negation, "There exists an irrational number x and a nonzero rational number y such that the product xy is rational," is false.
The incorrect "negation" given in the problem is also true.

Lo-2.7 There exists a computer program P such that P is correctly programmed and P compiles with warning messages. Which is true depends on your interpretation of "correctly programmed." Often a program will run just fine with warning messages. For example, many compilers give a warning message if a loop has no code in its body, but the empty body may be intentional.

Lo-2.8 The proposed negation is incorrect. A correct negation is "There exist real numbers x and y such that $x^2 = y^2$ but x does not equal y." The negation is true (take $x = -1$ and $y = 1$).
The incorrect "negation" has the contrapositive of the original statement inside the "for all" quantifier. Since a statement and its contrapositive are equivalent, this statement is also false (take $x = -1$ and $y = 1$ just as in the original).

Lo-2.9 "There exists $p \in \mathbb{P}$ such that p is even and $p \neq 2$." The original statement is true.

Lo-2.10 "There exists an animal x such that x is a tiger and either x has no stripes or x has no claws." There is probably a declawed captive tiger, in which case the negation is true.

Lo-2.11 (a) "$\forall x \in \mathbb{R}$, $\exists$ negative $y \in \mathbb{R}$, $x > y$." Both statements are true. For the original statement, take $x = 1$. For the statement here, take $y = -|x| - 1$.

(b) Applying Theorem 2, we move negation through the quantifiers one at a time. Let $\mathbb{R}^-$ be the negative reals. We have

$$\sim(\exists x \in \mathbb{R},\ \forall y \in \mathbb{R}^-,\ x > y) \quad \Leftrightarrow \quad \forall x \in \mathbb{R},\ \sim(\forall y \in \mathbb{R}^-,\ x > y)$$
$$\Leftrightarrow \quad \forall x \in \mathbb{R},\ \exists y \in \mathbb{R}^-,\ x \leq y.$$

This cannot be true since it is the negation of a true statement. Again, take $x = 1$.

Lo-2.12 Contrapositive: "For all computer programs P, if P compiles with error messages then P is incorrect."
Converse: "For all computer programs P, if P compiles without error messages then P is correct."
Inverse: "For all computer programs P, if P is incorrect then P compiles with error messages."

Lo-2.13 Contrapositive: "$\forall n \in \mathbb{N}$, if n is odd then its square is odd."
Converse: "$\forall n \in \mathbb{N}$, if n is even then its square is even."
Inverse: "$\forall n \in \mathbb{N}$, If n^2 is odd then n is odd."
All statements are true in this problem because the square of an integer is even if and only if the integer is even. The "if" part is proved using the contrapositive and the "only if" part is proved using the converse.

Lo-2.14 Contrapositive: "$\forall n \in N$, if n is even and not 2 then n is composite."
Converse: "$\forall n \in N$, if n is odd or equal to 2 then n is prime."
Inverse: "$\forall n \in N$, if n is composite then n is even and not equal to 2."
The statement and its contrapositive are true. The converse and the inverse are false.

Lo-2.15 (a) $\forall x \in P,\ H(x) \Rightarrow L(x)$.

(b) Everyone who is happy has a large income.

(c) We have

$$\sim\big(\forall x \in P,\ H(x) \Rightarrow L(x)\big)$$
$$\Leftrightarrow \quad \exists x \in P \sim\big(\sim H(x) \vee L(x)\big)$$
$$\Leftrightarrow \quad \exists x \in P\ \big(H(x) \wedge \sim L(x)\big).$$

(d) There is someone who is happy and does not have a large income.

Lo-2.16 (a) False: For $x = 1$ and $x = -1$, both x and $1/x$ are integers.

(b) True: If $x \in \mathbb{R}$ and $x + y = 0$, then $y = -x$ and so $y \in \mathbb{R}$ and y is unique. Alternatively for uniqueness: Suppose $x + y = 0$ and $x + z = 0$. Then $x + y = x + z$ and so $y = z$.

Lo-2.17 $\Big(\exists x \in D,\ S(x)\Big) \wedge \forall x, y \in D,\ \big(S(x) \wedge S(y)\big) \Rightarrow (x = y)\Big)$

Solutions for Logic

Lo-2.18 Both are true. Since there are infinitely many primes, there is $p \in \mathbb{P}$ with $p > m$ and p odd.

(a) Let $n = p + 3$ and $q = 3$.

(b) Let $n = p + 2$ and $q = 2$.

Lo-2.19 (a) Equivalent statements: If the first is true, then $\forall x \in D \ P(x)$. Likewise, $\forall x \in D \ P(x)$. Thus the second is true. Suppose the first is false, then there is $x \in D$ such that either $P(x)$ is false or $Q(x)$ is false. If $P(x)$ is false, then so is "$\forall x \in D, \ P(x)$" and hence the second statement is false. If $Q(x)$ is false, similar reasoning applies.

(b) Not equivalent statements: Let $D = \mathbb{Z}$, let $P(x)$ be "x is even" and let $Q(x)$ be "x is odd." Then the first statement is true and the second is false.

(c) Not equivalent statements: The example for (b) works here also.

(d) Equivalent: If the first is true, then there is some $x \in D$ such that either $P(x)$ is true or $Q(x)$ is true. Hence either "$\exists x \in D, \ P(x)$" is true or "$\exists x \in D, \ P(x)$" is true. Thus the second statement is true. Suppose the first is false, the for all $x \in D$, both $P(x)$ and $Q(x)$ are false. From this you can conclude that the second statement is false.

Note: You actually only need to do one of (a) and (d) and one of (b) and (c) because of negation. Negating both statements in (a) gives both statements in (d) with the predicates $\sim P$ and $\sim Q$ in place of the predicates P and Q. Likewise for (b) and (c).

Lo-2.20 Let $n = ab$. By the formula for summing a geometric series,

$$1 + 2^a + (2^a)^2 + \cdots + (2^a)^{b-1} = \frac{1 - (2^a)^b}{1 - 2^a} = \frac{2^n - 1}{2^a - 1}.$$

Multiplying by $2^a - 1$ gives us a factorization of $2^n - 1$.

Lo-2.21 Let $p = 2^n - 1$. The divisors of N are 2^k and $2^k p$ where $0 \le k \le n - 1$. Thus, the sum of the divisors, including N, is two geometric series:

$$(1 + 2 + 2^2 + \cdots + 2^{n-1}) + (p + 2p + 2^2 p + \cdots + 2^{n-1}p) = (1 + 2 + 2^2 + \cdots + 2^{n-1})(1 + p).$$

The sum of the geometric series is $2^n - 1$ and $p + 1 = 2^n$. Thus the sum of the divisors of N is $(2^n - 1)2^n = 2N$. Since we included the divisor N, the sum of the divisor of N that are less than N is $2N - N = N$.

Solutions for Number Theory and Cryptography

NT-1.1 (a) True. Assume, without loss of generality, that x is even and y is odd. Then $x = 2k$ and $y = 2j + 1$, whence $x + y = 2k + 2j + 1 = 2(k + j) + 1$, which is odd.

(b) True. Use the contrapositive. First suppose that both x and y are odd, say $x = 2k + 1$ and $y = 2j + 1$. Then $x + y = 2(k + j + 1)$ is even. Now suppose that both x and y are even, say $x = 2k$ and $y = 2j$. Then $x + y = 2(k + j)$ is even.

NT-1.2 (a) False. Counterexample: $5 - 3 = 2$.

(b) False. Counterexample: $1 + 3 = 4$.

NT-1.3 (a) True. Negating both A and B in "A if and only if B" gives an equivalent statement. (It's the contrapositive.) In this case, the result is "The product of two integers is odd if and only if neither of them is even." Since "neither of them is even" is the same thing as "both of them are odd," this is the closure property mentioned in Example 1. You could also prove it from scratch. We do that now. For the "if" part, assume that x and y are integers and $x = 2k$ is even. Then $xy = 2ky = 2(ky)$ is even. For the "only if" part, use the contrapositive: "If x and y are both odd, there product is odd." Assume that $x = 2k + 1$ and $y = 2j + 1$. Then $xy = 2(2kj + k + j) + 1$ is odd.

(b) False. Counterexample: $3 \times 2 = 6$.

NT-1.4 (a) True: One can write out various proofs, breaking things down into cases depending on whether m and n are even or odd. Alternatively, one can construct a table with four cases Here's the table:

m	n	$m - n$	m^3	n^3	$m^3 - n^3$
even	even	even	even	even	even
even	odd	odd	even	odd	odd
odd	even	odd	odd	even	odd
odd	odd	even	odd	odd	even

By comparing the $m - n$ and $m^3 - n^3$ columns, we see that the result is true. An alternative proof can be obtained by doing calculations modulo 2. This is because "even" corresponds to 0 (mod 2) and "odd" to 1 (mod 2). We have $m^3 = m$ (mod 2) and $n^3 = n$ (mod 2). Therefore $m^3 - n^3 = m - n$ (mod 2).

(b) True: In fact, this is equivalent to (a) because $A \Leftrightarrow B$ is true if and only if $\sim A \Leftrightarrow \sim B$ is true.

NT-1.5 (a) False: Try $n = 3$. (What is the answer for $n > 3$?)

(b) True: Note that $(-1)^2 = 1$. If $n = 2k$, then $(-1)^{2k} = ((-1)^2)^k = 1^k = 1$. If $n = 2k + 1$, then $(-1)^{2k+1} = -(-1)^{2k} = -1$.

NT-1.6 (a) True: One of n and $n + 1$ is even. Therefore $n(n + 1) = n^2 + n$ is even. Therefore, since 5 is odd, $(n^2 + n) + 5$ is odd.

(b) True: With a little algebra $6(n^2 + n + 1) - (5n^2 - 3) = (n + 3)^2$.

(c) False: For every $M > 0$, let $n = 11M > M$. Note that

$$n^2 - n + 11 = 11(11M^2 - M + 1),$$

which is composite because $11M^2 - M + 1 \geq M(11M - 1) > 1$.

(d) True: Factor $n^2 + 2n - 3$ to get $(n + 3)(n - 1)$. For this to be a prime, we must choose n so that one factor is ± 1 and the other is $\pm p$, where p is a prime and both factors have the same sign. Thus, either $n + 3 = p$ and $n - 1 = 1$ or $n + 3 = -1$ and $n - 1 = -p$. The first pair of equations give $n = 2$ and $p = 5$, which is a prime. The second pair of equations give $n = -4$ and $p = 5$, which is the same prime. You may ask about the choices $n + 3 = 1$ and $n - 1 = p$ or $n + 3 = 1$ and $n - 1 = p$. They lead to negative values for p and primes must be positive by definition.

NT-1.7 (a) False: You should be able to prove that 7 is a counterexample since the only possible values for x, y and z are 0, 1 and 4.

(b) True: We can write such a product as $N(k) = k(k + 1)(k + 2)(k + 3)$. Let's look at some values: $N(1) = 24 = 5^2 - 1$, $N(2) = 120 = 11^2 - 1$, $N(3) = 360 = 19^2 - 1$. It looks like N is always one less than a square. Since squares are not very close together this would mean that N is not a square. We have a plan. Let's write a proof. Since $k(k + 3)$ and $(k + 1)(k + 2)$ are close together, we rewrite the product as

$$N = \big(k(k + 3)\big)\big((k + 1)(k + 2)\big) = \big(k(k + 3)\big)\big(k(k + 3) + 2\big) = n(n + 2) > n^2,$$

where $n = k(k + 3)$. Now $n(n + 2) = (n + 1)^2 - 1$ — one less than a square as we conjectured. Thus $N < (n + 1)^2$. We've shown that N lies between two consecutive squares; that is, $n^2 < N < (n + 1)^2$. Hence N cannot be a square.

NT-1.8 (a) True: Let $x = m^{1/2} + n^{1/2}$ and $y = m^{1/2} - n^{1/2}$. Then $xy = m - n$ is a nonzero integer. Hence x and y are either both rational or both irrational. (See discussion in Example 3.)

(b) True: Define x and y as in (a). Since at least one is rational, they are both rational by (a). Thus $(x + y)/2 = m^{1/2}$ is rational and so m is a perfect square by Theorem 3. Likewise, $(x - y)/2 = n^{-1/2}$ is also a perfect square.

(c) True: Note that $m + 2m^{1/2}n^{1/2} + n = (m^{1/2} + n^{1/2})^2$. Clearly this is a perfect square if m and n are perfect squares. Conversely, suppose this is a perfect square. Then $m^{1/2} + n^{1/2}$ is rational and so m and n are perfect squares by (b).

(d) You should be able to see that (a) and (b) are false whenever $m = n$ is *not* a perfect square. If $m = n$, $m + 2m^{1/2}n^{1/2} + n = 4m$ and hence is a perfect square if and only if m is a perfect square. Thus (c) is true.

NT-1.9 The "if" part is obvious. We prove the "only if" part. Suppose n is composite. Let p be the smallest prime dividing n. Then $n = pm$ where $m > 1$ since n is composite. Let q be a prime dividing m. Then $q \leq m$. Since p is the smallest prime dividing n, $p \leq q$. Hence $n = pm \geq pq \geq p^2$ and so $p < n^{1/2}$.

NT-1.10 If x terminates with d digits after the decimal place, we can write it as $10^d x / 10^d$. For (a), $d = 4$ and so we can write it as $31415/10000$. This is not in lowest terms, but you need not reduce it.

We call a decimal in which a pattern repeats a *repeating decimal*. Thus (b) and (c) are repeating decimals. The *period* is the number of digits in the repeating pattern. In (b) the period is two because of the pattern 30. In (c) the period is three because of the pattern 215. If x is a repeating decimal with period k, then $10^k x - x$ will be a terminating decimal, which can be written as a rational number a/b by the previous discussion. Thus $x = a/(b(10^k - 1))$. We now apply this.

(b) Since $x = 0.303030\ldots$, $k = 2$, $10^2 x - x = 30$, and so $x = 30/99$.

(c) $x = 6.3215215215\ldots$, $k = 3$,

$$10^3 x - x = 6321.5215215\ldots - 6.3215215\ldots = 6315.2 = 63152/10,$$

and so $x = 63152/9990$.

NT-1.11 Yes: We can solve for x to obtain $x = \frac{d-b}{a-c}$. Fill in the missing steps.

NT-1.12 (a) True: Note that $(k - 1) + k + (k + 1) = 3k$ which equals 0 mod 3. Equivalently, for three consecutive integers, in some order, one is equal to 0 (mod 3), one is equal to 1 (mod 3), and one is equal to 2 (mod 3). The sum is equal to $0 + 1 + 2 = 3 = 0$ (mod 3).

(b) True: Let the even integers be $2k$ and $2j$. Then $2k \times 2j = 4kj = 0$ (mod 4). Equivalently, an even integer is equal to 0 (mod 4) or 2 (mod 4). Modulo 4, the product of two even integers is either 0 or $2 \times 2 = 4$. In both cases the product equals 0 (mod 4).

(c) True: If $n = 16k$ then $n = 8(2k)$, so n is divisible by 8.

(d) True: Let $n = 2k + 1$. Then $3n + 3 = 6k + 3 + 3 = 6(k + 1)$ is divisible by 6.

NT-1.13 (a) True: If $b = ak$ then $bc = akc$.

(b) True: Since $b \mid c$, we have $c = bd$ for some $d \in \mathbb{Z}$. Use (a) with $a = a$, $b = b$ and $c = d$.

(c) False: Let $a = 2$, $b = 3$, $c = 4$.

NT-1.14 (a) False: Let $a = 2$ and $b = c = 1$.

(b) False: Let $a = 6$, $b = 2$ and $c = 3$.

(c) True: If $a \mid b$, then $b = ka$ for some $k \in \mathbb{Z}$. Then $b^2 = (k^2)a^2$ and so $a^2 \mid b^2$.

(d) False: Let $a = 4$ and $b = 2$.

NT-1.15 (a) $1404 = 2^2\, 3^3\, 13$ (b) $9702 = 2\, 3^2\, 7^2\, 11$ (c) $89250 = 2\, 3\, 5^3\, 7\, 17$

NT-1.16 (a) $p_1^{me_1} \cdots p_k^{me_k}$.

(b) Yes. We give a proof by contradiction along similar lines to the proof that $n^{1/2}$ must be irrational when n is not a perfect square as done in Example 3. Suppose $s^{1/m}$ is rational and not an integer. Then $s^{1/m} = a/b$ for some integers a and b. We can suppose that a/b is in lowest terms and that $b > 1$. From $s^{1/m} = a/b$, we have $a^m = sb^m$. If $p \mid b$, then $p \mid a^m$. As in Example 3, $p \mid a$, a contradiction.

NT-1.17 We have

$$\begin{aligned}
20! =&(2^2 \times 5) \times (19) \times (2 \times 3^2) \times (17) \times (2^4) \times (5 \times 3) \\
&\times (2 \times 7) \times (13) \times (3 \times 2^2) \times (11) \times (2 \times 5) \times (3^3) \\
&\times (2^3) \times (7) \times (2 \times 3) \times (5) \times (2^2) \times (3) \times (2) \times (1) \\
=&19 \times 17 \times 13 \times 11 \times 7^2 \times 5^4 \times 3^8 \times 2^{18}.
\end{aligned}$$

Solutions for Number Theory and Cryptography

Every zero at the end of 20! corresponds to a factor of $10 = 2 \times 5$. Since we have 2^{18} and 5^4, there will be four zeroes at the end.

For (b) and (c), the powers in the prime factorization are doubled and tripled, respectively. Thus the same happens to the number of zeroes at the end, giving us eight and twelve.

NT-1.18 Suppose we are given a number $A = a_n a_{n-1} \ldots a_1 a_0$. The statement $3 \mid A$ is equivalent to the statement $A \pmod 3 = 0$. Note that $10 = 1 \pmod 3$ and so $10^k = 1 \pmod 3$ for all $k \in \mathbb{N}$. We have

$$A = a_n 10^n + a_{n-1} 10^{n-1} + \cdots + a_1 10 + a_0$$
$$= a_n + a_{n-1} + \cdots + a_1 + a_0 \pmod 3.$$

Thus A equals the sum of its digits modulo 3 and so we are done. Can you find similar results for divisibility by 5? by 9? by 11?

NT-1.19 We prove it. If you list the remainders of all nonnegative integers, $0, 1, 2, 3, 4, 5, 6, 7, 8, \ldots$ when divided by four you get $0, 1, 2, 3, 0, 1, 2, 3, 0, 1, 2, 3, \ldots$. Given any four consecutive integers, one, call it x, must be divisible by 4 (remainder 0) and one other, call it y has remainder 2 when divided by 4 and so is even. Thus $x = 4j$ and $y = 2k$ for some k and j and thus $xy = 8kj$ is divisible by 8. Since the product of all four consecutive integers includes xy as a factor, it is also divisible by 8.

NT-1.20 If n is even, n^2 is even and so $n^2 \neq 3 \pmod 4$. If n is odd, then $n = 2k + 1$ for some k. Then

$$n^2 = 4k^2 + 4k + 1 = 4k(k+1) + 1.$$

Thus $n = 1 \pmod 4$.

NT-1.21 Let $n = 2k + 1$ By the solution to the previous exercise, $n^2 = 4k(k+1) + 1$. One of k and $k + 1$ is even. Thus $k(k+1)$ is even and so $8 \mid 4k(k+1)$. Thus $n^2 = 8j + 1$ for some j. Hence $n^4 = 64j^2 + 16j + 1 = 1 \pmod{16}$.

NT-1.22 Yes. Since $m - n = 0 \pmod d$ and $n = n \pmod d$, it follows from Theorem 4 that $(m - n) + n = 0 + n \pmod d$. Thus $m = n \pmod d$. This is the same as saying m and n have the same remainder when divided by d.

NT-1.23 No. Let $m = n = a = b = 2$, $d = 3$. Then $(m + n) \bmod d = 1$.

NT-1.24 (a) Since $k = j \pmod{!d}$, $j - k = id$ for some $i \in \mathbb{Z}$ If $x \in d\mathbb{Z} + j$, we have $x = md + j$ for some $m \in \mathbb{Z}$. Then $x = md + k + (j - k) = (m + i)d + k$ and so $x \in d\mathbb{Z} + k$. Similarly, if $y \in d\mathbb{Z} + k$, then $y \in d\mathbb{Z} + j$.

(b) We give a proof by contradiction. Suppose $x \in (d\mathbb{Z}+j) \cap (d\mathbb{Z}+k)$. Then $x \in (d\mathbb{Z}+j)$ and $x \in (d\mathbb{Z} + k)$. Thus $x = j \pmod d$ and $x = k \pmod d$. It follows that $j = k \pmod d$, a contradiction since we are given that $j \neq k \pmod d$.

NT-1.25 (a) We give a proof by contradiction. Suppose $log_p(q) = a/b$. Then $q = p^{a/b}$ and so $q^b = p^a$. This is impossible by the uniqueness of prime factorization.

(b) It is not true. Let $q = p^2$.

(c) Suppose $\log_a(b) = k/m$. By the definition of $\log_a$, we have $b = a^{\log_a(b)} = a^{k/m}$. Taking the mth power of both sides, we have $b^m = a^k$.

Conversely, suppose $b^m = a^k$. Taking the mth root of both sides, we have $b = a^{k/m}$. By the definition of $\log_a$, we have $\log_a(b) = k/m$.

NT-1.26 (a) False: Let $x = 1.1$ and $y = 0.9$. Then $\lfloor 1.1 - 0.9 \rfloor = \lfloor 0.2 \rfloor = 0$, $\lfloor 1.1 \rfloor = 1$ and $\lfloor 0.9 \rfloor = 0$.

(b) True: Suppose $n \le x < n+1$. Then $\lfloor x \rfloor = n$ and, since $n - k \le x - k < (n-k)+1$, $\lfloor x - k \rfloor = n - k$. The same is true for the ceiling function. The statements are also true with "+" in place of "−."

(c) False: Let $k = 2$ and $x = 1.5$.

NT-1.27 (a) True: Let $n = kq + r$ where $0 \le r < k$. Then $\frac{n}{k} = q + \frac{r}{k}$ and so $q \le \frac{n}{k} < q+1$. Hence $\lfloor \frac{n}{k} \rfloor = q = \frac{n-r}{k}$.

(b) False: Take $a = b = 2$ and $x = 1/2$.

NT-1.28 We prove both of them. Let $x = n + r$ where $n \in \mathbb{Z}$ and $0 < r < 1$. Then $-x = (-n - 1) + (1 - r)$ where $0 < 1 - r < 1$ You should verify that

$$\lfloor x \rfloor = n \qquad \lceil x \rceil = n + 1 \qquad \lfloor -x \rfloor = -n - 1 \qquad \lceil -x \rceil = -n.$$

Now (a) and (b) follow easily.

NT-2.1 (a) The algorithm gives $1001 > 544 > 457 > 87 > 22 > 21 > 1 > 0$. Thus $\gcd(1001, 544) = 1$ (the numbers are "relatively prime") and 1 is the only common divisor.

(b) The algorithm gives $3510 > 672 > 150 > 72 > 6 > 0$. Thus $\gcd(3510, 652) = 6$ and the common divisors are the divisors of 6, namely 1, 2, 3 and 6.

Both answers can be checked by factoring the numbers looking at the result to find the gcd. For example $1001 = 7 \times 11 \times 13$ and $544 = 2^5 \times 17$, so they have no common factor.

NT-2.2 The algorithm gives $252 > 180 > 72 > 36 > 0$. The common divisors are the divisors of $\gcd(252, 180) = 36$. Since $36 = 2^2 \times 3^2$, the common divisors are found by multiplying one of $\{1, 2, 2\}$ by one of $\{1, 3, 3^2\}$. We obtain 1, 2, 3, 4, 6, 9, 12, 18, 36.

NT-2.3 The Euclidean algorithm gives $59400 > 16200 > 10800 > 5400 > 0$. Thus $\gcd(59400, 16200) = 5400$. Factoring: $5400 = 2^3 \times 3^3 \times 5^2$. The common divisors are of the form $2^a \times 3^b \times 5^c$ where $0 \le a \le 3$, $0 \le b \le 3$ and $0 \le c \le 2$. There are four choices for a, four for b, and three for c. Thus there are $4 \times 4 \times 3 = 48$ common divisors.

NT-2.4 We compute the remainders and quotients:

$$252 \; > \; 180 \; > \; 72 \; > \; 36 \; > \; 0$$
$$1 \qquad 2 \qquad 2$$

Thus $72 = 252 - 1 \times 180$ and $36 = 180 - 2 \times 72$ whence

$$36 = 180 - 2 \times (252 - 1 \times 180) = 180 - 2 \times 252 + 2 \times 180$$
$$= -2 \times 252 + 3 \times 180.$$

Hence $A = -2$ and $B = 3$.

Solutions for Number Theory and Cryptography

NT-2.5 We can proceed as in the previous exercise; however, to keep numbers smaller, we could first divide m and n by a common factor; that is any divisor of $\gcd(59400, 16200)$, which we found in Exercise 2.3 to be 5400. To keep it simple, we'll just divide by 100. Computing the remainders and quotients:

$$594 \underset{3}{>} 162 \underset{1}{>} 108 \underset{2}{>} 54 > 0$$

Thus $108 = 594 - 3 \times 162$ and $54 = 162 - 1 \times 108$ whence

$$54 = 162 - 1 \times 108 = 162 - 1 \times (594 - 3 \times 162)$$
$$= -1 \times 594 + 4 \times 162.$$

Hence $A = -1$ and $B = 4$.

NT-2.6 We compute the remainders and quotients:

$$163 \underset{1}{>} 86 \underset{1}{>} 77 \underset{8}{>} 9 \underset{1}{>} 5 \underset{1}{>} 4 \underset{4}{>} 1 > 0$$

We won't bother to list the equations for each remainder, but just use them as needed. We have

$$1 = 5 - 1 \times 4 = 5 - 1 \times (9 - 1 \times 5)$$
$$= -1 \times 9 + 2 \times 5 = -1 \times 9 + 2 \times (77 - 8 \times 9)$$
$$= 2 \times 77 - 17 \times 9 = 2 \times 77 - 17 \times (86 - 1 \times 77)$$
$$= -17 \times 86 + 19 \times 77 = -17 \times 86 + 19 \times (163 - 1 \times 86)$$
$$= 19 \times 163 - 36 \times 86.$$

NT-2.7 One way to do it is to note that, since $\gcd(a, b) \mid a$ and $a \mid \text{lcm}(a, b)$, it follows that $\gcd(a, b) \mid \text{lcm}(a, b)$ by Exercise 1.13(b).

Another way to do it is to use the prime factorizations in Theorem 6. The power of p_i in $\gcd(a, b)$ is $\min(e_i, f_i)$ and the power of p_i in $\text{lcm}(a, b)$ is $\max(e_i, f_i)$. Since $\min(e_i, f_i) \leq \max(e_i, f_i)$, divisibility follows.

NT-2.8 (a) $120 = 2^3 \times 3 \times 5$ and $108 = 2^2 \times 3^3$. Thus $\text{lcm}(120, 108) = 2^3 \times 3^3 \times 5 = 1080$.

(b) By the Euclidean algorithm (We omit the steps.), we have $\gcd(120, 108) = 12$. Since $\gcd(a, b)\text{lcm}(a, b) = ab$, we have $\text{lcm}(120, 108) = 120 \times 108/12 = 1080$.

NT-2.9 Since $a \mid b$ and $b \mid b$, b is a common multiple of a and b. It is clearly the *least* strictly positive multiple of b and so is the *least* common multiple of a and b.

NT-2.10 As in the example, A sends $11^{13} \% 163 = 19$ to B. However, B sends $11^{15} \% 163 = 17$ to A. Now A computes $17^{13} \% 163 = 142$ and B computes $19^{15} \% 163$ which, of course, is also 142. This is their shared key K.

NT-2.11 We use the notation of the example. Suppose computer A chooses $s = 1$. It then sends $S = b^1 = b$, so if Joe sees that b is sent by computer A, he knows that $s = 1$. Computer B sends T and the shared secret is $K = T^s = T$. Since Joe sees T, he knows K.

NT-2.12 Since N is a prime, $\phi(N) = N - 1$. Joe solves the equation $ed = 1 \pmod{(N - 1)}$ for d using the method in Euclidean 13. Since he has d, he can compute $C^d \% N$, which equals M for the same reason explained in Example 17.

NT-2.13 We need to compute $2^d \% N$; that is $2^{37} \% 77$. This can be done in a variety of ways. Here is one way to do it without even using a calculator, with all calculations modulo 77:

$$2^7 = 128 \equiv 51, \qquad 2^8 \equiv 2 \times 51 \equiv 25, \qquad 2^{10} \equiv 4 \times 25 \equiv 23,$$
$$2^{12} \equiv 4 \times 23 \equiv 15, \qquad 2^{24} \equiv 15 \times 15 = 225 \equiv -6,$$
$$2^{36} \equiv (-6) \times 15 = -13, \qquad 2^{37} \equiv 3 \times (-13) \equiv 38.$$

NT-2.14 Since $N = 5 \times 13$, $\phi(N) = 4 \times 12 = 48$. Thus we must solve $7d = 1$ mod 48. We can do this by writing $7d + 48x = 1$ and using Example 13. We omit details. In this case, we could also observe that $7 \times 7 = 49$ and so $d = 7$.

NT-2.15 Since $ed = 1 \pmod{\phi(N)}$, $\phi(N)$ must divide $ed - 1$. Since $\phi(N) = (p-1)(q-1)$, it is even and so $ed - 1$ is even. Thus ed is odd. Hence both d and e are odd.

Solutions for Sets and Functions

SF-1.1 (a) Yes.

(b) No. 2 is in $\{1, 2, 3, 4\}$ but $\{2\}$ is not.

(c) Yes.

(d) Yes. All elements of $\{1, 2\}$ (namely 1 and 2) are also elements of $\{1, 2, \{1, 2\}, \{3, 4\}\}$.

(e) No. $\{1\}$ is an element, but 1 is not.

(f) Yes. A set is always a subset of itself.

SF-1.2 Be sure to draw these diagrams in "most general form." In (b), for example, B and C are disjoint, but A and B should show no special relationship.

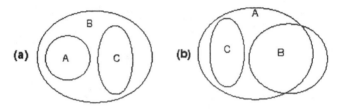

SF-1.3 (a) In lexicographic order, the elements of $A \times B$ are

$$(w, a), \ (w, b), \ (x, a), \ (x, b), \ (y, a), \ (y, b), \ (z, a), \ (z, b).$$

This is the order that the words wa, wb, xa, xb, etc., would appear in Webster's Baby-Talk Dictionary. The set $A \times B = \{(w, a), (w, b), (x, a), (x, b), (y, a), (y, b), (z, a), (z, b)\}$, when written in braces, does not formally imply any particular ordering of the elements. Generally though, any representation of a set will utilize some implicit ordering of the elements as a part of the "data structure" used.

(b) In lexicographic order (called lex order for short), the elements of $B \times A$ are

$$(a, w), \ (a, x), \ (a, y), \ (a, z), \ (b, w), \ (b, x), \ (b, y), \ (b, z).$$

(c) $(w, w), (w, x), (w, y), (w, z), (x, w), (x, x), (x, y), (x, z), (y, w), (y, x), (y, y), (y, z),$ $(z, w), (z, x), (z, y), (z, z).$

(d) $(a, a), (a, b), (b, a), (b, b).$

SF-1.4 (a) In lex order the list is $(1, (u, m)), \ (1, (u, n)), \ (1, (v, m)), \ (1, (v, n)), \ (2, (u, m)),$ $(2, (u, n)), (2, (v, m)), (2, (v, n)), (3, (u, m)), (3, (u, n)), (3, (v, m)), (3, (v, n)).$ This list has twelve elements, each a pair of elements. The first elements of the pair are ordered by the order on integers $(1, 2, 3)$. The second elements are ordered lexicographically based on the order of the alphabet. Lexicographic order as we are using it deals with "words over an alphabet," where the underlying alphabet is linearly ordered, or equivalently products of linearly ordered sets. Lexicographic order is itself a linear order.

(b) In lex order: $((1, u), m)$, $((1, u), n)$, $((1, v), m)$, $((1, v), n)$, $((2, u), m)$, $((2, u), n)$, $((2, v), m)$, $((2, v), n)$, $((3, u), m)$, $((3, u), n)$, $((3, v), m)$, $((3, v), n)$. The first components of these pairs are ordered lexicographically based on lex order of $A \times B$ (numerical in the first component and alphabetic in the second component).

(c) In lex order: $(1, u, m)$, $(1, u, n)$, $(1, v, m)$, $(1, v, n)$, $(2, u, m)$, $(2, u, n)$, $(2, v, m)$, $(2, v, n)$, $(3, u, m)$, $(3, u, n)$, $(3, v, m)$, $(3, v, n)$. This is lex order on $A \times B \times C$ based on numerical order in the first component and alphabetic order in each of the remaining two components. As a set,

$$A \times B \times C = \{(1, u, m), (1, u, n), (1, v, m), (1, v, n), (2, u, m), (2, u, n),$$
$$(2, v, m), (2, v, n), (3, u, m), (3, u, n), (3, v, m), (3, v, n)\}.$$

SF-1.5 (a) Here is the set of palindromes of length less than or equal to 4, listed in lex order: ϵ, x, xx, xxx, xxxx, xyx, xyyx, y, yxxy, yxy, yy, yyy, yyyy.

(b) In length-first lex order: x, xx, xy, xxx, xxy, xyx, xyy. In (ordinary) lex order: x, xx, xxx, xxy, xy, xyx, xyy. Do you see how to describe the difference? If a dictionary used length-first lex, zoo would come before able in the dictionary.

(c) xxxx, xxxy, xxyx, xxyy, xyxx, xyxy, xyyx, xyyy, yxxx, yxxy, yxyx, yxyy, yyxx, yyxy, yyyx, yyyy.

SF-1.6 We omit the Venn diagram. There can be more than one example in each case, so your examples may not be the same as the ones given here.

(a) Take A and B to be nonempty disjoint sets and take $C = A$.

(b) Take $C = \emptyset$ and $A \neq \emptyset$.

(c) Take $A = C \neq \emptyset$ and take $B = \emptyset$. The left hand side is A, the right hand side is $\emptyset$.

(d) Take $C = \emptyset$ and take A to be a proper subset of B; that is, $A \subseteq B$, but $A \neq B$.

(e) Take C to be nonempty, $A = C$, $B = \emptyset$.

(f) Take $A = B = C \neq \emptyset$. Then $A - (B - C) = A$, $(A - B) - C = \emptyset - C = \emptyset$.

SF-1.7 (a) Suppose $x \in A$. Since $A \subseteq B$ and $A \subseteq C$, $x \in B$ and $x \in C$. Thus $x \in B \cap C$ and so $A \subseteq B \cap C$.

(b) Suppose $x \in A \cup B$. Thus either $x \in A$ or $x \in B$. Since $A \subseteq C$ and $B \subseteq C$, $x \in C$. Thus $A \cup B \subseteq C$.

SF-1.8 If $x \in (A - B) \cap (C - B)$, then $x \in A - B$ and hence $x \in A$. Also, $x \in C - B$ and hence $x \in C$. Thus, $x \in A \cap C$. Also, $x \notin B$ because $x \in A - B$. Since $x \in A \cap C$ and $x \notin B$, $x \in (A \cap C) - B$. Thus $(A - B) \cap (C - B) \subseteq (A \cap C) - B$.

Conversely, suppose $x \in (A \cap C) - B$. Then $x \in A$ and $X \in C$ but $x \notin B$. Thus $x \in A - B$ and $x \in C - B$, and so $x \in (A - B) \cap (C - B)$. Thus $(A \cap C) - B \subseteq (A - B) \cap (C - B)$.

Thus, $(A - B) \cap (C - B) = (A \cap C) - B$. Note the general form of the element argument used to show two sets X and Y are equal. First assume x is in X and show it is in Y, then assume x is in Y and show it is in X. You must show both directions.

[203]

Solutions for Sets and Functions

SF-1.9 (a) Let U be the universal set. If $A \subseteq B$, then we show that $U - B \subseteq U - A$. Suppose $x \in U - B$. Then $x \notin B$. Since $A \subseteq B$, $x \notin A$. Thus $x \in U - A$. Thus $U - B \subseteq U - A$.

(b) Assume $x \in A \cap C$. Since $A \subseteq B$, $x \in B$. Thus $x \in B \cap C$ and so $A \cap C \subseteq B \cap C$.

(c) By (a), $A \subseteq B$ implies that $B^c \subseteq A^c$. With (b) applied to C^c, this implies that $B^c \cap C^c \subseteq A^c \cap C^c$. By (a), this implies that $(A^c \cap C^c)^c \subseteq (B^c \cap C^c)^c$. By DeMorgan's rule, this implies that $A \cup C \subseteq B \cup C$, which is what was to be shown.

SF-1.10 A mathematician would make use of "if and only if" statements used in succession here, with "iff" standing for "if and only if."

(a) $(x, y) \in A \times (B \cup C)$ iff $(x \in A)$ and $(y \in B \cup C)$
iff $(x \in A)$ and $(y \in B$ or $y \in C)$
iff $(x \in A$ and $y \in B)$ or $(x \in A$ and $y \in C)$
iff $\left((x, y) \in A \times B\right)$ or $(x, y) \in A \times C$ iff $(x, y) \in (A \times B) \cup (A \times C)$.
Thus $A \times (B \cup C) = (A \times C) \cup (A \times C)$.

(b) $(x, y) \in A \times (B \cap C)$ iff $(x \in A)$ and $(y \in B \cap C)$
iff $(x \in A)$ and $(y \in B$ and $y \in C)$
iff $(x \in A$ and $y \in B)$ and $(x \in A$ and $y \in C)$
iff $(x, y) \in A \times B$ and $(x, y) \in A \times C$ iff $(x, y) \in (A \times B) \cap (A \times C)$.
Thus $A \times (B \cap C) = (A \times C) \cap (A \times C)$.

Note: This exercise shows that Cartesian product $\times$ distributes over both set union and intersection. You should draw a picture that represents these identities. One way is to use the Cartesian plane $\mathbb{R}^2$ just like you use in high school math. Let $[s, t] = \{x \mid x \in \mathbb{R}$ and $s \leq x \leq t\}$. Let $A = [0, 3]$, $B = [0, 1.25]$ and $C = [.75, 2]$. Show the sets in the identities of (a) and (b) above.

SF-1.11 (a) Using $D - E = D \cap E^c$, we must prove that $(A \cap B^c) \cap C^c = A \cap (B \cup C)^c$. By the associative law and then DeMorgan's law,

$$(A \cap B^c) \cap C^c = A \cap (B^c \cap C^c) = A \cap (B \cup C)^c.$$

(b) Use (a) with the names of B and C interchanged to obtain $(A - C) - B = A - (C \cup B)$, which equals $A - (B \cup C)$ by the commutative law. By (a), this equals $(A - B) - C$.

(c) This is the same as proving $(A \cap B^c) \cup (B \cap A^c) = (A \cup B) \cap (A \cap B)^c$. By DeMorgan's law, the right side is $(A \cup B) \cap (A^c \cup B^c)$. Thus we want to prove $(A \cap B^c) \cup (B \cap A^c) = (A \cup B) \cap (A^c \cup B^c)$. The left side has $\cup$ as the outer operation and $\cap$ as the inner. The right side is just the reverse. Repeated use of the distributive, commutative and associative laws will convert one form to the other. We have

$$(A \cap B^c) \cup (B \cap A^c) = \left(A \cup (B \cap A^c)\right) \cap \left(B^c \cup (B \cap A^c)\right)$$
$$= (A \cup B) \cap (A \cup A^c) \cap (B^c \cup B) \cap (B^c \cup A^c).$$

Since $X \cup X^c$ is the universal set U and since $X \cap U = X$, this becomes

$$(A \cup B) \cap U \cap U \cap (B^c \cup A^c) = (A \cup B) \cap (B^c \cup A^c),$$

which is what we needed to prove.

SF-1.12 (a) True. Below is the Venn diagram with the regions numbered 1 to 8. $A-C$ consists of regions $\{2,3\}$, $B-C$ consists of regions $\{3,4\}$, and $A-B$ consists of regions $\{2,5\}$. There is no region common to all three of these sets, so $(A-C)\cap(B-C)\cap(A-B) = \emptyset$.

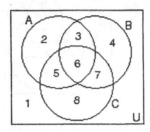

Here is an algebraic proof as well.

$$(A-C)\cap(B-C)\cap(A-B) = A\cap C^c\cap B\cap C^c\cap A\cap B^c = A\cap C^c\cap C^c\cap A\cap(B\cap B^c) = \emptyset$$

since $B\cap B^c = \emptyset$.

(b) True. Using the Venn diagram below, the proof that $A\cap(U-B) = \emptyset$ becomes $\{4,5\}\cap\{1,2\} = \emptyset$.

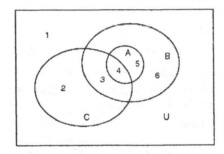

(c) False. Using the above Venn diagram: $A\cap(U-(B\cap C))$ becomes $\{4,5\}\cap\{1,2,5,6\} = \{5\}$ so the intersection is not empty. To construct a specific counterexample, we need to make sure that region 5 is not empty. That is, we want something in A that is not in C. We can take $A = B = \{a\}$ and $C = \emptyset$. Then $A\cap(U-(B\cap C)) = \{a\} \neq \emptyset$.

(d) False. To find a counterexample, we can go to the Venn diagram for (a) and cross off the numbers of the regions that must be empty because of the condition $(B\cap C) \subseteq A$. In this case, it is region 7. We drop this number for all our calculations. Now $A-B$ corresponds to $\{2,5\}$ and $A-C$ to $\{2,3\}$. The intersection is $\{2\}$. Thus we can take $A = \{a\}$ and $B = C = \emptyset$. Of course, you may have seen this without using the Venn diagram, which is fine.

(e) False. Counterexample: $A = \{a\}$ and $B = \{b\}$. Then $A \times B = \{(a,b)\}$.

SF-1.13 Note that $A\oplus B = (A\cap B^c)\cup(B\cap A^c)$. By Exercise 1.11(c), $A\oplus B = A\cup B-(A\cap B)$. In words, $A \oplus B$ consists of everything in A or B that is not in both A and B. For

[205]

Solutions for Sets and Functions

this problem, we refer to the following Venn diagram.

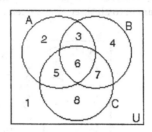

(a) $B \oplus C$ consists of regions $\{3,4,5,8\}$. $A \oplus (B \oplus C)$ consists of regions $\{2,6,4,8\}$.
$A \oplus B$ consists of regions $\{2,5,4,7\}$. $(A \oplus B) \oplus C$ consists of regions $\{2,4,6,8\}$.
The final set of regions, $\{2,4,6,8\}$, is the same in both cases. Thus, $A \oplus (B \oplus C) = (A \oplus B) \oplus C$.

Note: Since we have the associative and commutative laws we can combine a collection of sets using $\oplus$ in any order we wish and the answer will be the same. For example, $(A \oplus B) \oplus (C \oplus D) = D \oplus ((B \oplus C) \oplus A)$. Also note that $A \oplus (B \oplus C)$ consists of those elements that are in an *odd* number of A, B and C. This is true in general: $A_1 \oplus A_2 \oplus \cdots \oplus A_n$ consists of those elements that are in an odd number of $A_1, A_2, \ldots, A_n$. You can apply this fact to get alternate proofs for parts (b), (c) and (d) of this problem.

(b) $A \oplus \emptyset = (A \cup \emptyset) - (A \cap \emptyset) = A - \emptyset = A$.

(c) $A \oplus A^c = (A \cup A^c) - (A \cap A^c) = U - \emptyset = U$. (Note: $A \oplus U = A^c$.)

(d) $A \oplus A = (A \cup A) - (A \cap A) = A - A = \emptyset$.

(e) If $A \oplus C = B \oplus C$ then $(A \oplus C) \oplus C = (B \oplus C) \oplus C$. Thus $A \oplus (C \oplus C) = B \oplus (C \oplus C)$ and so $A \oplus \emptyset = B \oplus \emptyset$. Finally, $A = B$.

SF-1.14 Use the Venn diagram:

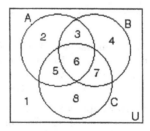

(a) Must be disjoint. $A - B$ is regions $\{2,5\}$ and $B - C$ is regions $\{3,4\}$, so there are no regions in common.

(b) May not be disjoint. $A - B$ is regions $\{2,5\}$ and $C - B$ is regions $\{5,8\}$. Region 5 is common to both and may be nonempty.

(c) Must be disjoint. $A - (B \cup C)$ is region 2 and $B - (A \cup C)$ is region 4. There are no regions in common.

(d) May not be disjoint. $A - (B \cap C)$ consists of regions $\{2,3,5\}$ and $B - (A \cap C)$ consists of regions $\{3,4,7\}$. Region 3 is common to both and may be nonempty.

SF-1.15 (a) No because 1 appears in $\{1,3,5\}$ and in $\{1,2,6\}$.

(b) Yes because every element in $\{1,2,\ldots,8\}$ appears in exactly one block.

(c) Yes because a partition is a *set* and so we can ignore the fact that $\{2,6\}$ was listed twice.

(d) No because 7 is missing.

SF-1.16 We can choose any refinement of $\{1,3,5\}$ (there are B_3), any refinement of $\{2,6\}$, and any refinement of $\{4,7,8,9\}$. The number of refinements is the product of the Bell numbers: $B_3 \times B_2 \times B_4 = 5 \times 2 \times 15 = 150$.

SF-1.17 (a) Suppose $x \in S \cup T$. Every element of $S \cup T$ appears in exactly one of S and T. If $x \in S$, then it appears in exactly one block σ. If $x \in T$, then it appears in exactly one block τ. Hence x appears in exactly one block of $\sigma \cup \tau$.

(b) We get each refinement of $\sigma \cup \tau$ by choosing a refinement of σ and choosing a refinement of τ. Thus there are $n_\sigma n_\tau$ refinements of $\sigma \cup \tau$.

SF-1.18 (a) $\{1,2,3\}$ has three elements. Here are the subsets and the characteristic functions with χ given as $\chi(1), \chi(2), \chi(3)$.

$\emptyset$	0,0,0	$\{1\}$	1,0,0	$\{2\}$	0,1,0
$\{1,2\}$	1,1,0	$\{3\}$	0,0,1	$\{1,3\}$	1,0,1
$\{2,3\}$	0,1,1	$\{1,2,3\}$	1,1,1		

(b) Since $X \times Y = \{(a,x),(a,y),(b,x),(b,y)\}$, we list χ in the order $\chi((a,x)), \chi((a,y)), \chi((b,x)), \chi((b,y))$.

$\emptyset$	0,0,0,0	$\{(a,x)\}$	1,0,0,0
$\{(a,y)\}$	0,1,0,0	$\{(a,x),(a,y)\}$	1,1,0,0
$\{(b,x)\}$	0,0,1,0	$\{(a,x),(b,x)\}$	1,0,1,0
$\cdots$		$\cdots$	
$\{(a,x),(a,y),(b,x)\}$	1,1,1,0	$\{(a,x),(a,y),(b,x),(b,y)\}$	1,1,1,1

SF-1.19 Remember that the power set of a set S contains the empty set $\emptyset$, the set S itself, and all proper, nonempty subsets of S.

(a) Here $S = \emptyset$, which has no nonempty, proper subsets. Thus $\mathcal{P}(\emptyset) = \{\emptyset\}$. This is *not* the empty set — it is a one-element set and its element is the empty set.

(b) $\mathcal{P}(\mathcal{P}(\emptyset)) = \mathcal{P}(\{\emptyset\})$, so $S = \{\emptyset\}$ is a set with one element. If we call this element a, then $\mathcal{P}(S) = \{\emptyset, \{a\}\}$. Replacing a with its value $\emptyset$, we have $\mathcal{P}(\mathcal{P}(\emptyset)) = \{\emptyset, \{\emptyset\}\}$.

(c) We are now starting with a two-element set, so it has four subsets. Thus

$$\mathcal{P}(\mathcal{P}(\mathcal{P}(\emptyset))) = \mathcal{P}(\{\emptyset, \{\emptyset\}\}) = \Big\{\emptyset, \{\emptyset\}, \{\{\emptyset\}\}, \{\emptyset, \{\emptyset\}\}\Big\}.$$

This looks confusing, but if you write down $\mathcal{P}(\{a,b\})$ and then replace a with $\emptyset$ and b with $\{\emptyset\}$, you should have no trouble.

Solutions for Sets and Functions

SF-1.20 (a) Since $A \subseteq A \cup B$, we have $\mathcal{P}(A) \subseteq \mathcal{P}(A \cup B)$. Similarly , $\mathcal{P}(B) \subseteq \mathcal{P}(A \cup B)$ and so $\mathcal{P}(A) \cup \mathcal{P}(B) \subseteq \mathcal{P}(A \cup B)$.

If $A \subseteq B$, then $A \cup B = B$ and $\mathcal{P}(A) \subseteq \mathcal{P}(B)$. Thus $\mathcal{P}(A \cup B) = \mathcal{P}(A) \cup \mathcal{P}(B)$. Similarly, they are equal if $B \subseteq A$.

Suppose this is not the case so that $a \in A - B$ and $b \in B - A$. Then $\{a, b\} \in \mathcal{P}(A \cup B)$ but $\{a, b\} \notin \mathcal{P}(A) \cup \mathcal{P}(B)$.

In summary, if either $A - B = \emptyset$ or $B - A = \emptyset$, then the given sets are equal. Otherwise, $\mathcal{P}(A) \cup \mathcal{P}(B)$ is a proper subset of $\mathcal{P}(A \cup B)$.

(b) They are equal. Proof: $X \in \mathcal{P}(A \cap B)$ iff $X \subseteq A \cap B$
iff $X \subseteq A$ and $X \subseteq B$ iff $X \in \mathcal{P}(A)$ and $X \in \mathcal{P}(B)$
iff $X \in \mathcal{P}(A) \cap \mathcal{P}(B)$. We have shown that $\mathcal{P}(A \cap B) = \mathcal{P}(A) \cap \mathcal{P}(B)$.

(c) For any sets A and B, $\mathcal{P}(A) \times \mathcal{P}(B)$ and $\mathcal{P}(A \times B)$ have no elements in common! The set $\mathcal{P}(A) \times \mathcal{P}(B)$ consists of pairs of subsets (S, T) where $S \subset A$ and $T \subset B$. The set $\mathcal{P}(A \times B)$ has elements W which are collections of pairs (x, y) where $x \in A$ and $y \in B$. The pair of sets (S, T) cannot equal the set of pairs W.

Let's count the number of elements. Recall that $|\mathcal{P}(C)| = 2^{|C|}$ and $|U \times V| = |U| \cdot |V|$. Thus

$$|\mathcal{P}(A) \times \mathcal{P}(B)| = |\mathcal{P}(A)| \cdot |\mathcal{P}(B)| = 2^{|A|} 2^{|B|} = 2^{|A|+|B|}$$

and

$$|\mathcal{P}(A \times B)| = 2^{|A \times B|} = 2^{|A| \cdot |B|}.$$

These two numbers will be equal if and only if $|A| + |B| = |A| \cdot |B|$. Thus we need to know what the solutions of $x + y = xy$ in nonnegative integers. We can rewrite this as $xy - x - y + 1 = 1$, which can be written $(x - 1)(y - 1) = 1$. The product of two integers is 1 if and only if both integers are $+1$ or both integers are -1. Thus the only solutions are $x - 1 = y - 1 = 1$ and $x - 1 = y - 1 = -1$. The first case says $|A| = |B| = 2$ and the second case says that $A = B = \emptyset$. In the first case, the two sets in the problem have $2^4 = 16$ elements and in the second case the two sets are $\{\emptyset\}$ and $\{(\emptyset, \emptyset)\}$. In all other cases, $|A| \cdot |B| > |A| + |B|$, which means that $\mathcal{P}(A \times B)$ has more elements than $\mathcal{P}(A) \times \mathcal{P}(B)$.

SF-1.21 We know that a set with m elements has 2^m subsets. Since T_1 is the set of all subsets of $\{2, \ldots, n\}$ (which has $n - 1$ elements), $|T_1| = 2^{n-1}$.

What about S_1? Here's one way. Every subset of S is either in S_1 (if it contains 1) or in T_1 (if it does not contain 1), but not both. Thus $|S_1| + |T_1| = 2^{|S|} = 2^n$ and so $|S_1| = 2^n - |T_1| = 2^n - 2^{n-1} = 2^{n-1}$.

Here's another way. We can remove 1 from a subset in S_1 and obtain a subset in T_1. This process is reversible. Thus $|S_1| = |T_1|$.

SF-2.1 In the figure below, the set $\mathbb{R} \times \mathbb{R}$ is represented in the usual manner by points in plane. Of course, the figure only shows a portion of $\mathbb{R} \times \mathbb{R}$. Points satisfying the relation are dark, forming a curve — the parabola.

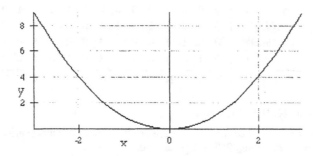

SF-2.2 (a) No (not true for empty set).   (b) Yes.   (c) No.

SF-2.3 (a) Pictorially, this relation S on B can be drawn as follows (there are many other ways ...):

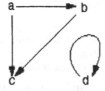

(b) Since the relation is symmetric, an arrow from a to b means there is also one from b to a. To avoid cluttering the figure, we draw this as an arrow with heads at both ends.

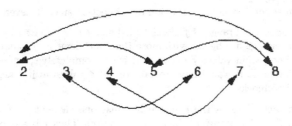

SF-2.4 Altogether, there are sixteen relations of which four are functional. We can list those that are *not* functional by listing all sixteen relations and then removing those that are functional. A subset S of $\{a, b\} \times \{x, y\}$ will be functional if exactly one of (a, x) and (a, y) is in the subset (defines $f(a)$) and exactly one of (b, x) and (b, y) is in the subset (defines $f(b)$). In the following table, each column gives a subset of $\{a, b\} \times \{x, y\}$ in characteristic function form. The last row indicates whether it is functional (Y) or not (N).

$\chi((a,x))$	0	0	0	0	0	0	0	0	1	1	1	1	1	1	1	1
$\chi((a,y))$	0	0	0	0	1	1	1	1	0	0	0	0	1	1	1	1
$\chi((b,x))$	0	0	1	1	0	0	1	1	0	0	1	1	0	0	1	1
$\chi((b,y))$	0	1	0	1	0	1	0	1	0	1	0	1	0	1	0	1
functional?	N	N	N	N	N	Y	Y	N	N	Y	Y	N	N	N	N	N

SF-2.5 $S = \{(3,6),(4,4),(5,5)\}$ and $S^{-1} = \{(6,3),(4,4),(5,5)\}$.

SF-2.6 (a) $|A \times B| = mn$. For any set S there are $2^{|S|}$ subsets. Thus there are 2^{mn} subsets of $A \times B$ or, same thing, 2^{mn} relations from A to B.

(b) Consider the definition of a function. Each $x \in A$ must be paired with exactly one $y \in B$. For each $x \in A$ there are $|B|$ choices for $y \in B$. List the elements of A as $a_1, a_2, \ldots, a_{|A|}$. There are $|B|$ choices to pair with a_1, $|B|$ choices to pair with a_2, etc. until finally $|B|$ choices to pair with $a_{|A|}$. The total number of choices is $|B| \cdot |B| \cdots |B| = |B|^{|A|}$.

SF-2.7 There are 17 edges in the digraph. To explain, list them all or draw a picture of the digraph.

SF-2.8 (a) No. Since no order is given for the domain, we cannot specify f in one-line form. Since f takes the value 3 two times, it is not an injection and hence not a bijection. Since we know the range, we can see that f is a surjection.

(b) We know the range and domain. Using the implicit order of the domain, we know f. Its two-line form is $\begin{pmatrix} 1 & 2 & 3 \\ ? & < & + \end{pmatrix}$. It is an injection but not a surjection or a bijection since it never takes the value $>$.

(c) We know the range and domain and are given the function values. The two-line form is $\begin{pmatrix} 1 & 2 & 3 \\ 4 & 3 & 2 \end{pmatrix}$. Using the implicit order of the domain, the one-line form $(4,3,2)$. It is an injection but not a surjection or a bijection since it never takes the value 1.

SF-2.9 (a) The domain and range of f are specified and f takes on exactly two distinct values. Since the coimage has blocks with more than one element, f is not an injection. Since we don't know what values f takes on, it is not completely specified; however, it is not a surjection since it would have to take on all 4 values in its range and the coimage has only two blocks.

(b) Since each block of the coimage has just one element, f is an injection. Since $|\text{Coimage}(f)| = 5 = |\text{range of } f|$, f is a surjection. Thus f is a bijection and, since the range and domain are the same, f is a permutation. In spite of all this, we don't know $f(x)$ for any $x \in \underline{5}$.

(c) We know the domain and range of f since $\{f^{-1}(2), f^{-1}(4)\}$ is a partition of the domain, we know $f(x)$ for all $x \in \underline{5}$. Thus we know f completely. It is neither a surjection nor an injection.

(d) We know that f is a surjection. It cannot be an injection because the domain is larger than the range. We cannot specify f.

(e) This specification is nonsense: Since Image(f) must be a subset of the range, it cannot have more than four elements.

(f) This specification is nonsense: Sinc each block of Coimage(f) corresponds to a different element of the image of f, it cannot have more than four blocks.

SF-2.10 (a) and (b) are contrapositives of each other and hence logically equivalent. Both are correct definitions of injective or one-to-one.

(c) No. What is being defined is a function and not all functions are one-to-one. For example, $A = \{a, b\}$, $B = \{c\}$, $f(a) = f(b) = c$ satisfies the definition.

(d) Correct. This is the definition of one-to-one.

SF-2.11 (a) g is one-to-on because $g(s) = g(t)$ means $3s - 1 = 3t - 1$ and so $s = t$.

(b) g is not onto. For example $g(s) = 0$ would mean $3s - 1 = 0$ and so $s = 1/3$, which is not an integer.

(c) g is onto in this case. Given $s \in \mathbb{R}$, we must find $x \in \mathbb{R}$ such that $g(x) = s$. In other words, we want $3x - 1 = s$, so let $x = (s+1)/3$. Then $g(x) = 3(s+1)/3 - 1 = s$.

SF-2.12 There are at least three ways to do this problem:

- Either find $x \neq y$ such that $f(x) = f(y)$ or prove that $f(x) = f(y)$ implies that $x = y$. This is the straightforward method, but it is not always convenient.

- Use calculus to show that $f(x)$ is strictly monotonic and hence one-to-one.

- Look at a carefully drawn graph of $f(x)$. This works if f is *not* one-to-one — you'll be able to see that there are x and y with $f(x) = f(y)$. You can't be sure when f is one-to-one because you can't graph $f(x)$ for *all* x in the domain. Calculus may help in this case.

We'll use various methods here.

(a) f is not one-to-one. Here is what the graph of $x/(x^2 + 1)$ looks like:

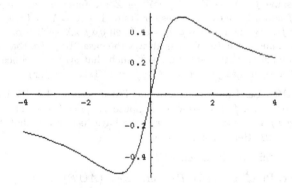

(b) f is one-to-one. It is easier to see if we write $f(x) = 2 + 1/x$. Suppose $f(s) = f(t)$. Then $1/s = 1/t$ and so $s = t$.

(c) f is one-to-one. $f(x) = 1 - 2/(x+1)$. Proceed as in (b).

Solutions for Sets and Functions

SF-2.13 It's easy to convert between one-line and two-line form: If the top line of two-line form is arranged in order $(1, 2, \ldots)$, then the bottom line is the one-line form. Hence we usually omit one or the other of these forms below. (a) For $(1,5,7,8)(2,3)(4)(6)$, the two-line form is $\begin{pmatrix} 1 & 2 & 3 & 4 & 5 & 6 & 7 & 8 \\ 5 & 3 & 2 & 4 & 7 & 6 & 8 & 1 \end{pmatrix}$. The inverse is $(1,8,7,5)(2,3)(4)(6)$ in cycle form, $(8,3,2,4,1,6,5,7)$ in one-line form.

(b) For $\begin{pmatrix} 1 & 2 & 3 & 4 & 5 & 6 & 7 & 8 \\ 8 & 3 & 7 & 2 & 6 & 4 & 5 & 1 \end{pmatrix}$, the cycle form is $(1,8)(2,3,7,5,6,4)$. The inverse is $(1,8)(2,4,6,5,7,3)$ in cycle form and $(8,4,2,6,7,5,3,1)$ in one-line form

(c) The one-line form $(5,4,3,2,1)$ has the cycle form $(1,5)(2,4)(3)$. The permutation is its own inverse.

(d) The cycle form $(5,4,3,2,1)$ is not in standard form. The standard form in $(1,5,4,3,2)$. Its the one-line form is $(5,1,2,3,4)$. Its inverse is $(1,2,3,4,5)$ in cycle form and $(2,3,4,5,1)$ in one-line form.

SF-2.14 Let $A = \{a_1, a_2, \ldots\}$.

(a) Note that $f(a_1) \in B$, $f(a_2) \in B - \{f(a_1)\}$ and $f(a_1) \in B - \{f(a_1), f(a_2)\}$. Thus there are $|B| = 3$ choices for $f(a_1)$, then $|B| - 1 = 2$ choices for $f(a_2)$ once $f(a_1)$ is chosen, etc. This gives $|S| = 3 \times 2 \times 1 = 6$.

(b) $5 \times 4 \times 3 = 60$ by the reasoning in (a).

(c) If $m > n$ the answer is zero. Otherwise, reasoning as in (a) the answer is

$$\overbrace{n \times (n-1) \times (n-2) \times \cdots \times (n-m+1)}^{m \text{ factors}}$$

This is called the "falling factorial" and written $(n)_m$.
Alternatively, one can choose the m elements of the image in $\binom{n}{m}$ ways and then write those m things in one-line form in $m!$ ways so the answer is $\binom{n}{m} m! = (n)_m$.

SF-2.15 (a) Assume $f : X \to Y$ and $g : Y \to Z$ are functions and $g \circ f : X \to Z$ is onto. Must f and g be onto? The answer is no. Let $X = Y = \{a, b\}$ and let $Z = \{c\}$. Let $f(a) = f(b) = b$ and $g(a) = g(b) = c$. Then $g \circ f : X \to Z$ is onto but f is not onto. In this example, g is onto. That is always the case. To prove that g is onto, we pick any $z \in Z$ and show that there is a $y \in Y$ such that $g(y) = z$. Since $g \circ f$ is onto, there is an $x \in X$ such that $g \circ f(x) = g(f(x)) = z$. Take $y = f(x)$.

(b) In this case, f must be one-to-one, but g need not be. We let you find an example for g. We prove f is one-to-one. Suppose $f(x_1) = f(x_2)$. Then $g(f(x_1)) = g(f(x_2))$. Since $g \circ f$ is one-to-one, $x_1 = x_2$. We have just shown that $f(x_1) = f(x_2)$ implies that $x_1 = x_2$. Hence f is one-to-one.

SF-2.16 We write "iff" for "if and only if."

(a) True. Proof: $y \in f(A \cup B)$ iff $\exists x \in (A \cup B),\ f(x) = y$
iff $\left(\exists x \in A,\ f(x) = y\right)$ or $\left(\exists x \in B,\ f(x) = y\right)$
iff $y \in f(A)$ or $y \in f(B)$ iff $y \in (f(A) \cup f(B))$.

(b) False. Counterexample: Let $A = \{1\}$, $B = \{2\}$ and $f(1) = f(2) = 3$. Then $f(A \cap B) = f(\emptyset) = \emptyset$ and $f(A) \cap f(B) = \{3\}$.

(c) False. Counterexample: Let $A = \{1\}$, $B = \{2\}$ and $f(1) = f(2) = 3$. Then $f(A - B) = f(A) = \{3\}$ and $f(A) - f(B) = \{3\} - \{3\} = \emptyset$.

(d) True. Proof: $x \in f^{-1}(C \cap D)$ iff $f(x) \in C \cap D$
iff $f(x) \in C$ and $f(x) \in D$ iff $x \in f^{-1}(C)$ and $x \in f^{-1}(D)$
iff $x \in f^{-1}(C) \cap f^{-1}(D)$.

SF-2.17 (a) False. Find the simplest counter example you can!

(b) False. Suppose $X = \{a\}$ and $Y = \{c, d\}$. Let $f(a) = c$ and take the set C of the problem to be Y. Is the statement true for this example? Remember, to show that $P = Q$ for two sets P and Q, you must show that P is a subset of Q and Q is a subset of P.

(c) True. Proof: $x \in (g \circ f)^{-1}(E)$ iff $g(f(x)) \in E$
iff $f(x) \in g^{-1}(E)$ iff $x \in f^{-1}(g^{-1}(E))$.

SF-2.18 (a) Let the elements of the domain be $\{a, b, c\}$ and the elements of the codomain be $\{u, v\}$. We can construct two onto functions f with a given coimage $\{S, T\}$:

- one by taking $S = f^{-1}(u)$ and $T = f^{-1}(v)$,
- one by taking $S = f^{-1}(v)$ and $T = f^{-1}(u)$.

The number of choices for $\{S, T\}$ is the number of partitions of $\{a, b, c\}$ into two blocks. There are three such partitions: $\{\{a, b\}, \{c\}\}$, $\{\{a, c\}, \{b\}\}$ and $\{\{b, c\}, \{a\}\}$. Thus the answer is $2 \times 3 = 6$.

(b) There no onto functions in this case. For an onto function $|\text{Image}(f)| = |\text{Range}(f)|$. For any function, $|\text{Image}(f)| \leq |\text{Domain}(f)|$. Thus $|\text{Range}(f)| \leq |\text{Domain}(f)|$ for an onto function. In other words $|B| \leq |A|$. In this case, that would mean $5 \leq 3$, which is not true.

(c) The number of possible coimages is equal to the number of partitions of a set of four elements into two blocks. Each such coimage gives rise to two onto functions, just like in part (a). Rather than list all partitions of four things into two blocks, we note that it is $S(4, 2)$ by definition and use the table in the text to see that $S(4, 2) = 7$. Hence there are $2 \times 7 = 14$ onto functions.

(d) If $m < n$ there are none because of the reasoning in part (b). Suppose that $m \geq n$. We proceed as in (c):

- Given a partition of A, we claim $n!$ onto functions have this partition as coimage. Why is this? Each block must be mapped to a *different* element of B by a function that has this partition as coimage. List the blocks in some order. There are n choices for the image of the first block. This leaves $n - 1$ choices for the image of the second block. This leaves $n - 2$ choices for the image of the third block, and so on. Thus we get $n(n - 1)(n - 2) \cdots = n!$ possible functions.

- Since the number of blocks in the coimage equals the size of the image, you can see that we must partition A into $|B|$ blocks. (Look at the discussion in part (b) if this is unclear.) Thus there are $S(m, n)$ possible partitions.

Since there are $n!$ onto functions for each of the $S(m, n)$ partitions, the answer is $n! \, S(m, n)$.

(e) Since we are dealing with onto functions, $k = n$. Now apply the formula and use the fact that $\binom{n}{n} = 1$.

SF-2.19 One way is to fill in the brief explanation given in the text. Here is another.

- Let $C = \text{Image}(f)$, the image of A. There are $\binom{n}{k}$ possible choices for C since it is a k-subset of B.

- How many functions are there with a given C? They are the onto functions from A to C. We counted them in part (d) of the previous exercise, except now the range is a k-set instead of an n-set. Thus there are $k!\, S(m, k)$ such functions.

Putting the two parts together gives the answer.

Solutions for Equivalence and Order

EO-1.1 Let $a : S \to \mathbb{N} \times \mathbb{N}$ be the function that assigns to each student x, the age of x paired with the years completed. The equivalence class partition is the coimage partition of the function a.

EO-1.2 There are d equivalence classes: $\{x, x+d, x-d, x+2d, x-2d, \ldots\}, x = 0, 1, 2, \ldots, d-1$. They form the coimage partition of the function $m(x) = x \pmod{d}$.

EO-1.3 Define $t(x)$ to be the truth table of x. The coimage partition of t is the set of equivalence classes. The equivalence classes correspond to sets of equivalent forms in the usual sense that they represent the same Boolean function. How many equivalence classes are there?

EO-1.4 Define a function $f(x) = x^k \pmod{d}$. The equivalence classes are the blocks of the Coimage(f). Can you describe the equivalence classes?

EO-1.5 This is a case where it is easy to show that the relation is reflexive, symmetric, and transitive. Defining the function f is not too bad either: $f(x) = x - \lfloor x \rfloor$. That is, $f(x)$ is x minus the least integer in x. We have reached a "cross over" point where proving directly that the relation is reflexive, symmetric, and transitive is easier (barely) than dealing with Coimage(f).

EO-1.6 There are $26 \times 26 = 676$ possible (first,last) letter pairs. Thus at least one block of the coimage of the function f that maps a name to its (first,last) letter pair must have more than one element.

EO-1.7 (a) Yes. The mapping from a set of k integers to remainder mod $k-1$ has $k-1$ blocks in its coimage and k elements in its domain.

(b) No. Take the set of integers to be $\{0, 1, 2, \ldots, k-1\}$. The mapping from integers to integers mod k is the identity mapping.

EO-1.8 Let's look at a particular case, say $n = 9$ so $S = \{1, 2, 3, 4, 5, 6, 7, 8, 9\}$. Choose $C = \{1, 2, 3, 4, 5\}$. No pair adds up to 10. Now suppose C has 6 elements. We claim there is a pair that adds to 10. To see why, let $f(x)$ be the set $\{x, (10 - x)\}$ for $x \in S$. Image(f) has exactly 5 elements and so Coimage(f) has exactly 5 blocks. Now restrict f to C. The domain has 6 elements but the coimage has at most 5 blocks. Therefore there must be two distinct elements x, y in C that have the same value. Thus $\{x, (10-x)\} = \{y, (10-y)\}$. and, since $x \neq y$, $x = 10-y$. Thus $x+y = 10$. For general n, $f(x) = \{x, (n+1 - x)\}$ and Coimage(f) has $\lceil n/2 \rceil$ blocks. Thus $k = \lceil n/2 \rceil + 1$.

EO-1.9 This is a more elementary idea than the pigeon-hole principle. Suppose n is even, say $n = 2j$. Note that there are $j+1$ even and j odd integers in S. To be sure of getting an odd integer you must pick at least $j+2$ elements of S. To be sure of getting an even integer, you must pick at least $j+1$. We leave it to you to do the case $n = 2j-1$.

EO-1.10 The primes between 1 and 50 are $P = \{2, 3, 5, 7, 11, 13, 17, 19, 23, 29, 31, 37, 41, 43, 47\}$. There are 15 of them. Thus $k = 15$ doesn't work: Take $S = P$. We now show that $k = 16$ works. Define a function f from the set S to P by

$$f(x) = (\text{smallest prime that divides } x).$$

Solutions for Equivalence and Order

If S has at least 16 elements then one block of the Coimage(f) must contain at least two elements. Say $f(s) = f(t) = p$. Then p is the smallest prime dividing s and also the smallest prime dividing t. Hence p divides $\gcd(s,t)$ and so $\gcd(s,t) > 1$.

EO-1.11 Think about the coimage partition of the function B that maps a person to his/her birth-month. This partition has as most twelve blocks. If each block had less than three elements there would be at most $12 \times 2 = 24$ elements (persons) in the domain of B. Thus, if the domain of B contains more than 24 elements (persons) there must be at least one block in Coimage(B) with more than two elements. In other words, if there are more than 24 in a group of people, there must be at least three with the same birth-month. Thus $k = 25$.

EO-1.12 The table below shows the structure of the possible coimage partitions of f, given that no block has more than three elements. The first row indicates the possible block sizes, 1, 2, or 3. The entries in the other rows indicate how many blocks there are of that size. Thus, in the row with entries 9, 1, 1, there are 9 blocks of size 3, 1 block of size 2, and 1 block of size 1. Our solution given in the statement of the problem corresponds to the first row.

3	2	1
10	0	0
9	1	1
9	0	3
8	3	0
8	2	2
7	4	1
6	6	0

EO-1.13 Let B denote the map from persons to birth ATCs. Coimage(B) can have at most 1461 blocks. If every block had at most 3 elements then the domain of B would be no larger than $3 \times 1461 = 4383$. Thus, if the domain of B has more than 4383 elements there must be at least one block of Coimage(B) with at least 4 elements. Thus $k = 4384$.

EO-1.14 Let f be function from the set of N students to their scores (in the range 27 to 94). The codomain of f has maximum size 65. We have $65 \times 2 = 130$. Thus, $N > 130$ guarantees that 3 students must have the same score.

EO-1.15 You choose x pennies and look at the map f from these pennies to their dates. There can be at most three blocks in the coimage. If each block had three elements, then $x = 9$. Thus, you had better pick more than nine to be sure of getting one block with at least four pennies. If you pick ten pennies, one block must contain at least four pennies. Thus $N_4 = 10$.

To find N_6 a little more care must be taken. There are only four 1971 pennies so $|f^{-1}(1971)| \leq 4$. To choose the most pennies and not get at least 6 with the same date, take all four 1971 pennies, five 1968 pennies, and five 1967 pennies for a total of 14 pennies. Thus, you can choose 14 pennies and still not have six pennies with the same date. One more penny forces you to take at least six pennies of the same date. Thus $N_6 = 15$.

By the same sort of reasoning, $N_8 = 19$. (You can take all the 1968 and 1971 pennies plus 7 of the 1967 pennies without getting eight pennies with the same date.)

EO-1.16 If $n \mid t_k$ for some k, we are done, so assume that n does not divide t_k for any k. Look at the set $R = \{t_k \% n \mid k = 1, 2, \ldots, n\}$ of remainders mod n. This set has cardinality at most $n - 1$, since $0 \notin R$ by assumption. Thus, by the pigeonhole principle, there must be at least two of the integers t_i and t_j that have the same value mod n. Thus, $n \mid (t_i - t_j)$.

EO-1.17 The approach is similar to previous problem. Let

$$S = \Big\{ \{0\}, \ \{1, n-1\}, \ \{2, n-2\}, \ \ldots, \ \{\lfloor n/2 \rfloor, \lceil n/2 \rceil\} \Big\},$$

a collection of $\lfloor n/2 \rfloor + 1$ sets, each with two elements except $\{0\}$ and, if n is even, $\{\lfloor n/2 \rfloor, \lceil n/2 \rceil\} = \{n/2\}$. Note that if any set in this collection has two elements, then the sum of those two elements is n. Let $R = \{t_i \mid i = 1, 2, \ldots, k\}$ be any collection of k integers and let $f : R \to S$ be a function from R to S. If $k > \lfloor n/2 \rfloor + 1$, then, by the pigeonhole principle, there must be two elements of R that are mapped to the same set of S.

Let $f(t_j)$ to be the set of S that contains the integer $t_j \% n$. If $f(t_i) = f(t_j)$, there are two possibilities:

(1) $t_i \% n = t_j \% n$ or

(2) $t_i + t_j = 0 \pmod{n}$. Thus, either $n \mid (t_i - t_j)$ or $n \mid (t_i + t_j)$. Hence $k \geq \lfloor n/2 \rfloor + 2$ will guarantee the condition of the problem. Why is this bound best possible? (That is, find an example that fails when $k = \lfloor n/2 \rfloor + 1$.)

EO-1.18 Consider an example. Suppose $n = 12$. The numbers are

$$D = \{1, 2, 3, 4, 5, 6, 7, 8, 9, 10, 11, 12\}.$$

Factoring out the highest power of two in each case, gives

$$2^0 \times 1, \ 2^1 \times 1, \ 2^0 \times 3, \ 2^2 \times 1, \ 2^0 \times 5, \ 2^3 \times 1, \ 2^0 \times 9, \ 2^1 \times 5, \ 2^0 \times 11, \ 2^2 \times 3,$$

where each number is written as $2^i \times x$ where x is odd. Define $f(2^i \times x) = x$. The function f give the "odd part" of a number. The domain of f is D and the codomain is $R = \{1, 3, 5, 7, 9, 11\}$, the odd integers in D. Note that $|R| = 6 = n/2$. If we take any subset S of D of size at least 7, then, by the pigeonhole principle, there are at least two different integers $p < q$ in S such that $f(p) = f(q)$. If $f(p) = f(q) = x$, we have that $p = 2^i \times x$ and $q = 2^j \times x$ and hence, since $p < q$, we must have $i < j$ and so $p \mid q$. This process works for any n. When n is even $|R| = n/2$ There are slight differences when n is odd, so you should check an odd case, say $n = 11$. When n is odd, $|R| = (n+1)/2$. In general, $|R| = \lfloor (n+1)/2 \rfloor = \lceil n/2 \rceil$. Thus, $m > \lceil n/2 \rceil$ works. Why is this value of m best possible? Look at the set of integers in D that are bigger than $n/2$. This set has $\lceil n/2 \rceil$ elements, and no two divide each other.

EO-1.19 (a) The sequence $1, 2, 3, \ldots, \iota$ obviously works.

(b) The sequence $2, 1, 4, 3, \ldots, (2k), (2k-1), \ldots, (2\iota), (2\iota-1)$ works. Why? It is clear that the longest decreasing subsequence has length 2. For each $j = 1, \ldots, \iota$, an increasing subsequence can contain at most one of $(2j)$ and $(2j-1)$.

Solutions for Equivalence and Order

(c) We use the idea in (b). Let S_k be the δ-long decreasing sequence

$$(\delta k),\ (\delta k - 1),\ \ldots,\ (\delta k - (\delta - 1)).$$

Note that its last term is $\delta(k - 1) + 1$. Our sequence is $S_1, S_2, \ldots, S_\iota$. A decreasing subsequence must take all of its elements from a single S_k and so has length at most δ. An increasing subsequence can take at most one element from each S_k and so has length at most ι.

EO-1.20 Since $m > q(p-1)$, by Example 11 there is either a $(q+1)$-long increasing subsequence or a p-long decreasing subsequence. We are told that the latter is not present. Thus there is a q-long increasing subsequence.

EO-1.21 (a) Since $(k - 1)^2 = n^4 < m$, there must be a k-long monotone subsequence by Example 11.

(b) Since $k > n^2$, there must be an $(n+1)$-long monotone subsequence by Example 11.

(c) By (b) the subsequence of b's is monotone. Since a subsequence of a monotone sequence is monotone, the subsequence $a_{t_1}, \ldots, a_{t_{n+1}}$ is monotone by (a).

EO-1.22 The fact that each client is assigned exactly one lawyer means that the assignment relation is a function $A : \{1, 2, \ldots, 15\} \to \{1, 2, 3, 4, 5\}$. Coimage($A$) has at exactly five blocks and each block has at most four elements. (The first condition is from the fact that each lawyer is to represent at least one client; i.e., A is onto. The second condition part of the statement of the problem.) Forget, for the moment, the question asked in the problem — a pretty good general approach in this type of problem. We need to make a table of possible assignments, without including unnecessary details, so that we can get a feel for the situation.

We consider assignments (functions) A where Coimage(A) has exactly five blocks. We make a "block-size chart" for possible coimages, with the top row given block sizes, from 1 to 4. The remaining four rows show how many blocks of that size can result from various assignments A. Thus, the first row shows 1 block of size 1 (one lawyer, one client), 1 block of size 2 (one lawyer, two clients), no blocks of size 3, and 3 blocks of size 4 (3 lawyers, 4 clients).

1	2	3	4
1	1	0	3
1	0	2	2
0	2	1	2
0	1	3	1
0	0	5	0

Now that we've got a general understanding, let's go back and find out what the question is! It was "Show that if two lawyers are assigned less than three clients, at least two must be assigned four clients." The truth of this is evident from our table. Rows one and two represent the situation where two lawyers are assigned less than three lawyers. In both cases, 2 or more (3 for row 1) are assigned four clients.

We're done but we could use the table to formulate other problems. For example

- Show that if less than two lawyers are assigned less than three clients then at least four lawyers must be assigned more than two clients.

- Show that at least three lawyers must be assigned three or more clients.

Can you do these problems by using the chart? The last question can be answered easily without the chart. Can you do it without the chart?

EO-2.1 (a) This relation is neither reflexive $(2 \not{R} 2)$, symmetric $(0\ R\ 3$ and $3\not{R} 0)$, nor transitive $(1\ R\ 0,\ 0\ R\ 3$ and $1 \not{R} 3)$.

(b) This relation is symmetric, but not reflexive and not transitive.

(c) This relation is transitive, but not reflexive and not symmetric.

(d) This relation is transitive and symmetric, but not reflexive.

(e) This relation is transitive and symmetric, but not reflexive.

EO-2.2 It is not reflexive $(1 \not{R} 1)$ and not transitive $(1\ R\ 0$ and $0\ R\ 1)$. It is symmetric since $x^2 + y^2 = n^2$ implies that $y^2 + x^2 = n^2$.

EO-2.3 It is reflexive by definition. It is symmetric since $x - y$ is an odd integer if and only if $y - x$ is an odd integer. (One difference is the negative of the other.) It is never satisfies the transitive condition for distinct x, y, z: If $x\ R\ y$ and $y\ R\ z$, then $x - y$ and $y - z$ are both odd integers and so $x - z$ is an **even** integer.

EO-2.4 It is reflexive, symmetric and transitive. One way to prove that it is an equivalence relation is to note that the equivalence classes are the blocks of the coimage partition of the function $f(x) = x^2$.

EO-2.5 It is symmetric since $\gcd(x, y) = \gcd(y, x)$. It fails to be reflexive because $\gcd(1, 1) = 1$. It is not transitive; for example $2\ R\ 10$ and $10\ R\ 5$ but $2 \not{R} 5$.

EO-2.6 It is reflexive and symmetric. (Why?) It is not transitive; for example, consider $\{1\}$, $\{1, 2\}$ and $\{2\}$.

EO-2.7 You should be able to explain why it is reflexive and symmetric. For transitive, consider $\{1\}$, $\{1, 2\}$ and $\{2\}$.

EO-2.8 To specify a relation R, we must make one of two choices (in R or not in R) for each pair (x, y). There are n^2 pairs. Thus there are 2^{n^2} relations — that's $2^{(n^2)}$, not $(2^n)^2$. (This was done in Example 14.)

A relation R is reflexive if and only if $x\ R\ x$ for all n values of x. Thus we are free to choose for only $n^2 - n$ of the pairs (x, y), namely those with $x \neq y$. Thus there are $2^{n^2 - n}$ reflexive relations. (This was done in Example 14.)

Subtracting the reflexive relations from all relations, we find that $2^{n^2} - 2^{n^2 - n}$ relations are not reflexive.

EO-2.9 For a symmetric relation, we can choose freely whether or not $(x, x) \in R$ for all n values of x; however, once we have made a choice for (x, y), we must make the same choice for (y, x). Thus we are free to choose for only half of those (x, y) with $x \neq y$. The number of such pairs is (x, y) is $n^2 - n$. Thus we are free to choose for $n + (n^2 - n)/2$ pairs. This number equals $(n^2 + n)/2 = n(n + 1)/2$. Hence there are $2^{n(n+1)/2}$ symmetric relations. Thus there are $2^{n^2} - 2^{n(n+1)/2}$ which are not symmetric.

EO-2.10 This is exactly the same as the previous exercise with one exception: We must have $(x, x) \in R$ for all R. You should be able to see that the answer is $2^{n(n-1)/2}$.

Solutions for Equivalence and Order

EO-2.11 The reasoning is exactly the same as in the two previous exercises. Also, half of this problem was done in Example 14.

EO-2.12 Computing the transitive closure "by inspection" of R can be tricky. It's much better to use either the incidence matrix or the directed graph diagram approach. We do the former. With rows and columns in the order 0, 1, 2, 3, the incidence matrix is

$$A = \begin{pmatrix} 1 & 0 & 0 & 1 \\ 1 & 0 & 1 & 0 \\ 1 & 0 & 0 & 0 \\ 0 & 0 & 1 & 0 \end{pmatrix}.$$

Using Boolean sum and product, we have

$$A^2 = \begin{pmatrix} 1 & 0 & 1 & 1 \\ 1 & 0 & 0 & 1 \\ 1 & 0 & 0 & 1 \\ 1 & 0 & 0 & 0 \end{pmatrix} \qquad S_2 = A^2 + A = \begin{pmatrix} 1 & 0 & 1 & 1 \\ 1 & 0 & 1 & 1 \\ 1 & 0 & 0 & 1 \\ 1 & 0 & 1 & 0 \end{pmatrix}$$

Next, we compute

$$S_3 = S_2 A + A = \begin{pmatrix} 1 & 0 & 1 & 1 \\ 1 & 0 & 1 & 1 \\ 1 & 0 & 1 & 1 \\ 1 & 0 & 1 & 1 \end{pmatrix} \qquad S_4 = S_3 A + A = \begin{pmatrix} 1 & 0 & 1 & 1 \\ 1 & 0 & 1 & 1 \\ 1 & 0 & 1 & 1 \\ 1 & 0 & 1 & 1 \end{pmatrix}.$$

Since $S_4 = S_3$, we are done.

EO-2.13 We leave experimentation to you. The results of the matrix calculations are

$$S_1 = A = \begin{pmatrix} 0 & 0 & 1 & 0 \\ 0 & 0 & 1 & 0 \\ 0 & 0 & 0 & 1 \\ 0 & 0 & 0 & 0 \end{pmatrix} \qquad S_2 = \begin{pmatrix} 0 & 0 & 1 & 1 \\ 0 & 0 & 1 & 1 \\ 0 & 0 & 0 & 1 \\ 0 & 0 & 0 & 0 \end{pmatrix} \qquad S_3 = S_2.$$

EO-2.14 The covering relation is

$$\{(4,8), (4,12), (4,20), (6,12), (6,18), (8,16), (9,18), (10,20)\}.$$

The minimal elements are 4, 6, 9, 10, 14 and 15. The maximal elements are 12, 16, 18, 20, 14 and 15. A chain of longest length is $4 \mid 8 \mid 16$. It has length two and is unique.

EO-2.15 The covering relation has cardinality 15 (draw the Hasse diagram). It consists of all pairs $(T, T \cup \{x\})$ where $T \cup \{x\}$ is an allowed subset of S and $x \notin T$. There are 6 chains of length three. The maximal elements are $\{1,3,5\}$, $\{1,4\}$, $\{2,4\}$ and $\{2,5\}$. The minimal element is $\emptyset$. Since there is more than one maximal element, there is no greatest element. Since there is only one minimal element, it is the least element.

EO-2.16 It can be shown that *every* finite poset has at least one maximal element, so your example must be infinite. One example is the integers with the usual order relation. Another example is the rationals (with the usual order relation). Yet another example is the rationals less than 1. Why are the rationals less than or equal to 1 not an example?

EO-2.17 The covering relation is

$$\{((0,0), (0,1)), ((0,0), (1,0)), ((0,1), (1,1)), ((1,0), (1,1))\}.$$

You can see this by drawing the Hasse diagram in the plane $\mathbb{R}^2$, using the elements of S_2 as coordinates. The picture is a square. The covering relation for the subsets of a two-element set under inclusion is essentially the same as this example. The correspondence is given by the isomorphism discussed in Example 17, obtained by using the characteristic function. As noted at the end of the example, the characteristic function provides an isomorphism for n-element sets. You could imagine drawing the Hasse diagram in n dimensions, obtaining an n-dimensional cube. Of course, we can't visualize this for $n > 3$.

EO-2.18 (a) In reverse order since $9 \mid 18$.

(b) In order.

(c) Incomparable since 16 and 6 are incomparable.

EO-2.19 (a) Let S consist of x and the $n - 1$ elements $s_2 \prec_C s_3 \prec_C \cdots \prec_C s_n$. There are three possible types of covering relations, depending on which of the s_i equals y.

- If $s_2 = y$, we have the additional relation $x \prec_C s_3$.
- If $s_n = y$, we have the additional relation $s_n \prec_C x$.
- If $s_k = y$ for $2 < k < n$, we have the additional relations $s_{k-1} \prec_C x \prec_C s_{k+1}$.

The Hasse diagram looks like the diagram for a chain except that one element of the chain has been split into two, namely x and y. You should draw the pictures.

(b) Consider two elements (a_1, a_2) and (b_1, b_2) of T. How can they be incomparable?

- We could have a_1 and b_1 incomparable. Thus one of a_1 and b_1 is x and the other is y. Since our pairs are not ordered, we can assume $a_1 = x$ and $b_1 = y$. The values of a_2 and b_2 can be anything. This gives us $n \times n = n^2$ pairs.
- If $a_1 = b_1$, then a_2 and b_2 must be incomparable and so must be x and y in some order. There are n choices for $a_1 = b_1$ and so this gives n pairs.
- If $a_1 \neq b_1$ and they are comparable, then (a_1, a_2) and (b_1, b_2) are also comparable, so this gives us nothing.

Adding up we obtain $n^2 + n$.

EO-2.20 There are 13 configurations. In lex order they are as follows: hhhhhh, hhvvhh, hvhvhh, hvvhhh, hvvvvh, vhhvhh, vhvhhh, vhvvvh, vvhhhh, vvhvvh, vvvhvh, vvvvhh, vvvvvv. You should draw pictures of some of them.

EO-2.21 As in the statement of the problem, we've omitted the commas and parentheses in the lists.

PASS 1
Bucket 1: 321, 441, 221, 311, 111
Bucket 2: 312, 422
Bucket 3: 143
Bucket 4: 214, 234

[221]

PASS 2
Bucket 1: 311, 111, 312, 214
Bucket 2: 321, 221, 422
Bucket 3: 234
Bucket 4: 441, 143

PASS 3
Bucket 1: 111, 143
Bucket 2: 214, 221, 234
Bucket 3: 311, 312, 321
Bucket 4: 422, 441

EO-2.22 The topological sort 15, 14, 10, 9, 6, 18, 4, 20, 12, 8, 16 has 26 in-order pairs.

EO-2.23 This poset has a least element, the empty set, and a greatest element, S. Thus every topological sort must start with $\emptyset$ and end with S. The three subsets $\{a\}$, $\{b\}$ and $\{c\}$ can be arranged in any order; that is in any of six ways. What about the three sets $\{a,b\}$, $\{a,c\}$ and $\{b,c\}$?

- They could be arranged in any manner after the three 1-element sets. Combining those six arrangements with the six 1-element set arrangements gives $6 \times 6 = 36$ topological sorts.

- If the one element sets are in the order $\{x\}$, $\{y\}$, $\{z\}$, then the set $\{x,y\}$ can be placed before $\{z\}$ and the other two 2-element sets can be placed after $\{z\}$ in either of two orders. This gives $6 \times 2 = 12$ topological sorts.

Adding the results gives us 48 topological sorts.

Solutions for Induction, Sequences and Series

IS-1.1 There is no single "right" answer. We illustrate this for (a) by giving some possible variations on the answer. In the other cases, we give just one answer.

(a) $\displaystyle\sum_{n=1}^{\infty} (-1)^{n-1} n^2$ or $\displaystyle -\sum_{k=1}^{\infty} (-1)^k k^2$ or $\displaystyle\sum_{n=0}^{\infty} (-1)^n (n+1)^2$

(b) $\displaystyle\sum_{k=1}^{\infty} (k^3 + (-1)^k)$

(c) $\displaystyle\prod_{k=2}^{\infty} (k^2 - (-1)^k)$

(d) $\displaystyle\prod_{n=0}^{\infty} (1 - r^{2n+1})$

(e) $\displaystyle\sum_{n=2}^{\infty} \frac{n-1}{n!}$

(f) $\displaystyle\sum_{k=0}^{\infty} \frac{n-k}{(k+1)!}$

IS-1.2 (a) For $n \geq 1$, $\frac{1}{n} - \frac{1}{n+1}$ — which equals $\frac{1}{n(n+1)}$.

(b) For $n \geq 1$, $\frac{n}{(n+1)^2}$ or, for $n \geq 2$, $\frac{n-1}{n^2}$.

(c) For $n \geq 1$, $\frac{(-1)^{n+1}n}{n+1}$ or, for $n \geq 2$, $\frac{(-1)^n(n-1)}{n}$.

(d) For $n \geq 1$, $n(n+1)$.

(e) For $n \geq 1$, $\lfloor (n-1)/2 \rfloor$ or, for $n \geq 0$, $\lfloor n/2 \rfloor$.

IS-1.3

(a) $\displaystyle\prod_{j=1}^{n} \frac{j^2}{j+1}$

(b) $\displaystyle\sum_{j=0}^{n-2} \frac{j+1}{(n-j-1)^2}$

(c) $\displaystyle\prod_{j=n-1}^{2n-1} \frac{n-j}{j+1}$

(d) $\displaystyle\prod_{j=0}^{n-1} \frac{j+1}{j+2} \prod_{j=0}^{n-1} \frac{j+2}{j+3}$

IS-1.4 Call the claim for n, $\mathcal{A}(n)$.
Base case ($n = 1$): $1^2 = 1(1+1)(2+1)/6$ proves it.

Solutions for Induction, Sequences and Series

Inductive step:

$$\sum_{k=1}^{n} k^2 = n^2 + \sum_{k=1}^{n-1} k^2 = n^2 + \frac{(n-1)n(2(n-1)+1)}{6} \qquad \text{by } \mathcal{A}(n-1)$$

$$= \frac{n(6n + (2n^2 - 3n + 1))}{6} = \frac{n(n+1)(2n+1)}{6} \qquad \text{by algebra.}$$

IS-1.5 By Theorem 2, the sum is a fourth degree polynomial with constant term 0 and lead coefficient 1/4. We could write down three equations for the three other coefficients by setting $n = 1$, $n = 2$ and $n = 3$. It is simpler to just verify that the given polynomial, $(n(n+1)/2)^2$ is fourth degree with no constant term and lead coefficient 1/4 and that it gives the correct answer for $n = 1$, $n = 2$ and $n = 3$.

For induction, the base case is trivial: $1^3 = (1(1+1)/2)^2$ and the inductive step is $n^3 + ((n-1)n/2)^2 = (n(n+1)/2)^2$, which is easily checked by algebra.

IS-1.6 Call the equation $\mathcal{A}(n)$. The base case $(n = 1)$ is simple. For the inductive step:

$$\sum_{i=1}^{n} \frac{1}{i(i+1)} = \frac{1}{n(n+1)} + \sum_{i=1}^{n-1} \frac{1}{i(i+1)} = \frac{1}{n(n+1)} + \frac{n-1}{n} = \frac{n}{n+1}.$$

IS-1.7 Call the equation $\mathcal{A}(n)$. The base case $(n = 0)$ is simple: $1 \times 2^1 = 0 + 2$. For the inductive step:

$$\sum_{i=1}^{n+1} i2^i = (n+1)2^{n+1} + \sum_{i=1}^{n} i2^i$$

$$= (n+1)2^{n+1} + (n-1)2^{n+1} + 2 \qquad \text{by } \mathcal{A}(n-1)$$

$$= 2n2^{n+1} + 2 = n2^{n+2} + 2.$$

IS-1.8 Call the equation $\mathcal{A}(n)$. The base case $(n = 2)$ is simple: $1 - 1/2^2 = (2+1)/2^2$. For the inductive step:

$$\prod_{i=2}^{n} \left(1 - \frac{1}{i^2}\right) = \left(1 - \frac{1}{n^2}\right) \prod_{i=2}^{n-1} \left(1 - \frac{1}{i^2}\right)$$

$$= \frac{n^2 - 1}{n^2} \frac{n}{2(n-1)} \qquad \text{by } \mathcal{A}(n-1)$$

$$= \frac{n+1}{2n} \qquad \text{by algebra}$$

IS-1.9 The base case $(n = 1)$ is simple. For the inductive step:

$$\sum_{i=1}^{n} i\,i! = n\,n! + \sum_{i=1}^{n-1} i\,i! = n\,n! + n! - 1 = (n+1)! - 1.$$

IS-1.10 The base case ($n = 0$) is simple. For the inductive step:

$$\prod_{i=0}^{n} \frac{1}{2i+1}\frac{1}{2i+2} = \frac{1}{(2n+1)(2n+2)} \prod_{i=0}^{n-1} \frac{1}{2i+1}\frac{1}{2i+2}$$

$$= \frac{1}{(2n+1)(2n+2)\,(2n)!} = \frac{1}{(2n+2)!}$$

IS-1.11 $\sum_{k=1}^{n} 5k = 5\sum_{k=1}^{n} k = 5n(n+1)/2$, by Example 2.

IS-1.12 There are various approaches. Here are some, with details omitted.

- If you know the sum of a geometric: $\sum_{k=0}^{N} Ar^k = \frac{A(r^{N+1}-1)}{r-1}$, you can use it with $A = 1$, $r = a$ and the two values $N = n$ and $N = t - 1$. Subtract the two results.

- As in the preceding, but with $A = a^t$, $r = a$ and $N = n - t$.

- Induction with $n = t$ the base case.

- Multiply both sides by $a - 1$ and note that

$$(a-1)\sum_{k=t}^{n} a^k = \sum_{k=t}^{n} a^{k+1} - \sum_{k=t}^{n} a^k = \sum_{j=t+1}^{n+1} a^j - \sum_{k=t}^{n} a^k = a^{n+1} - a^t.$$

IS-1.13 Without induction, one can do arithmetic mod 3. There are two ways to do this:

- $n^3 - 10n + 9 = n^3 - n = (n-1)n(n+1)$ mod 3. One of the three consecutive integers $n-1$, n and $n+1$ must be a multiple of 3 and so $n^3 - 10n + 9 = 0$ mod 3.

- n mod 3 is either 0, 1 or 2. In all three cases, one can check that $n^3 - 10n + 9 = 0$ mod 3.

Without induction, one can write $n^3 - 10n + 9 = (n-1)n(n+1) - 9(n-1)$ and continue; however, this is just a messier version of arithmetic mod 3.

By induction, it's true for $n = 0$. To keep the algebra simple, we prove the case $n + 1$ from the case n. We have

$$(n+1)^3 - 10(n+1) + 9 = n^3 + 3n^2 + 3n + 1 - 10n - 10 + 9 = (n^3 - 10n + 9) + 3(n^2 + n - 3).$$

By the induction assumption, $n^3 - 10n + 9$

IS-1.14 The base case ($n = 1$) is trivial $(x - y) \mid (x^1 - y^1)$.
For the inductive step, we have $(x - y) \mid (x^{n-1} - y^{n-1})$. To get x^n we could multiply by x to conclude that $(x - y) \mid (x^n - xy^{n-1})$. This is not quite right because we want $x^n - y^n$. The difference between what we want and what we have is

$$(x^n - y^n) - (x^n - xy^{n-1}) = (x - y)y^{n-1}.$$

How does this help? If $a \mid b$ and $a \mid c$, then $a \mid (b + c)$. Apply this with $a = x - y$, $b = x^n - xy^{n-1}$ and $c = (x - y)y^{n-1}$ and you are done.

Solutions for Induction, Sequences and Series

Another way to prove this is by the equation

$$\frac{x^n - y^n}{x - y} = x^{n-1} + x^{n-2}y + x^{n-3}y^2 + \cdots + xy^{n-1} + y^n.$$

Again, this can be proved by induction by using $x^n - y^n = x(x^{n-1} - y^{n-1}) - (x-y)y^{n-1}$. You should fill in the proof.

IS-1.15 Modulo 6 we have $n(n^2 + 5) = n(n^2 - 1) = (n-1)n(n+1)$, a product of three consecutive integers. One is divisible by 3 and at least one is divisible by 2.

For induction, the base case is trivial. We prove the case $n+1$ using the case n:

$$(n+1)((n+1)^2 + 5) = (n+1)(n^2 + 2n + 6) = n^3 + 3n^2 + 8n + 6 = n(n^2 + 5) + 3n(n+1) + 6.$$

By induction, $n(n^2 + 5)$ is divisible by 6. Since $n(n+1)$ is even, $3n(n+1)$ is divisible by 6. Thus we are done.

IS-1.16 The condition $n \neq 3$ looks strange — we haven't had anything like that before. To see what's going on, let's try the inductive step, proving the $n+1$ case. We want $(n+1)^2 \leq 2^{n+1}$ and we have $n^2 \leq 2^n$. The right sides are double each other, so we could do this if we had something similar for the left sides, namely $(n+1)^2 \leq 2n^2$ for then $(n+1)^2 \leq 2n^2 \leq 2 \times 2^n = 2^{n+1}$. We've reduced the inductive step to proving $(n+1)^2 \leq 2n^2$. This is true for $n \geq 3$. There are various ways to see that. Here's one:

$$2n^2 - (n+1)^2 = n^2 - 2n - 1 = (n-1)^2 - 2,$$

which is nonnegative when $n \geq 3$.

Where are we? We've shown that we can use $n^2 \leq 2^n$ to prove $(n+1)^2 \leq 2^{n+1}$ provided $n \geq 3$. Thus, we must verify $n = 0$, $n = 1$ and $n = 2$ separately as well as the base case $n = 4$ for the induction. Why $n = 4$ instead of $n = 3$? Because $n = 3$ was excluded in the problem — the inequality is not true when $n = 3$.

IS-1.17 For those who are wondering where this inequality comes from, it was inspired by the Riemann sum approximation to $\int x^{-1/2} dx$. However, you don't need to know this so we'll omit the details.

The base case ($n = 2$): We want $\sqrt{2} < 1 + 1/\sqrt{2}$, which you can check on a calculator. For the inductive step: We have

$$\sum_{i=1}^{n} \frac{1}{\sqrt{i}} = \frac{1}{\sqrt{n}} + \sum_{i=1}^{n-1} \frac{1}{\sqrt{i}} > \frac{1}{\sqrt{n}} + \sqrt{n-1},$$

where we used the inequality for $n - 1$.

To complete the proof we need to show that

$$\frac{1}{\sqrt{n}} + \sqrt{n-1} \geq \sqrt{n}.$$

The best way to deal with something like this probably to clear of fractions, so we multiply by $\sqrt{n}$ and see that we want to prove

$$1 + \sqrt{n(n-1)} \geq n.$$

If we move the 1 to the other side and square, we get rid of the annoying square root: We want $n(n-1) \geq (n-1)^2$, which is equivalent to $n^2 - n \geq n^2 - 2n + 1$. After some algebra, this is easily seen to be equivalent to $n \geq 1$.

This completes the proof of the inductive step, but it's rather awkward because it's all done backwards. If we reverse the steps, the result is "cleaner," but it looks more like magic. Let's do it.

Since $n > 1$, we have $n + (n^2 - 2n) > 1 + (n^2 - 2n)$. Factoring both sides and taking square roots: $\sqrt{n(n-1)} > n-1$. Adding 1 to both sides and dividing by $\sqrt{n}$, we have $\sqrt{n-1} + 1/\sqrt{n} > \sqrt{n}$, which is what we needed to prove.

IS-1.18 Let the assertion $\mathcal{A}(n)$ be "$3 \mid f_n$. Clearly $\mathcal{A}(0)$ and $\mathcal{A}(1)$ are true. For $n \geq 2$ we have $f_n = f_{n-2} + f_{n-1}$. By $\mathcal{A}(n-2)$ and $\mathcal{A}(n-1)$, we have $3 \mid f_{n-2}$ and $3 \mid f_{n-1}$. Hence there sum is a multiple of 3 and we are done.

Here's another proof. Let $F_0 = 1$, $F_1 = 2$ and $F_k = F_{k-2} + F_{k-1}$ for $k \geq 2$. Clearly $F_k \in \mathbb{Z}$ for $k \geq 0$. (Strictly speaking, this requires an inductive proof.) By an inductive proof, which we omit, $f_k = 3F_k$.

IS-1.19 We use induction on t. From the recursion $F_2 = 1$ and so the result is true for $t = 0$. For the inductive step:

$$F_{3t} = F_{3t-1} + F_{3t-2} = 1 + 1 = 0 \bmod 2,$$
$$F_{3t+1} = F_{3t} + F_{3t-1} = 0 + 1 = 1 \bmod 2,$$
$$F_{3t+2} = F_{3t+1} + F_{3t} = 1 + 0 = 1 \bmod 2.$$

IS-1.20 The base case ($k = 1$) is trivial. For the induction, we need to know the value of $\lfloor k/2 \rfloor$. If k is even, it is $k/2$. If k is odd, it is $(k-1)/2$. Thus, either by $\mathcal{A}(k/2)$ or $\mathcal{A}((k-1)/2)$, we have

$$f_{\lfloor \frac{k}{2} \rfloor} = \begin{cases} k/2, & \text{if } k \text{ is even,} \\ (k-1)/2, & \text{if } k \text{ is odd.} \end{cases}$$

In either case, $2f_{\lfloor \frac{k}{2} \rfloor} \leq k$.

IS-1.21 The general step assumes that there are two nonnegative integers, s and t, less than $k+1$ with $s+t = k+1$. This does not apply to $k = 0$, the base case. So we must take $k = 1$ as a base case, too, and check if $r^1 = 1$ for all real numbers r. Not so!

IS-1.22 If $p = 1$ and $q = 2$, then $p - 1 = 0$ is not a positive integer.

IS-1.23 To prove these results, we evaluate the functions at n. We've used lots of parentheses to try to make things clearer.

(a) $\quad (\Delta af)(n) = af(n+1) - af(n) = a\big(f(n+1) - f(n)\big) = a(\Delta f)(n)$

(b) $\quad (\Delta(f+g))(n) = (f+g)(n+1) - (f+g)(n)$
$$= f(n+1) + g(n+1) - \big(f(n) + g(n)\big)$$
$$= \big(f(n+1) - f(n)\big)\big(g(n+1) - g(n)\big)$$
$$= (\Delta f)(n)) + (\Delta g)(n) = (\Delta f + \Delta g)(n)$$

(c) $\quad (\Delta(fg))(n) = (fg)(n+1) - (fg)(n) = f(n+1)g(n+1) - f(n)g(n)$
$$= f(n)\big(g(n+1) - g(n)\big) + g(n)\big(f(n+1) - f(n)\big)$$
$$+ \big(f(n+1) - f(n)\big)\big(g(n+1) - g(n)\big)$$
$$= f(n)(\Delta g)(n) + g(n)(\Delta f)(n) + \big((\Delta f)(n)\big)\big((\Delta g)(n)\big)$$

Solutions for Induction, Sequences and Series

IS-1.24 We begin with the hint. Using $\binom{n}{i}\frac{n!}{i!\,(n-i)!}$, we have

$$\binom{k-1}{j-1}+\binom{k-1}{j}=\frac{(k-1)!}{(j-1)!\,(k-j)!}+\frac{(k-1)!}{j!\,(k-j-1)!}$$
$$=\frac{(k-1)!\,(j+(k-j))}{j!\,(k-j)!}=\frac{k!}{j!\,(k-j)!}=\binom{k}{j}.$$

Now for the $(\Delta^k f)(n)$ formula. The base case ($k=1$) is the definition of Δ.
We now do the inductive step. If you have difficulty with all the manipulations of
sums, try writing it out explicitly for $k=2$ and $k=3$.

$$(\Delta^k f)(n)=(\Delta^{k-1})(\Delta f)(n)=\sum_{i=0}^{k-1}\binom{k-1}{i}(-1)^{k-1-i}(\Delta f)(n+i)$$

$$=\sum_{i=0}^{k-1}\binom{k-1}{i}(-1)^{k-1-i}(f(n+i+1)-f(n+i))$$

$$=\sum_{i=0}^{k-1}\binom{k-1}{i}(-1)^{k-1-i}f(n+i+1)+\sum_{i=0}^{k-1}\binom{k-1}{i}(-1)^{k-i}f(n+i)$$

$$=\sum_{j=1}^{k}\binom{k-1}{j-1}(-1)^{k-j}f(n+j)+\sum_{j=0}^{k-1}\binom{k-1}{j}(-1)^{k-j}f(n+j)$$

$$=\sum_{j=1}^{k-1}\left(\binom{k-1}{j-1}+\binom{k-1}{j}\right)(-1)^{k-j}f(n+j)$$

$$+\binom{k-1}{k-1}(-1)^{k-k}f(n+k)+\binom{k-1}{0}(-1)^{k-0}f(n+0)$$

$$=\sum_{j=1}^{k-1}\binom{k}{j}(-1)^{k-j}f(n+j)+\binom{k}{k}(-1)^{k-k}f(n+k)+\binom{k}{0}(-1)^{k-0}f(n+0)$$

$$=\sum_{j=0}^{k}\binom{k}{j}(-1)^{k-j}f(n+j).$$

IS-2.1 *Bounded*: Only (f) is bounded.
Monotonic: Only (d) and (e) are not monotonic. You can check (d) and (e) by computing a few values. Perhaps (c) is a bit tricky; however, if you compute a few values
you should note that $a_{n+1}=a_n$ whenever n is even and $a_{n+1}=a_n+2$ whenever n is
odd.
Eventually monotonic: Since monotonic implies eventually monotonic, we only need to
check (d) and (e). After computing a few terms of (d), you should note that $a_{n+1}<a_n$
when n is even and $a_{n+1}>a_n$ when n is odd. Thus it is not eventually monotonic.
This leaves only (e). After some computation, it appears that the terms are eventually increasing. In other words $a_{n+1}-a_n$ is eventually non-negative. Compute the
difference:

$$a_{n+1}-a_n=\left(2^{n+1}-10(n+1)\right)-\left(2^n-10n\right)=(2^{n+1}-2^n)-10=2^n-10.$$

This is positive for all $n \geq 4$. Thus (e) is eventually monotonic.

IS-2.2 (a) Converges: $\lim_{n \to \infty} \frac{2n^3 + 3n + 1}{3n^3 + 2} = \lim_{n \to \infty} \frac{2 + 3/n + 1/n^3}{3 + 2/n^3} = \frac{2}{3}$

(b) Diverges to $-\infty$: $\lim_{n \to \infty} \frac{-n^3 + 1}{2n^2 + 3} = \lim_{n \to \infty} \frac{-n + 1/n^2}{2 + 3/n^2}$

(c) Diverges: $\lim_{n \to \infty} \frac{(-n)^n + 1}{n^n + 1} = \lim_{n \to \infty} \frac{(-1)^n + 1/n^n}{1 + 1/n^n}$

(d) Converges: $\lim_{n \to \infty} \frac{n^n}{(n/2)^{2n}} = \lim_{n \to \infty} \frac{2^{2n}}{n^n} = \lim_{n \to \infty} (4/n)^n = 0$

IS-2.3 (a) Converges: Recall the formula for changing bases in logarithms, namely $\log_a(x) = \log_a(b) \log_b(x)$. Thus $\log_2(n) = \log_3(n) \log_2 3$ so $\frac{\log_2(n)}{\log_3(n)} = \log_2 3$

(b) Converges: Think of $\log_2(n)$ as x. Then we have $\frac{\log_2(x)}{x}$, which converges to zero.

IS-3.1 Both series diverge because the terms do not approach zero. (See Theorem 10.) In fact, the terms in (a) diverge to $+\infty$ and those in (b) converge to $1/2$.

IS-3.2 In both cases, we apply Theorem 11.

(a) Let $a_n = (4/5)^n$, which gives a geometric series that converges (Example 11). Let $b_n = n^5/4^n$, which is bounded (Example 10).

(b) Let $a_n = 1/n^2$, which gives a convergent general harmonic series (Example 16). Let $b_n = n^2/(n^2 - 150)$, which is bounded.

IS-3.3 In both cases, we apply Theorem 11.

(a) Let $a_n = 1/n^{3/2}$, which gives a convergent general harmonic series (Example 16). Let $b_n = \left(\frac{n^3}{n^3 - n^2 - 1} \right)^{1/2}$, which is bounded.

(b) The problem here is estimating $(n + 1)^{1/2} - (n - 1)^{1/2}$. If it were a sum instead of a difference, it would be no problem to estimate, so we use a trick:

$$(n + 1)^{1/2} - (n - 1)^{1/2} = \frac{\left((n + 1)^{1/2} - (n - 1)^{1/2} \right)\left((n + 1)^{1/2} + (n - 1)^{1/2} \right)}{\left((n + 1)^{1/2} + (n - 1)^{1/2} \right)}$$

$$= \frac{(n + 1) - (n - 1)}{\left((n + 1)^{1/2} + (n - 1)^{1/2} \right)}$$

$$= \frac{2}{\left((n + 1)^{1/2} + (n - 1)^{1/2} \right)}.$$

Let $a_n = 1/n^{3/2}$ as in (a). Let $b_n = \frac{2n^{1/2}}{\left((n+1)^{1/2} + (n-1)^{1/2} \right)}$, which is bounded.

IS-3.4 These are both alternating series. Calculating some terms seems to indicate that they are monotonic decreasing to zero. How can we prove this? We need to show that a_n goes to zero and $|a_n|$ is eventually monotonic, so we need to estimate the sums in the parentheses somehow. Showing that a_n goes to zero is not too hard, but the monotonicity is a bit tricky.

In (a) we have a partial sum of the generalized harmonic series, which converges. In (b) we have a partial sum H_n of the harmonic series which behaves like $\ln(n)$ by Example 12. Thus $\lim_{n \to \infty} a_n = 0$ in both cases.

Solutions for Induction, Sequences and Series

Now we need to show that $|a_n|$ is eventually monotonic. Let p_n be the sum in parentheses and call its nth term b_n. We have

$$\left|\frac{a_{n+1}}{a_n}\right| = \frac{(p_n + b_{n+1})/(n+1)}{p_n/n} = \frac{n}{n+1}\frac{p_n + b_{n+1}}{p_n} = \frac{1 + b_{n+1}/p_n}{1 + 1/n}.$$

In both (a) and (b), $b_{n+1} < 1/n$ and $p_n > 1$ so that $1 + b_{n+1}/p_n < 1 + 1/n$ and we are done.

IS-3.5 (a) The only information we have on a series like this is that the series with terms $a_n = (\sin(n))/n$ converges (Example 14). The idea discussed there applies to the series in (a) because the terms $\frac{1}{|n-99.5|}$ are strictly decreasing for $n \geq 100$.

(b) In this case, the idea used in Examples 13 and 14 can be applied because the sequence $a_n = \frac{-9n^2-5}{n^3+1}$ is monotone and converges to zero.

Notation Index

(Page references herein refer to book's chapter numbering system.)

Index

Subject Index

(Page references herein refer to book's chapter numbering system.)

Index

Index

one-to-one (= injection) SF-18
one-way (= trapdoor) NT-21
onto (= surjection) SF-18
permutation SF-18
range (= codomain) of BF-1,
  SF-15
surjective SF-18
trapdoor NT-21
two-line notation for SF-20,
  SF-20
Functional relation SF-16

Gate BF-18

Geometric series IS-22

Goldbach's conjecture Lo-13

Graph diagrams, directed EO-26

Greatest common divisor
  (= gcd) NT-16
  Euclidean algorithm NT-18

Greatest element in poset EO-29

Greatest integer function NT-9

Half adder BF-18

Harmonic series IS-22
  alternating IS-23
  general IS-25

Hashing SF-19

Hasse diagram EO-28

Hexadecimal number BF-11

Idempotent rule BF-6, Lo-3, SF-3

If ... then Lo-5

If and only if (logic) Lo-7

Image of a function SF-15, SF-23

Implication Lo-5

Incidence matrix EO-14

Incomparable elements EO-14

Incomparable subsets EO-14

Increasing sequence IS-17

Induction terminology IS-1

Inductive step IS-1

Infinite sequence
  see Sequence

Infinite series
  see Series

Injective function SF-18

Integral test for series IS-24

Intersection of sets SF-2

Inverse Lo-6

Inverse function SF-18

Inverse relation SF-16

Irrationality of square root NT-4

Key (cryptography) NT-13
  Diffie-Hellman NT-22
  RSA and public NT-23
  trapdoor function and NT-21

Lattice of subsets EO-13

Least common multiple
  (= lcm) NT-16

Least element in poset EO-29

Least integer function NT-9

Length-first lex order EO-21

Lexicographic bucket sort EO-22

Lexicographic order (= lex
  order) SF-7, EO-19
  length-first (= short) EO-21

Limit
  of a sequence IS-13
  sum of infinite series IS-20

Linear extension EO-30

Linear order SF-1, EO-14

List (= ordered set) SF-1

Logarithm
  discrete and Diffie-
    Hellman NT-22

Logarithm, rate of growth of IS-18

Logic
  predicate Lo-12
  propositional BF-4, Lo-1

Index

Perfect
  number   Lo-17
Perfect square   NT-4
Permutation   SF-18
  cycle   SF-22
  cycle form   SF-22
  cycle length   SF-22
PGP (= Pretty Good
    Privacy)   NT-20,
    SF-19
Pigeonhole principle   EO-5
  extended   EO-7
Plaintext   NT-13
Polynomial, rate of growth of   IS-18
Poset   EO-13
  comparable elements   EO-14
  coordinate (= direct product)
    order   EO-17
  covering relation   EO-28
  direct product of   EO-17
  divisibility   EO-14, EO-19
  greatest element   EO-29
  incomparable elements   EO-14
  isomorphic   EO-18
  least element   EO-29
  lex order   EO-19
  linear (= total) order   EO-14
  maximal element   EO-30
  minimal element   EO-30
  restriction of (= subposet)   EO-17
  subset lattice   EO-13, EO-17
Power set   SF-9, EO-13
Powers
  sum of   IS-5
Predicate logic
  algebraic rules   Lo-19
  predicate   Lo-12
  quantifier   Lo-12
  truth set   Lo-12
Prime factorization   NT-3, IS-2
  uniqueness of   NT-3
Prime number   Lo-13, NT-2
  how common?   IS-28
  infinitely many   NT-4
  unique factorization into   NT-3

Prime Number Theorem   IS-28
Principle
  extended pigeonhole   EO-7
  pigeonhole   EO-5
Product of sets   SF-2
Propositional logic   BF-4, Lo-1
  algebraic rules   Lo-3
Public key cryptography   NT-21
  PGP   NT-20
  RSA protocol   NT-23

Quantifier
  existential (∃)   Lo-13
  negation of   Lo-15
  universal (∀)   Lo-12

Range of a function   BF-1, SF-15
Rate of growth   IS-18
Refinement of set partition   SF-11,
    EO-16
Reflexive relation   EO-3, EO-13
Relation   SF-16
  antisymmetric   EO-13
  binary   EO-3
  covering   EO-28
  equivalence   EO-1
  functional   SF-16
  inverse   SF-16
  number of   EO-15
  order   SF-8, EO-12
  reflexive   EO-3, EO-13
  symmetric   EO-3
  transitive   EO-3, EO-13
  transitive closure of   EO-26
Residue class (modular
    arithmetic)   NT-6
Restriction of a poset
    (= subposet)   EO-17
RSA protocol   NT-23

Index

Astronomy

CHARIOTS FOR APOLLO: The NASA History of Manned Lunar Spacecraft to 1969, Courtney G. Brooks, James M. Grimwood, and Loyd S. Swenson, Jr. This illustrated history by a trio of experts is the definitive reference on the Apollo spacecraft and lunar modules. It traces the vehicles' design, development, and operation in space. More than 100 photographs and illustrations. 576pp. 6 3/4 x 9 1/4. 0-486-46756-2

EXPLORING THE MOON THROUGH BINOCULARS AND SMALL TELESCOPES, Ernest H. Cherrington, Jr. Informative, profusely illustrated guide to locating and identifying craters, rills, seas, mountains, other lunar features. Newly revised and updated with special section of new photos. Over 100 photos and diagrams. 240pp. 8 1/4 x 11. 0-486-24491-1

WHERE NO MAN HAS GONE BEFORE: A History of NASA's Apollo Lunar Expeditions, William David Compton. Introduction by Paul Dickson. This official NASA history traces behind-the-scenes conflicts and cooperation between scientists and engineers. The first half concerns preparations for the Moon landings, and the second half documents the flights that followed Apollo 11. 1989 edition. 432pp. 7 x 10. 0-486-47888-2

APOLLO EXPEDITIONS TO THE MOON: The NASA History, Edited by Edgar M. Cortright. Official NASA publication marks the 40th anniversary of the first lunar landing and features essays by project participants recalling engineering and administrative challenges. Accessible, jargon-free accounts, highlighted by numerous illustrations. 336pp. 8 3/8 x 10 7/8. 0-486-47175-6

ON MARS: Exploration of the Red Planet, 1958-1978--The NASA History, Edward Clinton Ezell and Linda Neuman Ezell. NASA's official history chronicles the start of our explorations of our planetary neighbor. It recounts cooperation among government, industry, and academia, and it features dozens of photos from Viking cameras. 560pp. 6 3/4 x 9 1/4. 0-486-46757-0

ARISTARCHUS OF SAMOS: The Ancient Copernicus, Sir Thomas Heath. Heath's history of astronomy ranges from Homer and Hesiod to Aristarchus and includes quotes from numerous thinkers, compilers, and scholasticists from Thales and Anaximander through Pythagoras, Plato, Aristotle, and Heraclides. 34 figures. 448pp. 5 3/8 x 8 1/2. 0-486-43886-4

AN INTRODUCTION TO CELESTIAL MECHANICS, Forest Ray Moulton. Classic text still unsurpassed in presentation of fundamental principles. Covers rectilinear motion, central forces, problems of two and three bodies, much more. Includes over 200 problems, some with answers. 437pp. 5 3/8 x 8 1/2. 0-486-64687-4

BEYOND THE ATMOSPHERE: Early Years of Space Science, Homer E. Newell. This exciting survey is the work of a top NASA administrator who chronicles technological advances, the relationship of space science to general science, and the space program's social, political, and economic contexts. 528pp. 6 3/4 x 9 1/4. 0-486-47464-X

STAR LORE: Myths, Legends, and Facts, William Tyler Olcott. Captivating retellings of the origins and histories of ancient star groups include Pegasus, Ursa Major, Pleiades, signs of the zodiac, and other constellations. "Classic." – *Sky & Telescope.* 58 illustrations. 544pp. 5 3/8 x 8 1/2. 0-486-43581-4

A COMPLETE MANUAL OF AMATEUR ASTRONOMY: Tools and Techniques for Astronomical Observations, P. Clay Sherrod with Thomas L. Koed. Concise, highly readable book discusses the selection, set-up, and maintenance of a telescope; amateur studies of the sun; lunar topography and occultations; and more. 124 figures. 26 halftones. 37 tables. 335pp. 6 1/2 x 9 1/4. 0-486-42820-6

Chemistry

MOLECULAR COLLISION THEORY, M. S. Child. This high-level monograph offers an analytical treatment of classical scattering by a central force, quantum scattering by a central force, elastic scattering phase shifts, and semi-classical elastic scattering. 1974 edition. 310pp. 5 3/8 x 8 1/2. 0-486-69437-2

HANDBOOK OF COMPUTATIONAL QUANTUM CHEMISTRY, David B. Cook. This comprehensive text provides upper-level undergraduates and graduate students with an accessible introduction to the implementation of quantum ideas in molecular modeling, exploring practical applications alongside theoretical explanations. 1998 edition. 832pp. 5 3/8 x 8 1/2. 0-486-44307-8

RADIOACTIVE SUBSTANCES, Marie Curie. The celebrated scientist's thesis, which directly preceded her 1903 Nobel Prize, discusses establishing atomic character of radioactivity; extraction from pitchblende of polonium and radium; isolation of pure radium chloride; more. 96pp. 5 3/8 x 8 1/2. 0-486-42550-9

CHEMICAL MAGIC, Leonard A. Ford. Classic guide provides intriguing entertainment while elucidating sound scientific principles, with more than 100 unusual stunts: cold fire, dust explosions, a nylon rope trick, a disappearing beaker, much more. 128pp. 5 3/8 x 8 1/2. 0-486-67628-5

ALCHEMY, E. J. Holmyard. Classic study by noted authority covers 2,000 years of alchemical history: religious, mystical overtones; apparatus; signs, symbols, and secret terms; advent of scientific method, much more. Illustrated. 320pp. 5 3/8 x 8 1/2.
0-486-26298-7

CHEMICAL KINETICS AND REACTION DYNAMICS, Paul L. Houston. This text teaches the principles underlying modern chemical kinetics in a clear, direct fashion, using several examples to enhance basic understanding. Solutions to selected problems. 2001 edition. 352pp. 8 3/8 x 11. 0-486-45334-0

PROBLEMS AND SOLUTIONS IN QUANTUM CHEMISTRY AND PHYSICS, Charles S. Johnson and Lee G. Pedersen. Unusually varied problems, with detailed solutions, cover of quantum mechanics, wave mechanics, angular momentum, molecular spectroscopy, scattering theory, more. 280 problems, plus 139 supplementary exercises. 430pp. 6 1/2 x 9 1/4. 0-486-65236-X

ELEMENTS OF CHEMISTRY, Antoine Lavoisier. Monumental classic by the founder of modern chemistry features first explicit statement of law of conservation of matter in chemical change, and more. Facsimile reprint of original (1790) Kerr translation. 539pp. 5 3/8 x 8 1/2. 0-486-64624-6

MAGNETISM AND TRANSITION METAL COMPLEXES, F. E. Mabbs and D. J. Machin. A detailed view of the calculation methods involved in the magnetic properties of transition metal complexes, this volume offers sufficient background for original work in the field. 1973 edition. 240pp. 5 3/8 x 8 1/2. 0-486-46284-6

GENERAL CHEMISTRY, Linus Pauling. Revised third edition of classic first-year text by Nobel laureate. Atomic and molecular structure, quantum mechanics, statistical mechanics, thermodynamics correlated with descriptive chemistry. Problems. 992pp. 5 3/8 x 8 1/2. 0-486-65622-5

ELECTROLYTE SOLUTIONS: Second Revised Edition, R. A. Robinson and R. H. Stokes. Classic text deals primarily with measurement, interpretation of conductance, chemical potential, and diffusion in electrolyte solutions. Detailed theoretical interpretations, plus extensive tables of thermodynamic and transport properties. 1970 edition. 590pp. 5 3/8 x 8 1/2. 0-486-42225-9

Browse over 9,000 books at www.doverpublications.com